Strategic Imperatives and Core Competencies in the Era of Robotics and Artificial Intelligence

Roman Batko
Jagiellonian University, Poland

Anna Szopa
Jagiellonian University, Poland

A volume in the Advances in Computational Intelligence and Robotics (ACIR) Book Series

IGI GLOBAL
DISSEMINATOR OF KNOWLEDGE

www.igi-global.com

Published in the United States of America by
IGI Global
Business Science Reference (an imprint of IGI Global)
701 E. Chocolate Avenue
Hershey PA 17033
Tel: 717-533-8845
Fax: 717-533-8661
E-mail: cust@igi-global.com
Web site: http://www.igi-global.com

Library of Congress Cataloging-in-Publication Data

Names: Batko, Roman, 1966- editor. | Szopa, Anna, 1982- editor. | Information Resources Management Association.
Title: Strategic imperatives and core competencies in the era of robotics and artificial intelligence / Roman Batko, Anna Szopa, editors.
Description: Hershey : Engineering Science Reference, [2017] | Series: Advances in computational intelligence and robotic | Includes bibliographical references and index.
Identifiers: LCCN 2016041614| ISBN 9781522516569 (hardcover) | ISBN 9781522516576 (ebook)
Subjects: LCSH: Organizational effectiveness. | Core competencies. | Robotics. | Automation. | Artificial intelligence.
Classification: LCC HD58.9 .S767 2017 | DDC 658.4/012--dc23 LC record available at https://lccn.loc.gov/2016041614

This book is published in the IGI Global book series Advances in Computational Intelligence and Robotics (ACIR) (ISSN: 2327-0411; eISSN: 2327-042X)

British Cataloguing in Publication Data
A Cataloguing in Publication record for this book is available from the British Library.

Advances in Computational Intelligence and Robotics (ACIR) Book Series

ISSN:2327-0411
EISSN:2327-042X

Mission

While intelligence is traditionally a term applied to humans and human cognition, technology has progressed in such a way to allow for the development of intelligent systems able to simulate many human traits. With this new era of simulated and artificial intelligence, much research is needed in order to continue to advance the field and also to evaluate the ethical and societal concerns of the existence of artificial life and machine learning.

The **Advances in Computational Intelligence and Robotics (ACIR) Book Series** encourages scholarly discourse on all topics pertaining to evolutionary computing, artificial life, computational intelligence, machine learning, and robotics. ACIR presents the latest research being conducted on diverse topics in intelligence technologies with the goal of advancing knowledge and applications in this rapidly evolving field.

Coverage

- Robotics
- Brain Simulation
- Algorithmic Learning
- Heuristics
- Artificial Life
- Evolutionary Computing
- Agent technologies
- Automated Reasoning
- Synthetic Emotions
- Cognitive Informatics

IGI Global is currently accepting manuscripts for publication within this series. To submit a proposal for a volume in this series, please contact our Acquisition Editors at Acquisitions@igi-global.com or visit: http://www.igi-global.com/publish/.

The Advances in Computational Intelligence and Robotics (ACIR) Book Series (ISSN 2327-0411) is published by IGI Global, 701 E. Chocolate Avenue, Hershey, PA 17033-1240, USA, www.igi-global.com. This series is composed of titles available for purchase individually; each title is edited to be contextually exclusive from any other title within the series. For pricing and ordering information please visit http://www.igi-global.com/book-series/advances-computational-intelligence-robotics/73674. Postmaster: Send all address changes to above address.

Titles in this Series

For a list of additional titles in this series, please visit: www.igi-global.com

Multi-Core Computer Vision and Image Processing for Intelligent Applications
Mohan S. (Al Yamamah University, Saudi Arabia) and Vani V. (Al Yamamah University, Saudi Arabia)
Information Science Reference • copyright 2017 • 292pp • H/C (ISBN: 9781522508892) • US $210.00 (our price)

Developing and Applying Optoelectronics in Machine Vision
Oleg Sergiyenko (Autonomous University of Baja California, Mexico) and Julio C. Rodriguez-Quiñonez (Autonomous University of Baja California, Mexico)
Information Science Reference • copyright 2017 • 341pp • H/C (ISBN: 9781522506324) • US $205.00 (our price)

Pattern Recognition and Classification in Time Series Data
Eva Volna (University of Ostrava, Czech Republic) Martin Kotyrba (University of Ostrava, Czech Republic) and Michal Janosek (University of Ostrava, Czech Republic)
Information Science Reference • copyright 2017 • 282pp • H/C (ISBN: 9781522505655) • US $185.00 (our price)

Integrating Cognitive Architectures into Virtual Character Design
Jeremy Owen Turner (Simon Fraser University, Canada) Michael Nixon (Simon Fraser University, Canada) Ulysses Bernardet (Simon Fraser University, Canada) and Steve DiPaola (Simon Fraser University, Canada)
Information Science Reference • copyright 2016 • 346pp • H/C (ISBN: 9781522504542) • US $185.00 (our price)

Handbook of Research on Natural Computing for Optimization Problems
Jyotsna Kumar Mandal (University of Kalyani, India) Somnath Mukhopadhyay (Calcutta Business School, India) and Tandra Pal (National Institute of Technology Durgapur, India)
Information Science Reference • copyright 2016 • 1015pp • H/C (ISBN: 9781522500582) • US $465.00 (our price)

Applied Artificial Higher Order Neural Networks for Control and Recognition
Ming Zhang (Christopher Newport University, USA)
Information Science Reference • copyright 2016 • 511pp • H/C (ISBN: 9781522500636) • US $215.00 (our price)

Table of Contents

Detailed Table of Contents

The chapter proposes a critical analysis of the panopticon, the model of an ideal prison devised in the 18th century by the English philosopher Jeremy Bentham, and its function in modern management. The original prison was designed in such a way that it enabled a single prison guard to watch all the inmates at all times, while the latter could never tell whether they were currently being watched or not. The idea behind the panopticon gave rise to the concept of panopticism, a philosophy developed by the French thinker Michel Foucault, which he voiced in Discipline & Punish: The Birth of the Prison (1975/1995). Analyzing the mechanism of control through the ages, he ponders the resilience and adaptability of such concepts as discipline, training and hierarchical surveillance. I intend to provide a critical analysis of methods and techniques of organizational control applied today, in the era of liquid modernity. Since that control, more often than not, is executed nowadays by electronic surveillance equipment and computer programs, hence my term of choice – "cybercontrol".

We observe situation in which human computation processes leverage the abilities of people via the Web to perform complex computation. In this context human intelligence is harnessed to solve computational problems that are beyond the scope of existing Artificial Intelligence (AI) algorithms. The aim of this paper is to achieve an in-depth understanding of the influence of crowdfunding business process into Artificial Intelligence and what kind of computational problems can be efficiently

The paper is based on the research carried out into Living Labs in Canada. The aim of the paper is presenting the essence of Living Labs as a concept facilitating innovation generation in businesses thanks to the cooperation of various actors, e.g. producers with users, inspiring the process of the development of new goods and services. The research questions raised pertain to the clarification how Living Labs create innovation in businesses. The Living Labs functioning in the Ontario region were the subject matter of the research. The described case study is theorygenic in character because of the early development stage of the knowledge. During the research process the multi-directional nature and the impact dynamics of the idea of Living Labs among peer partners of innovative processes have been noted, emphasizing the prosumer idea as well as the possibility of businesses cooperating in Living Labs.

Organizations try to survive in a hypercompetitive, changing, and unpredictable environment. The form of this survival continuously changes and requires different tools, solutions, dynamics and drivers according to the actual time. Today organizations face with a big paradigm shift, the industrialization of information age. Organizations should find a new form on the basis of the new age requirements. Some authors have suggested some strategic prescriptions harmonized with the actual term such as McKinsey's 7S Model including strategy, structure, systems, skills, staff, style and shared goals. It is considered that there is a gap related to updating and upgrading these formulas by activating some new concepts such as morphing, organizational ambidexterity and so on. In this study, it is aimed to understand the organizations in the future with internal elements and outer changes which would affect them at the edge of the new era by using this model and offer many propositions.

The past two decades—and the technology advancements experienced throughout them—have left marketers with a new context that has provided new business opportunities. This new context has prompted a change in the focus of the marketing function and demanded a shift in marketing imperatives and competencies. This chapter provides a comprehensive review of the technological changes experienced by the marketing function in a company, as documented by both scholars and practitioners. It also provides a thorough discussion of the ongoing academic debate regarding the new set of technical skills that have defined employability in the marketing circles for the past couple of decades and the challenges ahead for future professionals and executives.

Plato's Demiurge is the quintessence of perfection and power. "Whatever comes from me is indestructible unless I, myself, wish it to be destroyed" - says the creator of the gods, speaking to them in Timaeus. The gods and Demiurge are believed to collaborate on the creation of people in accordance with the standard of excellent ideas; soon the world, as we know, will be created. Plato's Demiurge is also the good, and the platonic gods are righteous. Without Demiurge the world is a chaos, an environment of irrational chances. Nowadays, the myth of Demiurge can be related to the perfection of creativity. Demiurge becomes equal to the anticipated, all-powerful driving force. Omnipotent, yet tamed and friendly. Demiurge is also the leader in the tradition of social research, the "divine" constructor of the economy, the originator of development. In the new media environment, Demiurge is a convenient metaphor for the presentation of the algorithm: mysterious, error-free, resistant to influence, free from human weaknesses. A transcendent being. The aim of this publication is to present new concept, the core myth of new media organizations - the myth of Demiurge associated with the operation of algorithms and critical analysis of myths created around it, which accompany the social, political and business role of algorithms. Considering the aspect connected with the interpretation of digital media operation and their social and business role, algorithms have not been so far analysed in the context of the presence of myths in organisational functioning. The author believes, however, that the common factor in the perception of algorithms in new media is their mythical aureole and mythical thinking associated with them.

The authors of the chapter proved that the fundamental intellectual processes, which lie on the basis of decision-making behavior of the human-operator, could be identified on the bases on the analogies with the devices (elements). The basic intellectual processes of the Rational decision-making models can be adequately identified by the transient processes of the PID-controller; the intellectual processes of the Bounded Rationality and Intuitive decision-making models can be identified by the transient processes of the nonlinear elements. Taxonomy of the most typical automatic control theory elements and their compliance with certain decision-making models with a point of view of decision-making processes specificity and on human-operator behavior in the context of the socio-technical concept was obtained Authors of the chapter also accept the suggestion that the instruments of collective multi-criteria decision-making and social-network analysis theory have similar mathematical and methodological bases.

Aggregation of media content comes as a popular form of media market activity which is supposed to facilitate access to information in the conditions of its excessive amount. It can be defined as selection and hierarchization of information and determination of its value. Aggregation can be performed with the direct participation of editorial teams, or it can be performed automatically, with the use of algorithms. The research conducted in April 2014 – April 2015 was to indicate the sources of information published by the news portals dominating on the Polish market, namely: Onet.pl and WP. pl. The research referred to the main section Wiadomości (News). During the research 1829 sources of information were identified in Onet.pl and 2272 sources of information in WP.pl. Considering Onet.pl, there were 549 pieces of the portal news and 1239 pieces of news obtained from different sources. WP.pl presented 582 pieces of the portal news and 1690 pieces of news received from other sources. Thus, Onet.pl presented its original content more frequently. The scale of using external sources allows us to state that both portals are, first of all, the aggregators of medial contents, and they use external information sources, however the scale of the practice is differentiated. Generally, it is possible to state that only every third message placed in the most important section of both portals was created in their editorial offices.

Preface

Modern management has to deal with unprecedented changes affecting today organizational structure and technologies used by companies. The modern enterprise is already a cyberfactory (Czarniawska, 2012). One can say we are facing the cyborgization of companies, "The Second Machine Age" (Brynjolfsson & McAfee 2014), with robotics, automation taking over tasks previously performed by humans; the phenomenon requires entirely new skills and competences as well as building new relationships in the workplace not only between people, but also between man and the thinking machine equipped with artificial intelligence.

The publication focuses on contemporary organizations - already called organizations of the future due to the fact that they undertake innovative projects on a previously unknown and unprecedented scale. In this context, it aims to identify these new skills and competences and proposes how to best manage them. In addition, the publication will provide best practices for managing the new competences. We are now in the "second machine age" which will again fundamentally change the workplace. Machines are still unable to perform as well as humans at jobs which require non-routine functions or creativity, since they are designed to copy human thoughts and processes rather than create them independently. However, over the last few years we have observed a growing number of manual workers being replaced by increasingly advanced robots. In the new era of robotics and artificial intelligence there is a need to discuss new competences of humans.

Recent literature has documented the positive relationship between new workplace design, innovative human resource management, and organizational performance. There is a strong correlation between the level of competence and some specific work practices. Organizations of the future need to be complemented internally by bundles of new competences in order to be more successful.

The book discusses three components of competencies:

1. **The Knowledge:** Knowledge of the foundations of robots, information resources, information and knowledge management, information technology, information resource management, organizational management, research and user studies.

2. **The Skills:** Critical thinking, problem-solving skills, communication skills, collaboration skills, creativity and innovation skills, communication, team working, analytical thinking, decision making, management and planning, teaching and training, conceptual thinking.
3. **The Personal Attributes:** Including the leadership, service mentality, morals and professional ethics, achievement motivation, accountability, self-management, flexibility, and adaptability.

Those competencies are closely related to sustained competitive advantage of the future organizations in the era of robotics and artificial intelligence.

OBJECTIVE OF THE BOOK

Companies of the future learn more rapidly and increase flexibility in a world of growing complexity and change. The book shows how these companies manage their new competencies; it also reviews the broad background of the new competencies and examines assumptions, strategies and models, and other aspects of new competencies. It explains how the necessary new competences can be developed: first, by discussing and illustrating the kinds of managerial and organizational competencies that are needed to navigate and be effective in the future; and second, by illustrating how organizations can develop competence plans. This book ventures beyond the mere identification of new competencies and specifically discusses strategies a diverse group of prominent companies have used to bridge competency gaps. It presents some insights, projections, perspectives, and observations regarding the competencies required in the future. Moreover, it offers a set of testable ideas and practices suggesting how to manage new competencies.

TARGET AUDIENCE

We address this book to professionals, executives and administrators, as well as specialists in the fields of personnel training and human resources management. We believe that high-tech companies' founders and managers will benefit from our research findings, and that faculties, students, and researchers will find here some new food for thought.

Roman Batko in his chapter conducts a critical analysis of methods and techniques of organizational control applied today, in the era of liquid modernity. Since that control, more often than not, is executed nowadays by electronic surveillance equipment and computer programs, hence his term of choice – "cybercontrol". This kind of control exerts an overwhelming influence on work environment and as such demands new strategic competencies on the part of management. According to Anna Szopa, wisdom of the crowd has been applied successfully to a range of tasks, from translating text and annotating images, to collecting information and building complex software. While traditionally these tasks have been small and could be completed by non-professionals, organizations are now starting to crowdsource larger, more complex tasks to experts in their respective to build large databases and to solve problems that are difficult for machines to solve. In her chapter she presents hybrid solutions which can be viewed as optimization- leveraging machine intelligence to help improve the accuracy and efficiency of algorithms. Robert Niewiadomski and Dennis Anderson explore the impact of the rapid expansion of artificial intelligence (AI) in relations to the labor market. The authors argue that this rather optimistic, even naïve scenario, collapses while confronted with the exponential growth of AI; in particular, with the potential arrival of syneoids – robotic forms of "strong AI" possessing, or even exceeding, the full range of human cognitive abilities.

Kijpokin Kasemsap and Suan Sunandha argue that critical thinking (CT) is an important goal of modern education, therefore, it is required for the learning organizations to promote their CT and develop a learning plan to regularly check their practical advancements toward satisfying educator requirements. In their chapter they propose that promoting CT has the potential to enhance organizational performance and reach strategic goals in modern learning environments.

The Role of Living Labs in the Process of Creating Innovation is the subject of the chapter by Anna Maria Sabat and Anna Katarzyna Florek-Paszkowska. Based on the research carried out into Living Labs in Canada, it presents the essence of Living Labs as a concept facilitating generating innovation in businesses thanks to the cooperation of various actors, e.g. producers with users, inspiring the process of the development of new goods and services. The research questions raised pertain to the clarification how Living Labs create innovation in businesses.

Oya Zincir and Ayşegül Özbebek Tunç write that hypercompetitive, changing, and unpredictable environment requires different tools, solutions, dynamics and drivers according to the actual time. The Authors propose the rethinking of the McKinsey's 7S Model in the context of An Imagination of Organizations in the Future.

Mario Gonzalez-Fuentes' chapter on The Organization of the Future and the Marketing Function provides a comprehensive review of technological changes experienced by the marketing function in a company, as documented by both scholars and practitioners. It also features a thorough discussion of the ongoing academic debate regarding the new set of technical skills that have defined employability in the marketing circles for the past couple of decades and the challenges ahead for future professionals and executives.

The objective of Jan Kreft's chapter is to present the core myth of new media organizations - the myth of the Demiurge - associated with the operation of algorithms and critical analysis of myths created around it, which accompany the social, political and business role of algorithms. The myth of the Demiurge allows us to approach an organization from a broader perspective, it becomes a basis for defining particular behavior of a media organization. Such a complex myth favors the formation and supports other myths which accompany the functioning of algorithms: the myth of democratization, visibility, and accessibility.

Nina Rizun and Tatyana Shmelova argue that the appearance of the term Socio-Technical Systems (STS) is the result of a rapidly developing interaction between the social and technical factors that create a new, sometimes unpredictable, synergistic effect on the performance of the contemporary company. In their chapter they present the taxonomy of the most typical automatic control theory elements and their compliance with certain decision-making models. The authors also accept the suggestion that the instruments of collective multi-criteria decision-making and social-network analysis theory have similar mathematical and methodological bases.

Jan Kreft and Mariana Petrova write about new media space and users who are confronted with excessive contents, a phenomenon which is metaphorically referred to as information or content overload. Its manifestations are commonly known. The aggregation of media content appears to be a popular form of media market activity which is supposed to facilitate access to information in the conditions of its excessive amount. It can be defined as selection and hierarchization of information and determination of its value. The aggregation can be performed with the direct participation of editorial teams, or it can be performed automatically, with the use of algorithms.

Zofia Bednarowska and Michał Andrzej Chrzanowski regard Competences as a Core Factor Impacting Market Research Usage in Poland. The market research industry is widely recognized as an essential element in the management and development of enterprises. This article identifies a battery of concepts proving that the use of market research supports the decision-making process within enterprises, and, as a result – company development as a whole. Not only does it provide knowledge about the competitive environment and reduces uncertainty during decision-making processes, but also contributes to the knowledge of the management and the growth of the company.

In sum, we present to you the book on your future workplace. What kind of competencies will we need as workers to survive and succeed when surrounded by very smart robots? What kind of new management skills will managements need to develop and acquire in order to tame the new human-cyborg environment? And, are we still living in what we have so far considered to be Reality, or is it already the Matrix? Here you just might find suggestions, and perhaps even solutions and answers.

Roman Batko
Jagiellonian University, Poland

Anna Szopa
Jagiellonian University, Poland

REFERENCES

Brynjolfsson, E., & McAfee, A. (2014). *Second machine age: Work, progress, and prosperity in the time of brilliant technologies*. New York: W. W. Norton.

Czarniawska, B. (2013). *Cyberfactories: How news agencies produce news*. Cheltenham, UK: Edward Elgar.

Chapter 1
Panopticon – Cybercontrol in Liquid Modernity:
What Does Control Really Mean in Contemporary Management?

Roman Batko
Jagiellonian University, Poland

ABSTRACT

The chapter proposes a critical analysis of the panopticon, the model of an ideal prison devised in the 18th century by the English philosopher Jeremy Bentham, and its function in modern management. The original prison was designed in such a way that it enabled a single prison guard to watch all the inmates at all times, while the latter could never tell whether they were currently being watched or not. The idea behind the panopticon gave rise to the concept of panopticism, a philosophy developed by the French thinker Michel Foucault, which he voiced in Discipline & Punish: The Birth of the Prison (1975/1995). Analyzing the mechanism of control through the ages, he ponders the resilience and adaptability of such concepts as discipline, training and hierarchical surveillance. I intend to provide a critical analysis of methods and techniques of organizational control applied today, in the era of liquid modernity. Since that control, more often than not, is executed nowadays by electronic surveillance equipment and computer programs, hence my term of choice – "cybercontrol".

DOI: 10.4018/978-1-5225-1656-9.ch001

INTRODUCTION

Exercising control is one of the fundamental functions of management. Organizational structures, processes, and rules enable managers to control the chaos which could otherwise easily substitute cooperation. Over the recent decades, however, the control mechanisms have escalated beyond any reasonable proportions, becoming the modern embodiment of the original panopticon (Taylor, 2015). There is a growing tendency to monitor all employee activity (including non-work-related behavior) by means of oppressive calculating equipment which records every computer click, every second of a telephone conversation, and the exact time spent away from one's desk. Some companies take the control issue further still, introducing elements which can only be described as a prelude to total control: for instance, the Swedish firm Epicenter makes its employees wear microchip implants equipped with a radio locator, which – according to the company's rhetoric - are to better safeguard their privacy (Cellan-Jones, 2015). Last but not least, there are the global internet services and networks, such as Google, Facebook and numerous others, which exercise control over billions of users, gathering personal information about their age, gender and interests, and are even able to trace, by means of the GPS system, their exact location. The services also record all search histories, e-mails, calendars, photographs and other files which become, in fact, indelible traces billions of people leave behind on servers, for anyone to use and abuse. And is not the struggle man today undertakes – raging, in the name of reclaiming his freedom, against the machine, our desperate attempt to save the virtues of humanism in the face of the imminent mutation of man into the cyborg? Armin Grunwald (2014) conducted research regarding the future based on narrative methods; he noted that modern stories and fantasies about the future, if you apply the hermeneutical method, tell us not so much about the future, as about the present - our emotions, fears, and values, here and now. Hermeneutics – as postulated in Being and Time by Martin Heidegger (2008), then followed by Hans Georg Gadamer (2004) and Paul Ricoeur (1981) – sets a new ontological dimension where "being" means trying to understand oneself and the world. I too have decided to follow that track and through the hermeneutics of the term panopticon, entangled in a number of multifaceted contexts, arrive at a new translation of the word, more adequate perhaps in our era of liquid modernity - omnipotent cybercontrol. In this new context even certain serious and prestigious scientific projects such as the Global Brain, with its attributes of omniscience, omnipresence, omnipotence (Heylighen, 2013) may raise ethical doubts and be viewed as yet another tool encouraging omnipotent control.

THE PANOPTICON: THE HERMENEUTICS OF THE IDEA

When writing about the pandemic of the Black Plague in medieval and post-medieval Europe, Foucault (1995) noted that it had brought into being certain rituals of exclusion and disciplinary patterns. As a simultaneously real and symbolic embodiment of chaos, it demanded an immediate antidote in the form of order. The fear of "contagion" – with a disease, rebellion, crime, and any kind of deviation from social order, which in itself is a cultural construct and a narrative of power - - has always been a sufficient reason for implementing an extra measure of order, understood chiefly as discipline and oppression. One of the most effective methods in this respect was exclusion – the task fulfilled in society by mental institutions, correctional facilities, penal colonies, and even partly hospitals, which propagated a binary system based on opposites: normal vs. abnormal, safe vs. dangerous, etc. The panopticon is a model of an ideal prison devised in the 18th century by the English philosopher Jeremy Bentham, and constitutes an architectural, tangible rendition of that exclusion and discipline implementation. The original prison was designed in such a way that it enabled a single prison guard to watch all the inmates at all times, while the latter could never tell whether they were currently being watched or not. Bentham's key premise was as follows: authority must be felt and visible at all times, without ever being verifiable. The central situation of the watchtower, the lighting in the prison cells, all that was intended to make the prisoner aware of being continuously under surveillance, but he was never supposed to be able to tell whether he was, or was not, in fact being observed, at a given time. The idea behind the panopticon gave rise to the concept of panopticism, a philosophy developed by Michel Foucault which he voiced in Discipline & Punish: The Birth of the Prison (1995). Analyzing the mechanism of control through the ages he ponders the resilience and adaptability of such concepts as discipline, training and hierarchical surveillance. The idea of the panopticon is present in all crucial areas of our existence: in politics, where it describes the discourse of power; in ethics, where it raises the questions of freedom, privacy, and dignity; in social relations, where it alerts us to the issues of human rights and examines the responsibilities and limitations of the community. It is also present in economics, where it relates to the commodification of everything, and especially to the delusion of security and "communication," being online, in the center of events. This is strongly related to the issue of management: capitalism, commercialization, the construction and consolidation of a hierarchical structure and discourse of power, the predatory exploitation of resources (both human and natural).

In opposition to Foucault's panopticon stands the synopticon, the term introduced by Mathiesen (1997), which allows "the many to see and contemplate the few" (Mathiesen, 1997, p. 219). Since then the synopticon has evolved directly into the post-panopticon, as Bauman defined the phenomenon in his interview with David Lyon (Bauman & Lyon, 2013), and which I prefer to call cybercontrol. There is a fundamental difference between the panopticon and the post-panopticon. According to Foucault, as an architectural rendition of the idea of constant surveillance, with an overseer placed and functioning within the system, the panopticon can easily be verified and inspected from outside that system, by an independent inspector who represents the people, so in a broader perspective the panopticon will become a transparent institution whose actions will no longer be secret, but openly visible to the whole society; thus the panopticon may become a mechanism which supports the democratic process. In the post-panopticon the surveillance system is of global proportions, and the accumulation of data and the observation process are conducted by "overseers" who are not only invisible to us, but also anonymous and unknown (because they hide their identity behind a "top secret" label, or are too numerous and widespread to be tracked down) ; here we are no longer able to apply the same democratic measures of control by the people, as with the old-fashioned panopticon: the institutionalized surveillance is done through a Venetian mirror – we are being watched, but we have no idea by whom and when. If we accept Foulcault's idea (1995) that we the Antiquity was a society of the spectacle, in which the majority was able to observe the chosen few who were the actors in that society – in forums, temples, theaters, circuses, then the modern times have turned these proportions upside down – with the few being able to observe the lives of masses, any time and all the time. The citizens had withdrawn from the agoras and other public places to the seeming peace and privacy of their homes and the state had to find new mechanisms which would allow it to maintain control over its people and govern effectively in the name of the "higher good" – the integrity of the society. To that end new narratives and new rhetoric had to be born, in which individual freedom is pitted against public security, and where a permanent state of endangerment, constantly held over the citizens' heads, is a sufficient and powerful enough reason to justify invigilation. Foucault defines modern society as a society of surveillance, or a "disciplinary" society, governed by such methods and techniques which allow the authorities to control communities in the most "economical" way, i.e., attain the maximum of power with the minimum of expenditure and effort. The best evidence of the effectiveness of these methods is the barely audible voice of social protest when the state, one by one, usurps every area of our privacy.

LIQUID MODERNITY: LIQUID SURVEILLANCE

Understanding the concept of constant surveillance today requires visualising it as one of the leading phenomena of the contemporary times. Nowadays, the experience which touches and disturbs us the most is that of a constant motion, change, liquidity. Liquid modernity – the term Bauman (Bauman) proposes as a way of defining that condition of the contemporary world - points to an amorphous nature of reality. Dynamic processes may transform it into any arbitrary configuration. Communities and organizations based on social contract are becoming more and more unstable and already perceive their own identity as liquid. What is, then, liquid modernity and how do we recognize it? Andrzej Szachaj (2003) names several characteristics of postmodernity:

- A shift from the production paradigm to the consumption paradigm. Consumerist attitude is characterized by impatience and desire for instant gratification;
- A differentiation of lifestyles, subjectivism of convictions, syncretism evident in the approach to religion postulating tolerant and ecumenical attitudes;
- An aestheticization of life consisting in treating one's life and identity in terms of an experiment;
- A steadily increasing role of the broadcast and communication media which instead of presenting reality, create it on their own terms; and
- A dispersion, lack of an axiological center, a patchwork mosaic and internal incoherence of the Western culture.

In thus experienced liquid modernity, surveillance also becomes liquid. Surveillance procedures, such as tracking, controlling and constant observation have become a part of our everyday life to such an extent that we no longer notice the omnipresence of invigilation. In the preface to his interviews with Zygmunt Bauman, David Lyon (Bauman & Lyon, 2013) notes that due to the omnipresent "security requirements" and intrusive marketing of new technologies, invigilation has been spreading in all directions, infusing every area of our lives like a liquid which cannot be contained, therefore we can label it liquid surveillance. The post-panopticon no longer needs "old school" surveillance systems consisting of walls and barred windows. Today, the walls have become firewalls, and the windows are "Windows." The liquid nature of surveillance allows it to adopt different forms. What is more, by means of gigantic data bases, supercomputers endowed with calculating superpowers, advanced algorithms and statistical methods, surveillance attempts to exert control over events which have not happened yet, calling this type of activity risk management. In Philip K. Dick's 1956 science fiction short story "The Minority Report" we encounter a criminal justice agency – "Precrime" – whose task is to identify and eliminate persons who will commit crimes in the future; in today's world, such actors, operating on the premise of precognition, appear to be a very realistic and not fictional concept at all. As early as the 1980s, long before mass computerization and hyper-invigilation which followed the trauma of September 11, 2001, Shoshana Zuboff in her critically acclaimed book *In the Age of the Smart Machine: The Future of Work and Power* (Zuboff, 1988), had pointed out the connection between panoptic power and modern management strategies and the consequences ensuing from applying horizontal panopticism in an organization:

This rendering of panoptic power reflects an important evolution of the original concept. It rests on a new collectivism in which "the many" view themselves and each views "the other." Horizontal visibility is created even as vertical visibility is intensified. The model is less one of Big Brother than of a workplace in which each member is explicitly empowered as his or her fellow worker's keeper. Instead of a single omniscient overseer, this panopticon relies upon shared custodianship of data that reflect mutually enacted behavior. This new collectivism is an important antidote to the unilateral use of panoptic power, but it is not a trouble-free ideal. Horizontal transparency breeds new human dilemmas as well... (Zuboff, 1988, p. 351).

From a manager's perspective Zuboff recognizes a new danger here – when behavioral complexity and versatility of attitudes and values are replaced with „black-and white" or „zero-one" attitudes which become subject to the tyranny of gathered data interpreted by an algorithm, the focus shifts from human beings to numbers gauging every employee's efficiency. Kevin Haggerty and Richard Ericson note that "technological solutions have a strong cultural allure in Western societies and high technology has emerged as an important marker of progress and organizational reputation" (Ericson & Haggerty, 2006, p. 14), stating, in addition, that invigilation is becoming a tool which reinforces social differences, because its intensity and content differs depending on the social group it targets. Deriving from the Bentham-Foucault panopticon model is Didier Bigo's concept of ban-opticon which describes the ways of selection and exclusion of socially marginal groups, but on a global, not merely state level. States and global corporations supplying information and biometric technologies have come together united by a common cause – invigilation of individual cross-border traffic. Ban-opticon takes on a particularly sinister meaning in the face of recent migration issues in Europe. According to Bigo:

The Ban-opticon is then characterized by the exceptionalism of power (rules of emergency and their tendency to become permanent), by the way it excludes certain groups in the name of their future potential behaviour (profiling) and by the way it normalizes the non-excluded through its production of normative imperatives, the most important of which is free movement (the so-called four freedoms of circulation of the EU: concerning goods, capital, information, services and persons) (Bigo, 2008, p. 35).

Today's world - a "stock exchange of threats, fears and uneases" (Bigo, 2008, p. 45) – very readily justifies bringing to life such organizations as Europol – European police whose task is to manage the escalating sense of endangerment and risk. Ban-opticon enjoys a unique position in neoliberal societies which tend to exclude undesirable minorities, driven by the determination to protect and retain their security and peace. To this end these societies rely on profiling which allows them to exclude certain groups of people based on predictions regarding their future behavior; simultaneously they want to standardize the behavior of groups which are not subject to exclusion – by persuading them that a free, smooth, undisturbed transfer of goods, information and capital is still happening, the way it always has done. In this concept the most focus is on security, not, as in Foucault's panopticon, on discipline. The term ban-opticon also resonates with the thinking of Giorgio Agamben (1998), and especially his concept of homo sacer and disenfranchisement, however I do not intend to develop this path of thought here, since it would transcend the scope of this article.

My goal is not to provide a detailed map of state and corporate hyperactivity with regard to constant surveillance of its citizens, nor do I intend to copy here that overwhelming amount of data, emerging daily, to document this phenomenon. My wish is to capture the general picture by putting together various bits and pieces of information fabric, as if creating a bricolage, and hoping the image, when seen from a distance, will leave no doubt as to what it depicts.

The Internet is transparent –such is the official message. True - but it is the transparency of the Venetian mirror (Orliński, 2013). We are being watched all the time, but we can hardly learn anything – sometimes nothing at all - about those who are watching us. We live in an age of profound epistemic asymmetry. The hypervisibility of the individual citizen – trackable through all sorts of smart devices – is matched by the growing hyperinvisibility of all other players (Morozov, 2014). Global internet services and networks such as Google, Facebook and numerous others exercise control over billions of users, gathering personal information about their age, gender and interests, and are even able, by means of the GPS system, to trace their exact location. The services also record all search histories, e-mails, calendars, photographs and other files which become, in fact, indelible traces billions of people leave behind on servers, for anyone to use and abuse. 4 December, 2009: Google announces "personalized search". Now – entices us Google – if you are looking for something using us, we have a better opportunity to provide you with the most relevant results, based on the 180 days back search of your browser. The answer to your search is determined by your previous activity, location, and other indicators on which Google is not required to divulge information, because, according to Google, it is confidential. Google will show you the answers on websites which earn it the most money through advertising. It it obvious that in order to be able to activate this kind of „personal search" Google keeps constantly tracking our activity online (all activity) and in this way can put together a priceless, from the marketing perspective, profile of a consumer who, on top of the information he wants, also gets to see "personalized," tailor-made advertisements on the websites he visits.

What is the role of "cookies" and other Big (Brother) Data in all this? Maximilian Schrems, an Austrian lawyer, aware that Facebook headquarters are located in Ireland, has been trying since 2011 to make the corporation comply with the European data protection law. He and a group of colleagues have demanded that the company release to them all the data it has gathered on them. They received printouts of up to 1 222 pages per person, in 57 separate categories! What is more, they noticed that a number of data were missing – among them, the data which Facebook must undoubtedly process, such as facial recognition. They are still waiting for these data ("Europe versus Facebook," 2015). To demonstrate how easy it is to access confidential information let us just quote a report which shows that in the United States alone more than 850 000 employees have access to the "top clearance," while 1 300 government agencies and 2 000 private companies collect intelligence and also have access to the most confidential information (Orliński, 2013). The borders of Europe today are becoming more and more technological. Data bases and biometric profiles transform the bodies of migrants into sources of information, invading their privacy and objectifying them in the process (Dijstelbloem & Meijer, 2011). Recently, Europe on an unprecedented scale has been building a virtual fortress, pumping unimaginable amounts of money into electronic monitoring systems and cybercontrol devices and services of all kinds. It is just a business like any other, as the numbers below prove: the turnover in the security services market in the years 2001-2011 increased from 10 to more than 100 billion euros. In Europe alone, the company employs 180 000 people, and the companies' turnover amounts to 30 billion euros per year (Tan, 2014). And this seems a powerful enough argument to begin a discussion on the mutual dependence of management and cybercontrol.

CYBERCONTROL AND MANAGEMENT

According to Foucault, man is held captive by the control machine equipped with an apparatus for observation, recording and training. Today these very concepts, if slightly euphemized, constitute the cornerstone of scientific management. This, however, has no bearing on the fact that control is one of the fundamental functions of management. Organizational structures, processes and rules, competency profiles, job descriptions and work flow enable managers to control the chaos which could otherwise easily substitute cooperation. In the famous passage in Don Quixote where he fights windmills (an economic artifact) thinking that they were mythological evil giants, the machines defeated him, the representative of a medieval spirit and holder of an old-fashioned and cavalier code of honor (Czarniawska, 2005). Modern windmills are cybercontrol artifacts. Control, and synonymous expressions such as: check, superintend, supervise, counteract, monitor, regulate, verify, restrain, discipline, auditing, validate – are among the most commonly used by management science. The very fact that English language dictionaries as one of the synonyms to "control" give the word "manage" shows how close these concepts are on both semantic and pragmatic levels. One of the most overused terms and tools in management today is "audit" While Patricia Broadfoot has rightly noted that "evaluations, assessments, audits, accounts are part of everyday human interactions" (Broadfoot, 1996, p.3), Michael Power has described a phenomenon which he calls the audit society. It is a society which is "engaged in constant checking and verification …, in which a particular style of formalized accountability is the ruling principle" (Power, 1997, p.4). There is a growing tendency to monitor all employee activity (including non-work-related behavior) by means of oppressive calculating equipment which records every computer click, every second of a telephone conversation, and the exact time spent away from one's desk (Taylor, 2015). Some companies take the control issue further still, introducing elements which can only be described as a prelude to total control: for instance, the Swedish firm Epicenter makes its employees wear microchip implants equipped with a radio locator, which – according to the company's rhetoric – are there "to better safeguard their privacy" (Cellan-Jones, 2015). A question worth asking in this context is why employees agree to these perverse language games and let themselves be treated in this way by management? On the organizational level they do it out of fear – they are afraid of losing their job, however, not only in the traditional sense of being let go and replaced by someone else; they are afraid of becoming obsolete as humans and being replaced with machines – so much more efficient and complacent; presented with such alternative they do not mind wearing microchips and becoming a little bit like machines themselves. On the level of the state people are persuaded that sacrificing an inch of their privacy is a small price to pay for security. Governments argue that they need total control, because only total control guarantees complete protection and peace of mind. Give us your freedom, they say, and we will give you security. The EU Security Program – Horizon 2020, has been designed to exercise this kind of control over EU citizens on a gigantic scale. The European Commission activities on trust and security cover the areas of Cybersecurity and Online Privacy in order to, according to the document, fight crime, illegal trafficking and terrorism, including understanding and tackling terrorist ideas and beliefs, improve cyber security, ensure privacy and freedom (including the Internet environment), and enhance the societal legal and ethical understanding of all areas of security, risk and management. This is a rhetorical device, and it should be considered as a language game. One of the elements of the game is business strategy by fear, where the security of the citizens acts as an effective rhetorical ruse. By creating an illusion that Europe is in a state of war – and these are not the realistic fears of normal citizens, but a marketing device - it targets new customers by arousing fear in them in order to sell a product. Other clever control strategies employed by corporations have made us unwitting guardians of ourselves. One of them is self-management, a particularly sophisticated form of exercising control, which says: be the manager of your own life, and be your own overseer.

An interesting study, especially in the context of management, was conducted by Oscar Grandy in The Panoptic Sort: A Political Economy of Personal Information (1993). He draws attention to the panoptical character of the segregation mechanism which marketing departments in corporations use when profiling their data bases. Such segregation greatly facilitates the application of particular marketing strategies, designed to provoke desirable consumer behavior. The strategy reflects market rationalization and helps isolate certain consumer "targets" (the term routinely used in marketing instead of "clients," "consumers" or simply "people"). The mechanism relies on feedback and is expected, on the one hand, to target the most desirable group of consumers and mark it as an "elite" by providing it with certain unique and prestigious (described so in the marketing jargon) goods and services; on the other hand, the elitist status of the group is supposed to generate in its members an urge for self-verification (am I still worthy, do I still deserve the "elite" member status?) and, simultaneously, an urge to prevent non-elite consumers from joining the elite group – thus protecting their own elitist satisfaction at belonging to the elite.

In the context of management and control it seems imperative to mention the phenomenon which Bauman defines as adiaforisation (Bauman & Lyon, 2013), where systemic procedures and actions are implemented without the slightest reflection regarding the ensuing consequences. Thus responsibility becomes diffused, turning into what Hannah Arendt describes as a transferable responsibility, and, as such, no man's responsibility, a condition which may crucially affect decision-making. At the same time, modern technologies which have replaced man at data gathering and processing, have also replaced him as a decision-maker, "liberating" him from taking responsibility for his actions, and, subsequently, experiencing a sense of guilt. Unmanned drones which on their own, without human interference, mark targets for destruction, will not be held responsible for killing innocent people in case of a technological error. Simultaneously, the drone operator can easily justify the error and his lack of any moral qualms by quoting an overwhelming amount of data he must take in and process, the task he is incapable of performing faultlessly

(Bauman & Lyon, 2013), because he is only human. The guilt will be transferred from man to machine, and machines cannot be tried for homicide.

The removal, or transference of responsibility is characteristic of the new managerial revolution happening now: middle management is no longer held responsible for the employees' performance, it is the workers themselves who – through strict self-control mechanisms - become entirely responsible for it. On the one hand, we have these unbelievably advanced technological systems, electronic microchips registering employee comings and goings, employee activity logs in computer systems, automatically generated goal achievement, production and sales reports, automatic programs handling client correspondence, etc. On the other, perhaps more significant, the new self-motivation strategy has replaced repression, and it seems that temptation and seduction are more important in today's organizations than normative regulations, while the escalation of employee desires and expectations has supplanted the expensive and conflict-generating policy of surveillance. This "privatization" of surveillance makes employees build - of their own accord - their own personal panopticons, with thick walls and watchtowers from which they watch themselves constantly, waiting for fulfillment and prizes (Bauman & Lyon, 2013).

CONCLUSION

Without question, the phenomenon of cybercontrol raises in contemporary man immense doubts and stirs up anxiety regarding the preservation of our basic rights to freedom and privacy. Even though Kevin Haggarty claims the metaphor of the panopticon has lost its hermenutic significance today, I am more inclined to trust Bauman, who says the panopticon is doing better than ever, flexing its electronically enhanced, cyborgized muscles whose power and magnitude defies not only Bentham's imagination, which would seem obvious, but also Foucault's (Bauman & Lyon, 2013). However, if there is hope, it is to be found in liquidity – for even cybercontrol will not succeed at controlling and containing all forms of activity – both civil and consumer. In liquid modernity constant, never-ending reconfiguration of social order is the norm, and freedom, when suppressed in one place, without failing tends to resurface in another. This phenomenon is already very much present in our reality; one of its manifestations is "interactive democracy" (Rosanvallon, 2011) which is characterized not so much by lack of civil involvement as by its dispersion. New citizens are no longer satisfied with just being able to vote in an election, which is what participation in representative democracy until recently used to mean, but they wish to express themselves fuller and be heard better and to that end they choose to become involved in various "counter-democratic" activities:

They look for information, trying to compel the government to explain and justify its actions. They challenge it and watch its every move. The scope of interactive democracy is much vaster than that of the system of electoral representation … Informal networks create invisible institutions which play a significant part in the makeup of contemporary democracy (Rosanvallon, 2011, s. 22–23).

One of such "invisible institutions" is the movement called Avaaz, with over 42 million Internet users declaring themselves as members. Avaaz defines itself as "the campaigning community bringing people-powered politics to decision-making worldwide"(Avaaz, 2015). Some of the issues they campaign for, or against, are persecution of civilians in war-waging countries, destruction of natural environment, or corporate greed. Free and constant exchange of information and opinions, posting comments online as events take place, expressing approval or, more frequently, disapproval of politicians' decisions, and in some drastic cases - leaving the virtual world and taking to the streets in order to demonstrate their power and disorient "the powers that be" with their organizational speed and efficiency – these are the main characteristics of this phenomenon. "Interactive democracy" is making its presence felt also in Europe - a few years ago protests against ACTA spread like wildfire throughout the continent. "The development of the Internet - concludes Rosanvallon – has shaken the balance between the hidden and the visible. Today, everything is out in the open" (Rosanvallon, 2011, p. 26). When explaining the meaning of an –archy Levinas proposes that and suggests it – something - had existed prior to arche (Lévinas, 2008). A rebellion against arche, then, means an active protest against continuation. The purpose of the negation is transgression, a subversion of the established order. In this context, the opposition between (an) and (archy) may be interpreted not so much as a call to "set the world on fire," but a voice advocating a return to the source.

REFERENCES

Agamben, G. (1998). *Homo Sacer: Sovereign Power and Bare Life* (1st ed.). Stanford, CA: Stanford University Press.

Avaaz. (2015). *Avaaz*. Retrieved November 24, 2015, from http://www.avaaz.org/en/index.php

Bauman, Z. (2000). *Liquid Modernity*. Cambridge, UK: Polity Press.

Bauman, Z., & Lyon, D. (2013). *Liquid Surveillance: A Conversation*. Cambridge, UK: Polity.

Bigo, D. (2008). Globalized (In)Security: The field and the Ban-Opticon. In D. Bigo & A. Tsoukala (Eds.), *Terror, Insecurity and Liberty: Illiberal Practices of Liberal Regimes after 9/11* (pp. 10–48). London: Routledge. doi:10.4324/9780203926765.ch2

Broadfoot, P. M. (1996). *Education Assessment and Society*. Buckingham, UK: Open University Pres.

Cellan-Jones, R. (2015, January 29). *Office puts chips under staff's skin*. Retrieved November 20, 2015, from http://www.bbc.com/news/technology-31042477

Czarniawska, B. (2005). Don Quixote and Capitalism in Poland. On the Cultural Context of Organising. In B. Czarniawska-Joerges & P. de Monthoux (Eds.), *Good Novels, Better Management: Reading Organizational Realities in Fiction*. Routledge.

Dijstelbloem, H., & Meijer, A. (Eds.). (2011). *Migration and the New Technological Borders of Europe*. Houndmills, UK: Palgrave Macmillan. doi:10.1057/9780230299382

Ericson, R. V., & Haggerty, K. D. (2006). The New Politics of Surveillance and Visibility. In R. V. Ericson & K. D. Haggerty (Eds.), *The New Politics of Surveillance and Visibility*. Toronto: University of Toronto Press.

Europe versus Facebook. (2015, November 24). Retrieved November 24, 2015, from http://www.europe-v-facebook.org/

Foucault, M. (1995). *Discipline & Punish: The Birth of the Prison*. New York: Vintage Books.

Gadamer, H.-G. (2004). Truth and Method (2nd ed.). London: Bloomsbury Academic.

Gandy, O. H. (1993). *The Panoptic Sort: A Political Economy of Personal Information*. Boulder, CO: Westview Press.

Grunwald, A. (2014). The hermeneutic side of responsible research and innovation. *Journal of Responsible Innovation*, *1*(3), 274–291. doi:10.1080/23299460.2014.968437

Heidegger, M. (2008). *Being and Time (Reprint edition)*. New York: Harper Perennial Modern Classics.

Heylighen, F. (2013). Return to Eden? Promises and Perils on the Road to a Global Superintelligence. In B. Goertzel & T. Goertzel (Eds.), *The End of the Beginning: Life*. Society and Economy on the Brink of the Singularity.

Lévinas, E. (2008). *Bóg, śmierć i czas*. Kraków: Wydawnictwo Znak.

Mathiesen, T. (1997). The Viewer Society Michel Foucault's `Panopticon' Revisited. *Theoretical Criminology*, *1*(2), 215–234. doi:10.1177/1362480697001002003

Morozov, E. (2014, July 30). *Digital Surveillance. Like clueless guinea pigs*. Retrieved from http://www.faz.net/aktuell/feuilleton/debatten/the-digital-debate/digital-surveillance-like-clueless-guinea-pigs-13070758.html

Orliński, W. (2013). *Internet. Czas się bać*. Warszawa: Wydawnictwo Agora.

Power, M. (1997). *The Audit Society: Rituals of Verification*. Oxford, UK: Oxford University Press.

Ricoeur, P. (1981). *Hermeneutics and the Human Sciences: Essays on Language, Action and Interpretation*. Cambridge, UK: Cambridge University Press. doi:10.1017/CBO9781316534984

Rosanvallon, P. (2011). Demokracja interakcyjna. *Res Publica Nowa*, *16*(206), 18–29.

Szahaj, A. (2003). Co to jest postmodernizm? In M. A. Potocka (Ed.), Postmodernizm: teksty polskich autorów (pp. 41–52). Kraków: "Bunkier Sztuki".

Tan, S. (2014). *The European Security Complex*. documentary film.

Taylor, S. S. (2015). Controls and Constraints. *Organizational Aesthetics*, *4*(1), 1–3.

Zuboff, S. (1988). *In the Age of the Smart Machine: The Future of Work and Power*. New York: Basic Books.

Chapter 2
The Influence of Crowdsourcing Business Model into Artificial Intelligence

Anna Szopa
Jagiellonian University, Poland

ABSTRACT

We observe situation in which human computation processes leverage the abilities of people via the Web to perform complex computation. In this context human intelligence is harnessed to solve computational problems that are beyond the scope of existing Artificial Intelligence (AI) algorithms. The aim of this paper is to achieve an in-depth understanding of the influence of crowdfunding business process into Artificial Intelligence and what kind of computational problems can be efficiently answered using human computation. Wisdom of crowd has been applied successfully to a range of tasks, from translating text and marking images, collecting information to building large databases and to solve problems that are difficult for machines to solve. Presented hybrid solutions will be considered as optimization- leveraging machine intelligence to help improve the accuracy and efficiency of algorithms.

DOI: 10.4018/978-1-5225-1656-9.ch002

INTRODUCTION

Progress in information, communication and multimedia technologies and the increasing expansion and use of the internet, are generating gradual innovation in diverse areas, leading to the creation of new business processes. The innovation process is currently undergoing significant change in many industries, it very visible in Information Technology (IT) sector. Recent developments in technologies and the usage of Artificial Intelligence techniques make the IT more intelligent and provide higher-level services to its users. Therefore, building an appropriate infrastructure for intelligent agents to run around the IT performing complex actions for their users is the biggest challenge for organizations of the future (Hendler, 2001). An enormous potential exists in rich collaboration between humans and computers. Countless open computer science problems remaining in Artificial Intelligence, can be readily solved by humans with the usage of their creativity, planning abilities, visual processing, and verbal tasks (Quinn & Bederson, 2011).

The new computer world has created a virtual collective intelligence- helped large groups of people connect and collaborate. According to Von Hippel a large number of users of a given technology comes up with most innovative ideas (Von Hippel, 2005). A decentralized, disorganized group of people, thinking independently can use "crowd intelligence" to make judgments and solve problems that would be very difficult to make using only one person (Surowiecki, 2005). Crowdsourcing business model have had important impact into AI it has been applied to a range of tasks such as: collecting information, translating texts, annotating images, building complex software. While traditionally these tasks have been small and could be completed employees, organizations are now starting to crowdsource larger, more complex tasks to experts in their respective fields. Presented combination of crowd and computers offers another point in the tradeoff space where solutions are faster than individual human efforts due to diversity and network effects of human participation (Long Tran-Thanh, 2014).

The strategy of combining computers and humans by delegating parts of the problem to the crowd is conducted via the internet, in this process a computer system having global knowledge of the problem to be solved requests the crowd to resolve it then proposed results are usually checked and aggregated by the computer system in presented process humans help computers to do computer-type projects. Finally, group of these participating humans later became the first professional programmers (Malone et.al., 2009)

In most advance processes of artificial intelligence (AI) crowd builds large databases of common sense facts. That's possible due to the fact humans acquire a great deal of common sense knowledge. Large databases of spread humans' knowledge collect information into knowledge bases and ontologies using data mining. Presented process influences of recent innovations in the competitiveness of organizations (Zaidan & Callison-Burch, 2011).

Background

The idea of humans and computers collaboration was implemented in 1960 in the article, "Man-Computer Symbiosis," (Licklider, 1960). However recently field of human computation is being advanced by researchers who explore human intelligence to solve computational problems that are beyond the scope of existing Artificial Intelligence (AI) algorithms (Michelucci, 2011; Bederson, 2010; Brachman et.al, 2011; Quinn, 2011). According to researchers from University of Maryland Alexander J. Quinn and Benjamin B. Bederson human computation is related to, but not synonymous with terms such as collective intelligence, crowdsourcing, and social computing. The term human computation was primarily used in philosophy and psychology literature, as well as more recently in the context of computer science theory. And it refers to a set of operators in addition to traditional computation. Lav von and Ahn emphasize role of humans and define the process as "...a paradigm for utilizing human processing power to solve problems that computers cannot yet solve ... the idea of using human effort to perform tasks that computers cannot yet perform" (Law & von Ahn, 2009). Chandrasekar, Chi, Chickering, Ipeirotis, Mason, Provost and Tam take under consideration tasks that are completed, defining human computation as "...a new research area that studies the process of channeling the vast internet population to perform tasks or provide data towards solving difficult problems that no known efficient computer algorithms can yet solve" (Chandrasekar et.al., 2009). Quinn and Bederson underline collaboration "A computational process that involves humans in certain steps ... systems of computers and large numbers of humans that work together in order to solve problems that could not be solved by either computers or humans alone" (Quinn & Bederson, 2009). And Law highlights complexity of the process "...a new area of research that studies how to build systems, such as simple casual games, to collect annotations from human users" (Law et.al., 2009). Most comprehensive definition of human computation present Brachman, Cohen, and Dietterich defining it as "intelligent systems that organize humans to carry out the process of computation—whether it be performing the basic operations (or units of computation), taking charge of the control process itself (e.g., decide what operations to execute next or when to halt the program), or even synthesizing the program itself (e.g., by creating new operations and specifying how they are ordered)" According to authors still many problems are simple for humans to solve, but complicated for even the most sophisticated computer algorithms. Those problems are mainly referring to Artificial Intelligence (AI) and include perceptual tasks such as object recognition, classification, natural language analysis- language translation, sentiment analysis and complex cognitive tasks including planning, analyzing and reasoning. Presented process stands new opportunities for artificial intelligence leveraging the human computation systems, tackling problems that neither humans nor machines alone can resolve where crucial role plays collective intelligence (Brachman et. al., 2008). According to Levy knowledge communities voluntary and temporary, create collective intelligence, he claims that "thinking community search, inscribe, connect, consult, explore … Not only does the cosmopedia make available to the collective intellect all of the pertinent knowledge available to it at a given moment, but it also serves as a site of collective discussion, negotiation, and development. Unanswered questions will create tension within cosmopedic space, indicating regions where invention and innovation are required" (Levy, 1977).

Crowdsourcing Business Model

The term crowdsourcing, was first used in a Wired magazine article by Jeff Howe (Hove, 2006) and the subject of his book (Hove, 2008), and it was derived from outsourcing. Howe defines crowdsourcing as: "the act of taking a job traditionally performed by a designated agent (usually an employee) and outsourcing it to an undefined, generally large group of people in the form of an open call." (Hove, 2008). The transformation of the customer from a passive client into a highly active one who wants to participate in all production and development process. The term is also defined also as "the outsourcing of tasks to the general Internet public" (Kleemann et al.2008), "describes a new web-based business model" (Brabham, 2008). In recent years, a wide range of organizations, have turned to a new emerging labour market to achieve their operating objectives. Using the internet, they announce request to a global audience this so-called crowdsourcing promises considerable flexibility, as it quickly connects employers and workers. The development of crowdsourcing has been leveraged for working with people in a larger geographic scope, as well as task demands for accomplishing human intelligence tasks (Doan et.al, 2011).

According to Brabham crowdsourcing harnesses the creative solutions of a distributed network of individuals through open call for proposals. He points out that the public can help design products, produce memorable commercials and images, and that it outperforms the industry faster and cheaper than ever (Brabham, 2008). It includes customers, considered as "co-workers", and various stakeholders who are not employees of the organization, in the production process (Rieder & Voß, 2010). Crowdsourcing processes includes four dimensions: pre-selection of contributors, accessibility of peer contributions, aggregation of contributions, and remuneration for contributions. The classification scheme focuses exclusively on an organizational perspective and on the mechanisms available to these organizations (Geiger et.al, 2011)

Hove in his publication "Why the power of the crowd is driving the future of business" classifies the applications of crowdsourcing as (Howe, 2009):

1. Collective intelligence when people solve problems and provide new insights and ideas.
2. Crowd creation when people create various types of content and share it with others.
3. Crowd voting when people give their opinion and ratings on ideas, products, or services, as well as evaluate information.
4. Crowdfunding when people raise money for investment, donations, or for micro-lending of funds.

Crowdsourcing process is very broad and includes everything from the design of a product or process, creating content, product development, solving problems, advertising, quality monitoring, etc. (Howe, 2006; Brabham, 2008; Kleemann et al., 2008). Crowdfunding has developed mainly through the expansion of social networks, which has allowed labor to be outsourced to the public with very diverse kinds of remuneration and motivation mechanisms for contributions. (Corney et al., 2009). Crowdsourcing business model is related to a few concepts that may be used to render similar concepts, including: open source, open innovation, user innovation, and co-creation. Open source software allows users to freely access, use, modify and redistribute software products and their source code under a public license (Howe, 2006). The idea of open innovation assumes that companies should use external ideas to create innovations, share risks, and improve productivity and competitiveness. Chesbrough indicates several methods for importing external ideas: consulting, employing focus groups, conducting joint design sessions and using other collaboration methods that were in use for decades to facilitate open innovation (Chesbrough, 2011). The last concept- co-creation describes a joint effort of producers and consumers aiming at developing new products or services and significates working collaboratively with a group of people with specialized skills or talents to come out with new ideas (Ramaswamy, 2000).

There is no doubt that crowdsourcing business model is a phenomenon involving the use of volunteers to accomplish complex tasks however it has couple drawbacks that should be taken under consideration in context of Artificial Intelligence. The first problem that may occur is the lack of contributors. Crowdsourcing business model relies on voluntary participation and benefits from network externalities and reaching of essential crowd workers not always is guaranteed. Second issue is the knowledge correctness- the model is relevant in fields where knowledge is highly codified - as it is in case of artificial knowledge The third problem that can occur while implementing crowdsourcing model is the statement formulation. When addressing a request to the crowd, the firm seeks to obtain a number or candidate solutions to its problem. If the request to the crowd is wrong defined, the process is very likely to lead to non-satisfactory contributions as result feedback cannot be implemented (Teece, Pisano, 1994). As stated above the problem and requested definitions should be major concerns for companies that implement crowdsourcing business model.

THE IMPORTANCE OF CROWDSOURCING BUSINESS MODEL FOR ARTIFICIAL INTELLIGENCE

The potential of intelligent machines associated with human minds, such as learning and problem solving cannot be conceived without the participation of people. In this sense, the source of competitive advantage of artificial intelligence can be found in crowd wisdom- where knowledge is the principal feature.

A growing number of businesses implementing Artificial Intelligence projects crowdsource work on large-scale that require many hours of activity by experts in a particular field. In this context expert crowdsourcing is used for the creation and development of large software applications. Accordin to Long Tran-Thanh, Sebastian Stein, Alex Rogers and Nicholas R. Jennings researchers who investigate the problem of expert crowdsourcing we can observe growing popularity of this approach. According to their data in 2013 oDesk platform had 2.5 million registered workers, while Freelancer had 6.7 million. Presented expert crowdsourcing instead of doing small units of work rises new challenges (Tran-Thanh, 2011).

Crowdsourcing business model enables knowledge creation, sharing and learning (Shiffman, 2008), therefore is able to facilitate central problems of artificial intelligence that include (Luger, Stubblefield, 2004):

1. **Deduction, Reasoning, Problem Solving:** Developed algorithms that imitate reasoning that humans use when they make logical deductions,
2. **Knowledge Representation:** The problems that are expected to solve require extensive knowledge about the world,
3. **Planning:** Intelligent machines should be able to set goals and achieve them. Therefore, need a way to project the future and be able to make choices that maximize the of the available choices,
4. **Learning:** Studying of computer algorithms that improve automatically through experience,
5. **Natural Language Processing:** The ability to read and understand the languages that humans speak. A sufficiently powerful natural language process-

ing system enables natural language user interfaces and the acquisition of knowledge directly from human,
6. **Social Intelligence:** Recognizing, interpretation, processing and stimulation of human affects. The machines interpret the emotional state of humans and adapt its behavior to them, giving an appropriate response for those emotions,
7. **Creativity:** Addressed theoretically (from a philosophical and psychological perspective) and practically (via specific implementations of systems that generate outputs that are considered creative),
8. **General Intelligence:** Combining all the skills presented above.

Artificial intelligence is associated with a great number of problems however those stated above can and even should be solved with the usage of crowdsourcing business model. Fits of all because AI requires human knowledge from the fields of computer science, mathematics, psychology, linguistics, philosophy, neuroscience and artificial psychology (Russell, Stuart, Norvig, 2003) and crowdsourcing is associated with board knowledge. In addition, creates environment where the knowledge is complex expanding and located in networks rather than individual activities (Kodama, 2005). Presented concept o is essential since it makes information more meaningful to crowd by making it more understandable to machines (Feigenbaum et al., 2007). Furthermore, possessing information from crowd intelligent machines read, understand, interrelate, and can manipulate data. Gathered information is essential for adapting and personalizing products, or services. The concept means a radical transformation with critical changes in the technology itself (Garrigos et al., 2011). Such collaboration with crowd represent a flexible source of information and knowledge. Strategically, the most important change in presented relation is the importance of meaning in the new environment. It stands essential tool for finding out what customers are demanding, discovering technological trends, innovations and expert opinions. Furthermore, crowdsourcing in presented model promotes innovations "drives all of society and corporations to work faster, create and manage more interdependencies, and operate on global markets" (Kalpic, Bernus, 2006).

Practical Aspect of Crowdfunding Business Model and Its Impact into Artificial Intelligence Algorithms

To outline practical aspect of crowdfunding business model and its impact into Artificial Intelligence algorithms this part will present three platforms Amazon's Mechanical Turk that supports the large-scale distribution of micro-tasks to human workers, Ushahidi which provides software for collecting information from the public, mainly during crisis situations, and Zooniverse that hosts a range of large citizen science projects.

Amazon's Mechanical Turk

Amazon's Mechanical Turk is a platform dedicated to create annotated data for natural language processing application including human-mediated translations, editing rates (HTER) and creation multiple reference translations- humans perform manual evaluations of machine translation quality using the web service. Tasks dedicated to humans at the moment cannot be completed by computers and include from data extraction from images, filtering adult content, audio transcription, filtering adult content, moderating blog comments to providing feedback on relevance of results for a search query (Huang et.al, 2010). Mechanical Turk mechanism consists of two types of players: requester and workers- the independent contractors. The jobs offered on Mechanical Turk refer to Human Intelligence Tasks (HITs). Amazon's Mechanical Turk, online labors are paid small sums of money to complete tasks (Winter, Siddharth, 2012).

Presented platform stands an inexpensive way of gathering human judgments and annotations for a wide variety of tasks- which can be utilized to ensure translation quality. According to Chris Callison-Burch from Center for Language and Speech Processing, Johns Hopkins University Baltimore by combining the judgments of many non-experts the system can achieve the equivalent quality of experts. According to the researcher, the platform is very useful for a wide variety of machine translated tasks like: word creation, verifying the accuracy of documents especially sentence alignments, performing non-simulated active learning experiments for statistical machine translation, and even collecting training data for low resource languages (Callison-Burch, 2009). Rion Snow, Brendan O'Connor, Daniel Jurafsky, and Andrew Y. Ng. investigating certain tasks including word sense disambiguation, word similarity, textual entailment, and temporal ordering of events, observed that a correlation between non-expert and expert annotators can be achieved by combining the judgments of multiple non-experts, example by voting (Snow et.al., 2008). To provide high quality services Amazon's Mechanical Turk provides special mechanisms that include option of rejecting the work of individual employee and blocking from doing future work for a requester. Requesters can require a particular set of qualifications, such as sufficient accuracy on a small test set or a minimum percentage of previously accepted submissions. Furthermore, requesters can design custom qualification tests that must be completed before working on a particular project. To secure employees the platform established a Firefox plug in called Turkopticon that provides ratings of how good the requesters are in terms of communicating with workers, being generous, and paying (Amazon Mechanical Turk).

Ushahidi

Ushahidi stands an open source crisis-mapping application, and is being used for humanitarian purposes around the world. Ushahidi supports and trains community based organizations to engage with local governments. The key component of Ushahidi is the ability to use mobile phones as a means of sending crisis incidents and receiving updates. The platform, Crowdmap, is amongst the first Web 3.0 technologies using both mechanical and human means to order information and present it in a layered mode. The platform specializes at turning small parts of data into information by using mapping. Ushahidi is an open source technology, which is linked to detailed reports, including the level of validity or authentication. In this process collected information can be view online by anyone. The platform exploits new forms of encoding, which in turn afford new outcomes and have material impacts. Human vision can compare small and large-scale structures at the same time, deal with layers and separations and intuitively find patterns. The platform has its qualification system- the information is verified before plotting on the map - decision makers don't receive unproven crowdsourced or web based information. Then checked information is visualized. Open modification of the system allows for multiple types of visualizations of this data (Ushahidi).

Zooniverse

Zooniverse is a research platform in a range of subjects from astronomy to zoology where human-based analysis still exceeds that of machine intelligence). Zooniverse stands a cluster of projects that source volunteer contributors to analyze and interpret huge datasets that are too complex to interpret using computer algorithms (Zooniverse). The platform engages workers without having a specialist knowledge or background in science to complete sufficiently simple tasks (Silvertown, 2009). Zooniverse projects invite the crowd to participate in genuine data analysis at a scale that researchers cannot accomplish. The whole process consists of two teams: the first the core team of developers and data scientists (has headquarters at Oxford University and the Adler Planetarium) the second crowd workers that implement the most rudimentary algorithms for functionality (Smith et.al, 2009). Crowd workers are shown how to perform required analysis via a tutorial to be able to identify, classify, mark, and label them as researchers would do. Research data is shown to the crowd in the form of images, audio and video via one of the Zooniverse websites (Simpson et.al, 2014).

CONCLUSION

This paper has attempted to analyze the influence of crowdsourcing business model into Artificial Intelligence, and the development of human-computer interaction. The study examined literature and describes the transformation of technologies, and the new business model emerging in the new context of Artificial Intelligence. The chapter also stressed the importance not only of the technology, but of the essential use of crowdsourcing and the management of cooperation. The principle for human computation is that given conditions, an unprecedented number of human workers can be mobilized to solve problems that are difficult for machines to solve. However, since human computers are still more costly than machines, hybrid solutions, involving both human and machine intelligence, may advance future human computation systems. The crowdsourcing of people in artificial intelligence processes has proven to be a technique capable of making a valuable contribution. The crowd combination of intuition with abilities in pattern recognition and analysis can outstrip the performance of sophisticated automatic systems. Furthermore, presented process evaluates and is becoming very popular presented above Zooniverse platform have collected more than 300 million data analyses from over 1 million volunteers providing rich datasets for not only the individuals working to produce research from their projects but also the machine learning and computer vision research communities the platform is evolving into a system capable of integrating human and machine intelligence in a live environment. Author is aware that this chapter is merely a first step in analyzing the impact of crowdsourcing on artificial intelligence and recognizes the limits of this exploratory analysis and the fact that all of these transformations require further analysis. For instance, future research should concentrate on the impact specific functions, or the most appropriate skills for developing. In addition, the work can open up new possibilities for the use of crowdsourcing business model by all types of organization, and considers this to be a crucial step toward the success of crowdsourcing business model by effectively utilizing the collaborative intelligence of internal and external stakeholders, as well as the general public.

REFERENCES

Amazon Mechanical Turk. (2016). Available at https://www.mturk.com/mturk

Brabham, D. (2008). Crowdsourcing as a model for problem solving. an introduction and cases. *Convergence (London)*, *14*(1), 75–90. doi:10.1177/1354856507084420

Brabham, D. (2013). *Crowdsourcing*. Boston, MA: The MIT Press.

Brachman, R., Cohen, W., & Dietterich, T. (2011). *Synthesis Lectures on Artificial Intelligence and Machine Learning*. Morgan Claypool Publishers.

Brooks, F. P. (1995). *The Mythical Man-Month: Essays on Software Engineering*. Addison-Wesley Pub. Co.

Callison-Burch, C. (2009). *Fast, Cheap, and Creative: Evaluating Translation Quality Using Amazon's Mechanical Turk. In Proceedings of the 2009 Conference on Empirical Methods in Natural Language Processing* (pp. 286–295). Singapore: Association for Computational Linguistics. doi:10.3115/1699510.1699548

Chandrasekar, R., Chi, E., Chickering, M., Ipeirotis, P. G., Mason, W., Provost, F., & Tam, J. (2010). Front matter.*Proc. SIGKDD HCOMP*.

Chesbrough, H. (2011). Bringing open innovation to services. *MIT Sloan Manag. Rev.*, *52*(2), 85–90.

Corney, J., Torres-Sanchez, C., Jagadeesan, A., & Regli, W. (2009). Outsourcing labour to the cloud. *International Journal of Innovation and Sustainable Development*, *4*(4), 294–301. doi:10.1504/IJISD.2009.033083

Doan, A., Ramakrishnan, R., & Halevy, A. (2011). Crowdsourcing systems on the world-wide web. *Communications of the ACM*, *54*(4), 86–96. doi:10.1145/1924421.1924442

Feigenbaum, L., Herman, I., Hongsermeier, T., Neuman, E., & Stephens, S. (2007). The semantic web in action. *Scientific American*, *297*(6), 90–97. doi:10.1038/scientificamerican1207-90 PMID:18237102

Garrigos, F., Gil, I., & Narangajavana, Y. (2011). The impact of social networks in the competitiveness of the firms. In A. M. Beckford & J. P. Larsen (Eds.), *Competitiveness: Psychology, Production, Impact and Global Trends*. Hauppauge, NY: Nova Science Publishers.

Geiger, D., Seedorf, T., Schulze, C., & Nickerson, M. (2011). Managing the crowd: towards a taxonomy of crowdsourcing processes. *Proceedings of the Seventeenth Americas Conference on Information Systems*.

Hendler, J. (2001). Agents and the Semantic Web. *IEEE Intelligent Systems*, *16*(2), 30–37. doi:10.1109/5254.920597

Howe, J. (2006). The rise of crowdsourcing. *Wired*, *14*(6), 176–180.

Howe, J. (2009). *Crowdsourcing: Why the power of the crowd is driving the future of business*. New York: Crown Business.

Huang, E., Zhang, H., Parkes, D. C., Gajos, K. Z., & Chen, Y. (2010). Toward automatic task design: A progress report. In *Proceedings of the ACM SIGKDD Workshop on Human Computation*. New York: ACM. doi:10.1145/1837885.1837908

Kalpic, B., & Bernus, P. (2006). Business process modeling through the knowledge management perspective. *Journal of Knowledge Management, 10*(3), 40–56. doi:10.1108/13673270610670849

Kleemann, F., Voß, G., & Rieder, K. (2008). Un(der)paid innovators: The commercial utilization of consumer work through crowdsourcing. *Science, Technology & Innovation Studies, 4*(1), 5–26.

Kodama, M. (2005). New knowledge creation through leadership-based strategic community –a case of new product development in IT and multimedia business fields. *Technovation, 25*(8), 895–908. doi:10.1016/j.technovation.2004.02.016

Law, E., & von Ahn, L. (2009). Input-agreement: a new mechanism for collecting data using human computation games.*Proc. CHI*, (pp. 1197-1206). doi:10.1145/1518701.1518881

Law, E., West, K., Mandel, M., Bay, M., & Downie, J. S. (2009). Evaluation of algorithms using games: the case of music tagging.*Proc. ISMIR*2009, (pp. 387-392).

Levy, P. (1997). *Collective intelligence: Mankind's emerging world in cyberspace.* Cambridge, MA: Perseus.

Licklider, J. C. R. (1960). Man-computer symbiosis.IRE Transactions on Human Factors in Electronics, 1, 4-11.

Luger, G., & Stubblefield, W. (2004). *Artificial Intelligence: Structures and Strategies for Complex Problem Solving* (5th ed.). Benjamin/Cummings.

Malone, R., Laubacher, C., & Dellarocas. (2009). *Harnessing Crowds: Mapping the Genome of Collective Intelligence*. MIT Sloan Research Paper, 4732-09.

Malone, T. W., Laubacher, R., & Dellarocas, C. N. (2009). *Harnessing crowds: Mapping the genome of collective intelligence*. MIT Sloan Research Paper 4732-09.

Michelucci, P. (2011). *Handbook on human computation*. Springer.

Parameswaran, M., & Whinston, A.B. (2007). Social Computing: An Overview. *CAIS, 19*(37), 762-780.

Quinn, A., & Bederson, B. (2009). *A Taxonomy of Distributed Human Computation*. Tech. Rep. HCIL-2009-23. University of Maryland.

Ramaswamy, V. (2010). The Power of Co-Creation: Build It with Them To Boost Growth. Productivity, and Profits, Gouillart, Francis.

Rieder, K., & Voß, G. (2010). The working customer- an emerging new type of consumer. *Psychology of Everyday Activity, 3*(2), 2–10.

Russell, S., & Norvig, P. (2003). *Artificial Intelligence: A Modern Approach* (2nd ed.). Upper Saddle River, NJ: Prentice Hall.

Schall, D., Truong, H., & Dustdar, S. (2008). The Human Provided Services Framework Science Platform. In *Zooniverse: Observing the World's Largest Citizen Science Platform*. Academic Press.

Shiffman, D. (2008). *The Age of Engage: Reinventing Marketing for Today's Connected, Collaborative, and Hyper-Interactive Culture*. Ladera, CA: Hunt Street Press.

Silvertown, J. (2009). A new dawn for citizen science. *Trends in Ecology & Evolution, 24*(9), 467–471. doi:10.1016/j.tree.2009.03.017 PMID:19586682

Simpson, R., Page, K., & De Roure, D. (2014). *Zooniverse: Observing the World's Largest Citizen*. Academic Press.

Smith, A., Lynn, S., & Lintott, S. (2013). *Introduction to the Zooniverse*. AAAI Workshop - Technical Report WS-13-18.

Snow, R., O'Connor, B., Jurafsky, D., & Ng, A. (2008). Cheap and fast - but is it good? Evaluating non-expert annotations for natural language tasks.*Proceedings of the 2008 Conference on Empirical Methods in Natural Language Processing*.

Surowiecki, J. (2005). *The Wisdom of Crowds*. New York: Anchor Books.

Teece, D., & Pisano, G. (1994). The dynamic capabilities of firms: An introduction. *Industrial and Corporate Change, 3*(3), 537–556. doi:10.1093/icc/3.3.537-a

Tran-Thanh, L., Stein, S., Rogers, A., & Jennings, N. (2014). Efficient Crowdsourcing of Unknown Experts using Bounded Multi–Armed Bandits. *Artificial Intelligence, 214*, 89–111. doi:10.1016/j.artint.2014.04.005

Ushahidi. (2016). Available at https://www.ushahidi.com

von Ahn, L. (2005). *Human Computation*. (Doctoral Thesis). CMU.

Von Hippel, E. (2005). *Democratizing innovation*. Cambridge, MA: MIT Press.

Winter, M., & Siddharth, S. (2012). Conducting Behavioral Research on Amazon's Mechanical Turk. *Behavior Research Methods, 44*(1), 1–23. doi:10.3758/s13428-011-0124-6 PMID:21717266

Yang, Y., Zhu, B. B., Guo, R., Yang, L., Li, S., & Yu, N. (2008). A comprehensive human computation framework: with application to image labeling.*Proc. MM*2008. doi:10.1145/1459359.1459423

Yuen, M., Chen, L., & King, I. (2009). A Survey of Human Computation Systems. *Proc CSE.*

Zaidan, C. (2011). Crowdsourcing translation: professional quality from non-professionals. In *Proceedings of the 49th Annual Meeting of the Association for Computational Linguistics: Human Language Technologies.*

Zooniverse. (2016). Available at https://www.zooniverse.org

KEY TERMS AND DEFINITIONS

Artificial Intelligence: Intelligence exhibited by machines. In computer science, an ideal "intelligent" machine is a flexible rational agent that perceives its environment and takes actions that maximize its chance of success at some goal.

Collective Intelligence: Shared or group intelligence that emerges from the collaboration, collective efforts, and competition of many individuals and appears in consensus decision making.

Crowdsourcing: Process of obtaining needed services, ideas, or content by soliciting contributions from a large group of people, especially an online community.

Human Computation: A computer science technique in which a machine performs its function by outsourcing certain steps to humans, usually as microwork. This approach uses differences in abilities and alternative costs between humans and computer agents to achieve symbiotic human-computer interaction.

Innovation: As the application of better solutions that meet new requirements, unarticulated needs, or existing market needs. This is accomplished through more-effective products, processes, services, technologies, or business models that are readily available to markets, governments and society.

Chapter 3
The Rise of Artificial Intelligence:
Its Impact on Labor Market and Beyond

Robert Niewiadomski
NYCDOE, USA

Dennis Anderson
St. Francis College, USA

ABSTRACT

Our inventions defined the work we engaged in for centuries; created new industries and employment opportunities around them. They, however, had often unforeseen consequences that affected the way we lived, interacted with each other, and re-defined our societal rules. The established narration portrays the impact of major technological leaps in civilization on employment as temporary disruptions: Many finds themselves without employment taken away from them by efficient, laborsaving inventions, but, in the long run, through gradual adaptations, improved education and gaining higher qualifications, everyone benefits. In this chapter, the authors explore the impact of the rapid expansion of artificial intelligence (AI) in relations to the labor market. The authors argue that this rather optimistic, even naïve scenario, collapses while confronted with the exponential growth of AI; in particular, with the potential arrival of syneoids – robotic forms of "strong AI" possessing, or even exceeding, the full range of human cognitive abilities.

DOI: 10.4018/978-1-5225-1656-9.ch003

FROM GALATEA AND GOLEMS TO FUTURE SYNEOIDS

We know that blind evolutionary processes can produce human-level general intelligence, since they have already done so at least once. Evolutionary processes with foresight—that is, genetic programs designed and guided by an intelligent human programmer—should be able to achieve a similar outcome with far greater efficiency. (Nick Bostrom, Superintelligence, Paths, Dangers, Strategies)

The unique urge to create an *artificial man* has a long tradition dating back to the dawn of Western intellectual history, "Hephaestus, the master craftsman, grants a human voice (…) to his golden mechanical handmaidens (Gera, 2003, p. 114)" and Jewish folklore includes many versions of golems –anthropomorphic beings created from clay. Myths, religions, and popular culture displayed this obsession with varying intensity thorough history. Pamela McCorduck (1979), who investigated this phenomenon, concludes that, as humans, we have been engaged in this peculiar form of self-reproduction by attempting to fulfill the urgent desire, bypassing the ordinary means, in order to recreate what is the essential to us. Contemporary incarnation of this ancient desire can be observed in robotics.

Currently, we are on the verge of an unprecedented technological revolution involving intelligent robots powered by artificial intelligence (AI). It is important to note that what we are experiencing at this stage – Google self-driving cars, Apple Siri, Google Photo Search, robots exhibiting behavior similar to human- are examples of so called "weak AI (WAI)." The concept asserts that machines could act *as if* they were intelligent. It is a sort of limited intelligence. To the contrary, "strong AI" (SAI) is the higher level of AI, often referred to as "artificial general intelligence" (AGI) –it entails the possibility of machines *actually* thinking (Russell & Norvig, 1995). According to this view SAI would possess the full range of human cognitive abilities. Such AI, would, as predicted, experience exponential growth, and quite swiftly reach a level exceeding human capacities – a point often referred to as "singularity." If we assume a physicalist position regarding the nature of human consciousness, we ought to also seriously consider the possibility that SAI would be, at some point, able to genuinely experience subjective mental states such as consciousness. The emergence of consciousness in SAI could be perhaps an incremental process, similar to varying degrees of mental capacities in animals. The terminology attempting to capture this phenomenon varies and it does not quite reflect its full nature and scope. For the purpose of our investigation, we will refer to it as the *syneoid* –a term derived from Greek *συνειδητός* (corresponding to English *conscious*) and the suffix *–oid* suggesting "likeness" or "form of" as well as bearing resemblance to android (human-like robot).

Throughout history, technological advancements have impacted how we worked, lived, and died. The canonical narration pertaining to the impact of major technological leaps in civilization on employment, particularly the paradigmatic Industrial Revolution, follows the familiar pattern: they lead to substantial shifts in types of jobs, affecting perhaps a large but limited area of employment. Certainly, the total number of jobs was not affected and, if anything, technological revolutions expanded employment opportunities by opening new areas of exploration. In a relatively short term, this was a messy process; many lost jobs taken away from them by efficient, laborsaving inventions, but, in the long run, through gradual adaptations, improved education and higher qualification, everyone benefited. Thus, in essence, a technological revolution has been typically painted, in spite of its shortcomings and damages, as a disruptor and the true engine of human progress.

The arrival of the syneoid, however, would have staggering consequences for the future of human employment and beyond. In this chapter, we intend to explore the impact of the rapid expansion of SAI in relation to the labor market. We argue that this rather optimistic, even naïve scenario, collapses while confronted with the explosion of SAI, as none of the historical extrapolations will apply to the processes that are currently shaping up. The history might have had repeated itself with several technological revolutionary cycles but, this time, we will be confronted with a radical new factor, for the arrival of which we are utterly unprepared, and as we are about to see, in principle, we cannot.

Interestingly, although the research on SAI is growing fast, it did not capture the interest of general public beyond some portrayals in science fiction. This does not translate, however, into a serious public debate on the speed and consequences of this technological revolution *in statu nascendi*. Popular imagination seems to be captivated by other occurrences. The general public and both established and aspiring politicians are typically more concerned, and not without a good reason, with the issues of climate change, the rise of jihadist extremism, and ever-widening income gap. Politicians and experts turn to history looking for solutions, and while some of them proved to yield positive results in the past, none will work to alleviate the incoming next wave of unemployment due to automation, since its causes are drastically different from anything from what we have experienced so far.

Will History Repeat Itself? Not This Time

In the past, the slower pace of transformations allowed enough time to adapt and balance the labor supply and demand. By the dawn of 19th century, technological advancements decreased the need for agricultural labor drastically. The excess labor supply was shifted to manufacturing and commerce. This transfer first happened on a small scale, typically in small workshops or homes. It is interesting to note that this was the beginning of wealth transfer from agriculture to industry. Most these workshops disappeared with the arrival of steam engine brought about by the Industrial Revolution, which marked a dramatic change in process of production, particularly -a division of labor, mechanization, and specialized workforce. The driving force behind automation has been, and still is, the desire to improve the profit margin by cutting the workforce perceived as cumbersome and inefficient.

The modern factory also created a new type of workers -precisely defined components in steam-driven machinery. As industrial instruments became more sophisticated, the demand for more skilled labor force increased. The low-skilled ones were pushed out yet again. The investment in education in the 19th century provided a supply of qualified labor. This trend continued into the 20th century as post-secondary education became increasingly common. WWII fueled and accelerated the development of complex machines and systems as created the global supply chain management. After the war, workers in the West enjoyed unprecedented prosperity that provided them with a secure pathway to middle class. Even the initial introduction of computers had no immediate impact on job market. Obviously, mathematical calculations done by hand by numerous people became obsolete but, in time, the invention greatly increased the productivity of others and created a plethora of computer-related jobs. Many other technical innovations had similar effects. New machinery displaced handicraft producers across numerous industries, from textiles to metalworking.

Naturally, the mere possibility of automation didn't mean that it always made sense to implement it. Often, it was more profitable to outsource jobs and take advantage of cheap labor in developing counties. This was accelerated by globalization. Western workers found themselves struggling yet again, facing increasing competition from both machines and cheap labor overseas. Since the 1950's employment in manufacturing in the U.S. declined sharply due to cheap imports and rising cost of labor. It was inevitable that most available jobs moved to service industry. Eventually, we transitioned into a consumption-based economy where many jobs shifted to service industries selling goods and delivering services at a massive scale. However, service industry already displays similar techniques of maximization of output on the expense of labor force as we have seen in manufacturing.

Eventually, the deterioration of employment opportunities expanded to highly skilled college educated workforce as well. As Martin Ford argues, AI ultimately will lead to evaporation of most blue and white collar jobs alike leading to further acceleration of the process of middle-class shrinking. Middle and lower-middle classes are already experiencing tremendous costs of health care and education. Ford scenarios predict massive unemployment, and a complete breakdown of consumer-based economy itself (Ford, 2009).

As of now, WAI led to the emergence of a radically different sort of technologies capable of performing tasks previously done by humans: quite sophisticated and speedy reasoning, judgment, and perception. One would be perhaps tempted to describing theses capacities as constituting a sort of agency. As Jeremy Howard (2014) points out, "Computers right now can do the things that humans spend most of their time being paid to do, so now's the time to start thinking about how we're going to adjust our social structures and economic structures to be aware of this new reality." This rapid automation of physical, mechanical or repetitive jobs has already resulted in the elimination of various occupations and the process only just started. Job market needs a balance of supply and demand but when human labor is no longer required, this balance no longer exists as humans are no longer part of the labor market. At this historical junction, the fundamental question then is: Will artificial intelligence reduce the need for human toil and, as a result, cause technological unemployment? There are still two opposing main answers to this question as articulated by Nils Nilsson (1984, p. 5).

Some claim that AI is not really very different from other technologies that have supported automation and increased productivity (…) Like them, AI may also lead ultimately to an expanding economy with a concomitant expansion of employment opportunities. At worst, (…) there will be some, perhaps even substantial shifts in the types of jobs, but certainly no overall reduction in the total number of jobs. In my opinion, however, such an outcome is based on an overly conservative appraisal of the real potential of artificial intelligence. Others accept a rather strong hypothesis with regard to AI -one that sets AI far apart from previous labor-saving technologies. Quite simply, this hypothesis affirms that anything people can do, AI can do as well. Certainly AI has not yet achieved human-level performance in many important functions, but many AI scientists believe that artificial intelligence inevitably will equal and surpass human mental abilities.

The "optimistic scenario" as articulated above, is based on the naïve extrapolation of historical experience. In a nutshell, it basically claims that AI is no different from other technological advances we have witnessed in the past. It will, just like them, lead to the expansion of employment opportunities in other fields. This shift might be quite unpleasant in a short run, the human cost conceivably high, but, in the end, people, who are pushed out from their jobs by automation will eventually find employment elsewhere. To believe otherwise –according to the optimists – is erroneous. Keynes (1931) also wrote about the causes of widespread technological unemployment occurring due to discoveries capable of economizing the use of labor that are surpassing the pace at which we can find new uses for labor. Nevertheless, he concluded that technological unemployment is only, as he put it, a temporary phase of maladjustment that will be solved in a long run.

Those, who support this view, argue that those concerns pertaining to long–lasting or even permanent technological unemployment result from the "Luddite fallacy" chiefly because they fail to recognize compensation effects. The term "Luddite fallacy" was coined by drawing a comparison to the 19th century reasoning of the English textile workers, who were concerned with the machinery threatening their livelihood, introduced during the Industrial Revolution, and resorted to violent uprisings. The optimists believed that textile workers were fundamentally mistaken, since the technological progress will eventually lead to the increase of wealth of the society as a whole, and that the advances, as mentioned above, will lead to the creation of new employment opportunities. In other words, the "Luddite fallacy" proponents also tend to point out that the worries about long-lasting or permanent technological unemployment are unwarranted because there is no such thing as a finite amount of work available to us and new occupational opportunities will always emerge that only humans will be able to perform.

On the surface, this assumption might seem justified and intuitive, since we tend to extrapolate from the past. Historical events, however, are not necessary a solid predictor of the future, particularly when we are confronted with a radically novel phenomenon. The belief in "Luddite fallacy" constituted a prevalent view of mainstream economists on the problem of technological unemployment until recently and many still subscribed to it.

Nevertheless, some cracks in this reasoning are easy to spot. Assuming, for the sake of the argument, that the technological progress, in fact, will increase the aggregate wealth of the society, it is not obvious that the distribution of the wealth will be equitable. It goes without saying that distribution of wealth matters a great deal. For example, in a technocratic society, such as ours, an enormous amount of wealth is concentrated in the hands of an extremely small group. This does not mean, however, that everyone else benefits through mechanisms of job-creation. We can see this in similar mechanism – so-called trickle down economy. The correlation between lowering taxes for the wealthiest and job -creation is highly arguable. Thus, supposed increase in overall wealth does not automatically translate into a lower rate of unemployment –as enthusiasts of trickle down economy would like to believe.

In addition, according to the optimistic narrative, long-lasting technological unemployment typically requires a massive effort to educate the population. Again, it is a lesson learned from the Industrial Revolution. In short, education is supposed to be the key to solving this issue. As Paul Krugman (2013, para. 4) puts it by revisiting the problems faced by, the previously ridiculed, 19th century textile workers:

Until recently, the conventional wisdom about the effects of technology on workers was, in a way, comforting. Clearly, many workers weren't sharing fully — or, in many cases, at all — in the benefits of rising productivity; instead, the bulk of the gains were going to a minority of the work force. But this, the story went, was because modern technology was raising the demand for highly educated workers while reducing the demand for less educated workers. And the solution was more education.

Indeed, mostly free and compulsory public education emerged in the West as a result of ideas of the Enlightenment a feeding pipe for factories of the industrial age. This model of "keeping up "with the technology seemed to work fairly well for some portions of the population, at least until recently. Several factors are at play here: the outdated model of education that is simply too specialized and often provides training that quickly lacks relevance on the labor market. After all, it is difficult to blame institution of higher education for not predicting what skills will be desirable in the rapidly shifting labor landscape even in not-to-distant future. The second factor pertains to the massive democratization of higher education experienced virtually in all developed countries. This educational inflation, occasionally leading to outburst of frustration comparable to those of mentioned textile workers, effectively cancels out any competitive edge previously enjoyed by college degree holders in the past,

Today (...) a much darker picture of the effects of technology on labor is emerging. In this picture, highly educated workers are as likely as less educated workers to find themselves displaced and devalued, and pushing for more education may create as many problems as it solves. (...) So should workers simply be prepared to acquire new skills? The woolworkers of 18th-century Leeds addressed this issue back in 1786: "Who will maintain our families, whilst we undertake the arduous task" of learning a new trade? Also, they asked, what will happen if the new trade, in turn, gets devalued by further technological advance?
And the modern counterparts of those woolworkers might well ask further, what will happen to us if, like so many students, we go deep into debt to acquire the skills we're told we need, only to learn that the economy no longer wants those skills? Education, then, is no longer the answer to rising inequality, if it ever was (Krugman, 2013, para. 6).

Putting aside the question of what kind of training and qualifications would be needed to access these emerging employment opportunities and how many individual would have an actual chance to do so etc., there are more serious problems with this line of reasoning. Even individuals, who are flexible and well adjusted to the job market might not be able to keep up with the speed of technological acceleration. This is how Krugman (2013, para. 1-2) explain this phenomenon:

In 1786, the cloth workers of Leeds, a wool-industry center in northern England, issued a protest against the growing use of "scribbling" machines, which were taking over a task formerly performed by skilled labor. "How are those men, thus thrown out of employ to provide for their families?" (...) Those weren't foolish questions. Mechanization eventually — that is, after a couple of generations — led to a broad rise in British living standards. But it's far from clear whether typical workers reaped any benefits during the early stages of the Industrial Revolution; many workers were clearly hurt. And often the workers hurt most were those who had, with effort, acquired valuable skills — only to find those skills suddenly devalued.

More fundamentally, there is another crucial point that gets lost in the current debates. Even if some number of very highly specialized occupations will be created, with the arrival of syneoids, "this kind of work can also be performed by AI devices without necessarily implying more jobs for humans (Nilsson, Ibid.)."

The above set of issues gradually begins to gain traction in the high strata of the more forward-thinking individuals. According to the posted agenda of the Davos World Economic Forum in 2016 the incoming shift, framed as the Fourth Industrial Revolution, is recognized a paradigmatic change that will affect virtually every sphere of human life. To the question why this revolution would be any different from the previous ones we experienced, Klaus Schwab delivers the answer with blunt honesty,

The first industrial revolution began in the latter half of the 18th century, and brought us mechanical production through the steam engine, cotton spinning and then railroads. The second industrial revolution took place around the turn of the 20th century and brought mass production through assembly lines and electrification. The third industrial revolution was the computer revolution, starting in the 1960s that brought us the mainframe and then personal computing, as well as the Internet. Today we are in the midst of the Fourth Industrial Revolution, which will affect governments, businesses and economies in very substantial ways. We should not underestimate the change ahead of us, as there are at least three differences between this revolution and the previous ones (Schwab, 2015, para. 4-5).

Schwab articulates three major components that differentiate this new fundamental change from previous technological revolutions. In a nutshell, the first one has to do with its speed. The current change is happening with an unprecedented acceleration, as opposed to the previous ones that took decades or more to fully unfold. Secondly, the Fourth Industrial Revolution is not limited only to one area.

It incorporates such elements as mobile networks, nanotechnology, brain research, 3D printing, materials science, computing, networks, and "the interplay between all of these. (…) In addition, the accessibility and affordability of complex technologies will spread them farther and faster (Ibid. para. 6-8)." The third difference, according to Schwab, pertains to the innovation of the entire system, as opposed to one product or area. The phenomenon of "sharing economy" is only one of many manifestations of that aspect.

As mentioned, the impact of this paradigmatic shift will be profound, and will affect the business model, the ways governments operate, and last but not least, the future usefulness of skills and labor. As Fon Mathuros (2015, para. 2) points out,

Earlier Industrial Revolutions advanced human progress through new forms of power generation, mass production and information processing. Building on a ubiquitous and mobile internet, smaller, cheaper and more powerful sensors, as well as artificial intelligence and machine learning, the Fourth Industrial Revolution is distinct in the speed, scale and force at which it transforms entire systems of production, distribution, consumption – and possibly the very essence of human nature.

The Davos Forum organizers, although recognizing the magnitude of imminent shift, tend to portray the Fourth Industrial Revolution in cautiously optimistic terms as perhaps one of the greatest challenges we will encounter, and yet, a possible opportunity. Nevertheless, they cannot disregard that two aspects of human life will be affected most significantly by the future changes –the value and significance of human work and profound changes in social structures.

THE NEW CAMBRIAN EXPLOSION

In a peculiar way, some aspects of the development of robotics mimic the natural processes of evolution. Gill Pratt (2015, p. 51) draws an intriguing comparison between the two.

About half a billion years ago, life on earth experienced a short period of very rapid diversification called the "Cambrian Explosion." Many theories have been proposed for the cause of the Cambrian Explosion, with one of the most provocative being the evolution of vision, which allowed animals to dramatically increase their ability to hunt and find mates (…) Today, technological developments on several fronts are fomenting a similar explosion in the diversification and applicability of robotics. Many of the base hardware technologies on which robots depend—particularly computing, data storage, and communications—have been improving at exponential growth rates. Two newly blossoming technologies—"Cloud Robotics" and "Deep Learning"—could leverage these base technologies in a virtuous cycle of explosive growth. In Cloud Robotics (...) every robot learns from the experiences of all robots, which leads to rapid growth of robot competence, particularly as the number of robots grows. Deep Learning algorithms are a method for robots to learn and generalize their associations based on very large (and often cloud-based) "training sets" that typically include millions of examples. (...) One of the robotic capabilities recently enabled by these combined technologies is vision—the same capability that may have played a leading role in the Cambrian Explosion.

Initially, AI and robotics were separate fields, -computer science and engineering, respectively. Currently, it is the precisely the combination of the two -the robotic embodiment of AI- that sets the trend for the future. Particularly, Deep Learning (DL), which is an algorithm-based method of "teaching a machine" how to learn from massive sample sets, is one on the major mechanism driving forces. Current applications of DL already include automatic speech and image recognition, natural language processing, transforming images into description, discovery of drugs and toxicity, medical diagnostics, among others. DL certainly has the potential to become a stepping-stone to creating syneoids, which through exponential growth would rapidly exceed any human faculties to a degree unimaginable. Ray Kurzweil, who shares Searle's view that an appropriately programmed computer is a mind, makes a prediction that the equivalent capacity of one human brain will be available on desktop computers by 2020. Moreover, when AI starts to outstrip the collective total of all human intelligence, Kurzweil argues, humanity will enter, what he calls, the Singularity, marking a point at which change is so radical that it is no longer predictable (as cited in Wallach, p. 57).

The kinds of operations that WAI is capable of performing now cannot be called thinking in the traditional sense, provided that what we mean by thinking, by definition, only takes place inside the human cranium (McCorduck, 2004) Nevertheless, many consider "making difficult judgments, the kind usually left to experts, choosing among plausible alternatives, and acting on those choices (Ibid.)" a form of thinking. McCorduck further elaborates, "Along with most people in AI, I consider what artificial intelligences do as a form of thinking, though I agree that these programs don't think just like human beings do, for the most part. I'm not sure that's even desirable. Why would we want AIs if all we want is human-level intelligence? There are plenty of humans on the planet. The field's big project is to make intelligence that exceed our own. As these programs come into our lives in more ways, we'll need programs that can explain their reasoning to us before we accept their decisions (Ibid.)."

Certainly, it is fair to say that AI and robotics have not crossed yet the level of singularity and thus reached the level of syneoids. However, this could be just a matter of time. Once they reach that stage, all jobs can be automated including key decisions. No human will be needed for manufacturing, services, knowledge, or even creative processes. Thus, "Luddite fallacy", described before, may not be, in the end, a fallacy at all. This time around we simply might not be able to "catch up." It is precisely because human gradual learning cannot possibly be a competition to exponential growth of Deep Learning as Jeremy Howard (2014) explains,

80 percent of the world's employment in the developed world is stuff that computers have just learned how to do. What does that mean? Well, it'll be fine. They'll be replaced by other jobs. For example, there will be more jobs for data scientists. Well, not really. It doesn't take data scientists very long to build these things. For example, these four algorithms were all built by the same guy. So if you think, oh, it's all happened before, we've seen the results in the past of when new things come along and they get replaced by new jobs, what are these new jobs going to be? It's very hard for us to estimate this, because human performance grows at this gradual rate, but we now have a system, deep learning, that we know actually grows in capability exponentially.

In light of the inevitable systemic technological unemployment, experts are constantly considering different solutions ranging from quite unrealistic ideas of banning innovation to various forms of wealth redistribution. It is not our intention here to examine the feasibility of these proposals. The reasons for that choice are quite simple: stopping the entire process is practically impossible, and once syneoids arrive, humans won't make any decisions regarding that matter alone, as we will have to work it out with "others", presumably far more intelligent and capable than us. They might be very well able to colonize places inaccessible to human because of our limitations such us the need of oxygen, water, and food.

Even though, we are not there yet, the loss of jobs due to the already available WAI will be staggering in a relatively short time. A very conservative study conducted at Oxford University by Frey and Osborne (2013) examined the effects of computerization alone on 702 detailed occupations. The authors used a Gaussian process classifier and determined that "about 47 percent of total U.S. employment is at risk (Ibid. p.1)" of computerization in a decade or two. As they point out, "The impact of computerisation on labour market outcomes is well-established in the literature, documenting the decline of employment in routine intensive occupations – i.e. occupations mainly consisting of tasks following well-defined procedures that can easily be performed by sophisticated algorithms. (…) Following recent technological advances, however, computerisation is now spreading to domains commonly defined as non-routine (Ibid. pp.2-15)." Frey and Osborne further argue that even though Machine Learning (ML) and Mobile Robotics (MR) enable computerization across large areas of non-routine tasks, there exist some "engendering bottlenecks" inhabiting total de-humanization of labor.

Beyond these bottlenecks, however, we argue that it is largely already technologically possible to automate almost any task, provided that sufficient amounts of data are gathered for pattern recognition. Our model thus predicts that the pace at which these bottlenecks can be overcome will determine the extent of computerisation in the twenty-first century. Hence, in short, while the task model predicts that computers for labour substitution will be confined to routine tasks, our model predicts that computerisation can be extended to any non-routine task that is not subject to any engineering bottlenecks to computerisation. These bottlenecks thus set the boundaries for the computerisation of non-routine tasks (Ibid. p.23).

They identify three major "bottlenecks." The first one, perception and manipulation, in essence, pertains to the abilities to make precisely coordinated movements. This "bottleneck" seems to pose the least of the problem for computerization and robotization, since human body itself can be seen as a biological machine that had a very long to evolve and adapt to perform very complex ways of manipulating the environment. In principle, however, there is nothing, besides technological and perhaps financial limitations, that would prevent replication of these abilities. The authors admit that, "with incremental technological improvements, the comparative advantage of human labour in perception and manipulation tasks could eventually diminish (Ibid. p. 39)." The second "bottleneck" indeed poses a serious challenge for AI at this stage. It has to do with creative intelligence –faculties we typically associate with solving unusual problems, artistic production etc. The third one, social intelligence, pertains to the ability of being aware of other's reactions and understanding why they react in a given way; the ability to persuade others, and, finally, with a broadly understood, emotional support (Ibid. p. 31).

This last obstacle is indeed unlikely to be solved by WAI at its present stage. Thus, the study reports that, "generalist occupations requiring knowledge of human heuristics, and specialist occupations involving the development of novel ideas and artifacts, are the least susceptible to computerization (Ibid. p. 40)." Based on these technological "bottlenecks", their model assigns even specific probabilities that computerization will lead to job losses within about the next two decades. And, so they estimate that that probability would be 0.99 for telemarketers, 0.55 for commercial pilots, and only 0.003 for recreational therapists, presumably because the latter one would require technological overcoming of the social intelligence "bottleneck."

It is important to keep in mind, that Fry and Osborne's analysis does rely on the technological "bottlenecks" as they present themselves today. The history of technological advances, however, can be largely perceived as a struggle, astonishingly successful, of overcoming such "bottlenecks." Thus, there is no reason to think that these current, undoubtedly serious "bottlenecks," will become obsolete once the technology moves beyond the stage of WAI towards syneoids.

Again, skeptics find it difficult to accept that syneoids could completely fill every available niche in the labor market and make human skills virtually obsolete. They believe that there is something uniquely special about human capacities, whether be intellectual or emotional that there will be always a margin of tasks syneoids simply won't be able to do, or people would prefer them being done by other humans. Those occupations would require human sensitivity that can satisfy deep interpersonal needs. This would involve such activities as making decisions in a court of law because, supposedly, we would prefer a person to be accountable for such decisions. The same would apply to CEO's, generals, military and government leaders at every level.

This could be the case if syneoids would exceed human capacities only intellectually, but as we stipulated, they would also become decision making fully autonomous agents, with whom we would have to engage in interactions involving making moral choices.

Schwab (2015, para. 13-15) gestures towards this issue of syneoids and humans living side by side by sketching two compelling scenarios,

Very importantly, will the Fourth Industrial Revolution have a human heart and soul? Again, imagine living in the future, when robots and humans live side by side. Who would you trust, human or robot? Concretely:

- *If you were told you had a life-threatening illness and a human doctor prescribed treatment regime A, while an artificially intelligent robot prescribed treatment regime B, which treatment regime would you follow?*
- *If you were falsely accused of a crime, would you rather be tried by a human judge or by an AI judge?*

It is obvious that Schwab scratches just the surface of the possible issues of the moral aspect of interactions with syneoids. For some this very prospect of such interactions and the moral uncertainty connected to it are frightening ideas. Perhaps, as some prominent figures such as Stephen Hawking and Elon Musk, argue, it would be better for us to stop the expansion of AI altogether, particularly with respect to autonomous warfare agents (Musk et al.). Some postulate that our efforts should be concentrated on designing these agents in such way "so that they honor the broader set of values and laws humans demand of human moral (Wallach, p. 6)."

However, at this point, it is far too late to ask the question whether the humanity needs intelligent machines making moral decisions. Machines are already making decisions with moral consequences. As mentioned, machines, whether intelligent or not, even tools, had a moral impact on our lives and, in fact, they always had. James Moor proposed a hierarchical model that categorizes different levels of machine's ethical agency. At the very bottom are, what he refers to, "ethical impact agents." This category includes "any machine that can be evaluated for its ethical consequences (as cited in Wallach, p. 33-34)." The next level Moore calls "implicit ethical agents" – "machines whose designers have made an effort to design them so that they don't have negative ethical effects, by addressing safety and critical reliability concerns during the design process (Ibid.)." "Explicit ethical agents" occupy the third level of Moor's hierarchy. These are machines that can reason about ethics using ethical categories as part of their internal programming. Setting proper ramifications for the design of such "implicit ethical agents" and establishing an appropriate ethical code for their decision-making is currently the main concern of the field of computer ethics. Everything that lies beyond these three levels Moor considers "full ethical agents" –machines that can make "explicit moral judgments and are generally quite competent in justifying such decisions. This level of performance is often presumed to require a capacity for consciousness, intentionality, and free will. If any of these three is lacking in a human context, then the person's moral agency and legal culpability comes into question (Ibid)." This highest level of full ethical agents matches our definition of a syneoid. Naturally, it would be quite naïve to think that syneoids would necessarily share our moral standards and sensitivities. Syneoid's morality could go much further beyond anything we can envision based on our anthropological imagination and knowledge of varying moral standards in human cultures; it could be a set of norms radically alien in general, based on different preference and goals.

WHY CONSCIOUSNESS MATTERS?

Some philosophers and scientists doubt the very possibility of artificial "full ethical agents" precisely because the idea of them possessing consciousness, intentionality, and free will seem uniquely human. From all three –consciousness is probably considered to be at the core of human personhood, which is the only point of reference of personhood we have available. In order to understand why consciousness matters in the context of syneoids, we should go back to the18th century materialist philosophy. In reaction to the previous substance dualism, La Mettrie, in his *L'homme Machine*, extended to human beings Descartes' argument that animals are automatons. Karl Popper (1978, pp. 224-225) was convinced that La Mettrie's idea gained a greater reinforcement by the theory of evolution.

... The doctrine that man is a machine was argued most forcefully and seriously in 1751, long before the theory of evolution became generally accepted, by de Lamettrie; and the theory of evolution gave the problem an even sharper edge, by suggesting that there may be no clear distinction between living matter and dead matter.

In principle, the emergence of conscious syneoid is plausible provided that few premises that we would need to commit to are true. One of them is an ontological assumption, rooted in neuroscience, concerning the nature of consciousness. The brain itself –as the argument runs- is a naturally occurring material object that emerged through evolutionary processes and as such, at least in principle, is replicable with sufficiently advanced technology. Secondly, mental states are essentially brain states. Thus, consciousness supervenes on the brain and, just as the brain, it is a natural phenomenon that is a product of a very long incremental evolutionary process. Thirdly, we have to commit to yet another premise that neurons are not the only material that can create a mental state (Dennett, 1991). Therefore, it is just a matter of time that we, as conscious beings, will be able to imitate it by some means that might not be necessarily biological in nature.

This argument is further supported by "Artificial Life" projects that essentially see biology as a naturally occurring information technology. That "technology" could be reversed-engineered and synthesized in other kinds of technologies, again, not necessarily biological ones (Sullins, 2014). Consciousness would, therefore, be a product of this naturally occurring information technology. It is important to make another observation –consciousness, after all, is present, to varying degrees, in different species. In short, if we, as humans, are naturally occurring robots, and just so happened we possess consciousness, then there is nothing, in principle, preventing replication of this natural phenomenon artificially.

There is obviously the question of how would we even know that we have created a conscious machine, or that a conscious machine emerged through, for example, processes similar to Deep Learning? The famous Turing Test cannot answer this question as demonstrated by Searle in his "Chinese room" thought experiment. One might even argue that perhaps a consciousness test, in principle, cannot be done if we set the testing standard unreasonably high. After all, how can we be absolutely sure that other human fellows have mental states or just behave as if they have them? Such high requirements might lead us towards metaphysical solipsism – a doctrine with very vague practical consequences, and as such not very useful in our investigation.

One can obviously stipulate that consciousness is a something that occurs in the brain alone, and then, by analogy, for example, make inferences that other humans possess that property as well. It seems fair to say that most people use this analogy or symmetry principle, in their common sense approach to the issue of consciousness of others. This can be further supported by MRI imaging showing a correlation between changes in brain activity with reports of subjective mental states by the subject. In fact, similar methods can be used to investigate the mental states of individuals who are unable to communicate in any way. These are obviously well established and perfectly sound scientific practices, and yet strictly speaking, they can only provide indirect proofs that others possess consciousness. Another approach to solving the problem is based on common ancestry –the offspring, as a matter of rule, should display capabilities similar to those of those of its ancestors. This line of reasoning is just another way of making inferences by analogy. However, it is easy to spot a weak point of these common sense based reasoning. Provided that consciousness develops artificially and will be neither brain-based, not biologically related to us, none of these common sense reasoning by analogy will apply. In case there will be no way to discern if the intelligent machine *really* possesses consciousness, or is simulating it, one can obviously stick to the believe that only humans have it by definition, but that would be terribly arbitrary, and ultimately unjustified. Ultimately, if syneoids displayed mental and cognitive capabilities that are indiscernible from those displayed by humans that we "know" that they possess consciousness, it would be unclear, to say at least, on what basis we could conclude that syneoids do not have consciousness after all.

The more fundamental question is obviously: Why would it even matter to know whether syneoids possess consciousness or they do not? With regards to the issue of technological unemployment the answer is – it matters. One could obviously argue that such question is superfluous since super intelligent "zombies" could replace us easily, and the issue whether they are conscious has nothing to do with that. However, going back to the previous issue that people might prefer for certain services or occupations to be performed by human persons only precisely because they are conscious beings, it becomes clear how artificially generated consciousness would eliminate this need as well. Thus, there would be virtually nothing left that "only humans could do."

The answer to the consciousness question also becomes extremely relevant when we consider the issue of interaction with syneoids. Again, one can dodge the bullet and insist that although we are dealing with beings displaying intelligence similar or superior to ours, nevertheless, they are qualitatively infinitely different; they don't have some sort of "essence" that we have and, as such, they should be excluded from our moral considerations. Needless to say, there would be something incredibly odd in the argument that a syneoid capable of performing human functions, or even exceeding them -intellectual, cognitive, and emotional- is not treated as a person.

Obviously, it goes without saying, that the notion of personhood is an incredibly complex one. Nevertheless, the most important components of the notion that are crucial for the purpose of the argument are: consciousness, agency, self-awareness, and possession of rights and duties (Taylor, 1985). It is also important to realize that personhood could be a matter of degree; it could be gained and lost incrementally, and that does not need to be limited to humans. The latter statement has remarkable consequence, namely, being a member of the human species is neither necessary, nor sufficient to be, at least to some degree, a person. It is currently commonly accepted that at least primates and some other mammals possess a degree of personhood far beyond a rudimentary one, that qualifies them to be included in the circle of our moral considerations. Since the Oxford Group (Finsen & Finsen, 2009) the movement of animal rights gained traction in the mainstream intellectual community. It drew its conclusions from Darwinism and consequentialism, and basically stated that there is no mysterious essential difference between humans and other animals biologically speaking. Thus, the same type of moral continuum should be applied to other beings. It would be expected that a syneoid, conscious by definition, would realize its own personhood. Thus, if someone or something is considered to be a person, we ought to have moral obligations towards it. The fact of its origin, the fact that it is neither a human person, nor a biological being is utterly irrelevant. It would be preposterous to deny such syneoids moral consideration based on the claim that it does not belong to human species. In fact, it could be argued that such attitude is similar to speciesism –a type of prejudice such as sexism or racism.

Moreover, it could be argued that, as along as a syneoid behaves as a person, it is actually irrelevant if it actually possesses consciousness the way we conceive it at all, in order for us to have moral obligations towards it. After all, many of us commit to the notion that we have moral considerations for certain animals even if, for what we know, they do not possess consciousness, or display it to a certain, typically to a lesser degree than ours. The consequences of committing to the notion that syneoids should be included in the area of our moral concern are staggering. It would be incomprehensible that we could treat such AI as merely our tools, an extension of our bodies, something that we can use, control, and manipulate. It would have to rule out any possibility of the master-slave relationship. We would have no choice but to concede that they are entitled to all rights that we reserve for persons. We would have to concede that they have freedom of choice according to their preferences and needs, etc. the notion of our "uniqueness" would evaporate in an instant.

Co-Existence, Competition, Convergence

Job, occupation, profession, career –they are not merely sources of economic sustenance; they are far more than a mere necessity or burden. In many cases, occupations define who we are. Work, whether paid or pursued for passion, whether fairly simple or highly sophisticated, provides fulfillment, sense of community, exploration, discovery, an invention. Professionalism, craftsmanship, refinement, and artistry have been truly the driving forces of our civilization; for many –a fulfillment of an existential void, a deep sense of participation in a process of creation, mastery over nature, an instrument of control, and, ultimately, power.

Our inventions defined the work we engaged in for centuries from melding to typing; created new industries and employment opportunities around them. These inventions, intending to control the environment, had often unforeseen consequences that reshaped the way we lived, interacted with each other, and redefined our societal rules. To echo the famous maxim attributed to McLuhan that encapsulates this phenomenon: "We shape our tools and thereafter our tools shape us." Occasionally, an emergence of new technological advances such as, domestication of animals, creation of printing press, or invention of computers had a paradigm-changing effect leading to major disintegration in the status quo, shifts and disruptions in the existing power structures. We were adapting, not without struggle, failures, and suffering, to these changes. The unstoppable progress of AI, however, and the eventual arrival of syneoids will confront us with the challenge to radically renegotiate norms of coexistence with them without any guarantee of success.

In the context of a workplace, but not limited to it, possibly the best-case scenario would involve a collaborative co-existence. But it might likely turn into a combative competition over resources and power–something we are quite familiar with already. The Darwinian principle of the survival of the fittest could very well swing the odds to the advantage of syneoids. The end result of such competition could very well be the end of our kind, or perhaps, on the more optimistic note, as Buchannan argues: "humans will go the way bacteria did. They existed before the Cambrian explosion, and they still exist today. In fact, they continue to thrive. They just don't hold the dominant position they once did (Buchannan, 2015)." This perceived disadvantage of our species prompted some to entertaining the idea that we ought to use the ever-expanding information technology to transform ourselves into a new post-human species, as a part of adaptive response to environmental pressures. Preserving the natural state of a human body -claim the proponents of transhumanism- is an obstacle to progress in the self-directed evolution (Bostrom, 2005).

The arrival of syneoids will confront us on so many levels with the type of radical "other" - so different from us and yet a former "tool" "created in our image." It will launch a new stage of our journey -without precedence, unpredictable, unforeseeable, and volatile. We will be faces with many questions. Perhaps, to echo Pamela McCorduck (2004), "at that point, we turn to our own smart machines for advice on what the best next move is for the human race." And just like the well-being of most species currently existing is greatly affected by our decisions and action, so we might find ourselves in a similar position upon the arrival of syneoids.

REFERENCES

Bostrom, N. (2005). A history of transhumanist thought. *Journal of Evolution and Technology.*

Bostrom, N. (2014). *Superintelligence: Paths, dangers, strategies.* Oxford, UK: Oxford University Press.

Buchanan, M. (2015, August 13). *Get Ready for the Robot Explosion.* Retrieved November 13, 2015, from http://www.bloombergview.com/articles/2015-08-13/get-ready-for-the-robot-explosion

Davis, N. (2015, November 10). *5 ways of understanding the Fourth Industrial Revolution.* Retrieved November 12, 2015, from https://agenda.weforum.org/2015/11/5-ways-of-understanding-the-fourth-industrial-revolution/

Dennett, D. (1991). *Consciousness Imagined. In Consciousness Explained* (pp. 431–455). Back Bay Books.

Finsen, S., & Finsen, L. (2009). Animal rights movement. In M. Bekoff (Ed.), *The Encyclopedia of Animal Rights and Animal Welfare.* Greenwood.

Ford, M. (2009). *The lights in the tunnel: Automation, accelerating technology and the economy of the future.* Acculant Publishing.

Frey, C., & Osborne, M. (2013, September 17). *The Future of Employment: How Susceptible are Jobs to Computerisation?* Retrieved from http://www.oxfordmartin.ox.ac.uk/downloads/academic/The_Future_of_Employment.pdf

Gera, D. (2003). *Ancient Greek ideas on speech, language, and civilization.* Oxford, UK: Oxford University Press. doi:10.1093/acprof:oso/9780199256167.001.0001

Howard, J. (2014, December). *The wonderful and terrifying implications of computers that can learn.* Lecture presented at TED.

Keynes, J. M. (1931). Economic possibilities for our grandchildren (1930). In Essays in persuasion, (pp. 358–73). Academic Press.

Krugman, P. (2013, June 13). *Sympathy for the Luddites.* Retrieved November 13, 2015, from http://www.nytimes.com/2013/06/14/opinion/krugman-sympathy-for-the-luddites.html

Mathuros, F. (2015, November 10). *What is the theme of Davos 2016?* Retrieved November 12, 2015, from https://agenda.weforum.org/2015/11/what-is-the-theme-of-davos-2016/

McCorduck, P. (1979). *Machines who think: A personal inquiry into the history and prospects of artificial intelligence.* San Francisco: W.H. Freeman.

McCorduck, P. (2004). *FAQ answered by Pamela McCorduck.* Retrieved November 7, 2015, from http://www.pamelamc.com/html/machines_who_think.html

Musk, Wozniak and Hawking urge ban on warfare AI and autonomous weapons. (n.d.). Retrieved November 5, 2015, from http://www.msn.com/en-gb/news/other/musk-wozniak-and-hawking-urge-ban-on-warfare-ai-and-autonomous-weapons/ar-AAdxy1e

Nilsson, N. (1984). Artificial Intelligence, Employment, and Income. *AI Magazine*, 5–14.

Popper, K. (1973). Of Clouds and Clocks. In Objective Knowledge (pp. 206–55). Oxford, UK: Oxford University Press.

Pratt, G. (2015). Is a Cambrian Explosion Coming for Robotics? *The Journal of Economic Perspectives*, *29*(3), 51–60. doi:10.1257/jep.29.3.51

Russell, S., & Norvig, P. (1995). *Artificial intelligence: A modern approach.* Englewood Cliffs, NJ: Prentice Hall.

Schwab, K. (2015, October 27). *Will the Fourth Industrial Revolution have a human heart?* Retrieved November 12, 2015, from https://agenda.weforum.org/2015/10/will-the-fourth-industrial-revolution-have-a-human-heart-and-soul/

Sullins, J. (2014). Information Technology and Moral Values. In *The Stanford Encyclopedia of Philosophy*. Retrieved from http://plato.stanford.edu/archives/spr2014/entries/it-moral-values/

Taylor, C. (1985). The concept of a person. *Philosophical Papers Human Agency and Language*, *1*, 97–114. doi:10.1017/CBO9781139173483.005

Wallach, W. (2009). *Moral machines: Teaching robots right from wrong*. Oxford, UK: Oxford University Press. doi:10.1093/acprof:oso/9780195374049.001.0001

Chapter 4
Promoting Critical Thinking in the Modern Learning Environments

Kijpokin Kasemsap
Suan Sunandha Rajabhat University, Thailand

ABSTRACT

This chapter explains the overview of critical thinking (CT); the emerging trends of CT in the modern learning environments; CT in nursing education and health care; reflective judgment and CT; constructivist learning and CT; metacognitive strategies, cognitive strategies, and CT; and the promotion of CT in the modern learning environments. The promotion of CT is essential for the learning organizations that seek to serve educators, increase educational performance, strengthen competitiveness, and achieve continuous success in the modern learning environments. Therefore, it is required for the learning organizations to promote their CT and develop a learning plan to regularly check their practical advancements toward satisfying educator requirements. The chapter argues that promoting CT has the potential to enhance organizational performance and reach strategic goals in the modern learning environments.

DOI: 10.4018/978-1-5225-1656-9.ch004

INTRODUCTION

Critical thinking (CT) is an important goal of modern education (Ku, Ho, Hau, & Lai, 2014). CT is one of the most important skills deemed necessary for college graduates to become the effective contributors in the modern learning environments (Liu, Frankel, & Roohr, 2014). CT is an ability to utilize reason to move beyond the acquisition of facts to reveal the deep meaning (Weissberg, 2013). CT is an essential competency that is a priority for teaching and learning in an increasingly digital learning environment (Yang, Gamble, Hung, & Lin, 2014). The combination of human capital and competency is crucial for modern business to achieve better organizational performance (Kasemsap, 2016a). CT skills are required in order to contribute as a member of society, operate effectively in post-secondary institutions, and be competitive in the global marketplace (Carlgren, 2013).

CT is a metacognitive process, consisting of a number of sub-skills and dispositions, that, when used appropriately, increases the chances of producing a logical solution to a problem or a valid conclusion to an argument (Dwyer, Hogan, Harney, & O'Reilly, 2014). CT is the intellectually disciplined process of active conceptualization (Martincová & Lukešová, 2015). CT skills are essential to meeting the economic, sustainable, and social challenges of the future (Carmel-Gilfilen & Portillo, 2010). Leading critically is defined as applying CT skills to decisions about leadership actions in different situations and is a challenge for leaders, followers, and educators (Jenkins, 2012). Student perceptions of organized instruction are positively associated with gains in CT (Loes, Salisbury, & Pascarella, 2015).

Educational robotics is introduced as a transformational tool for learning, which promotes the learning of computational thinking, coding, and engineering, all viewed as the critical elements of STEM (Science, Technology, Engineering, and Mathematics) learning in K-12 education (Eguchi, 2016). Regarding the development of CT skills, robotics programs in K-12 settings can build the habits of mind (e.g., educational and social perspectives) that are valued in both formal and informal learning contexts (Gomez et al., 2016). Robot platform and a web-based cyber-infrastructure delivery system provide teachers with a flexible curriculum structure (Grandgenett, Ostler, Topp, & Goeman, 2014). Nowadays, young people enjoy robotics activities, appreciate the collaboration with others, and become deeply engaged in design experiences regarding artificial intelligence toward enhancing CT skills (Bernstein & Crowley, 2008).

Generating knowledge-based structural metadata can be done through using natural language processing (NLP)-based machine learning and artificial intelligence (Liu, Guo, & Zhang, 2014). The data becomes an asset that requires the cost-effective innovations in information processing that enable process automation, enhanced insight, and decision making (Kasemsap, 2016b). The analysis of educational results and the comparison of behavior displayed by human and artificial agents regarding artificial intelligence allow students to identify the significant role played by the perspectives affecting the agent/environment interaction, the relationship between category and action development, and the role of cognitive biases related to CT skills from previous knowledge (Morlino, Gianelli, Borghi, & Nolfi, 2015).

This chapter aims to bridge the gap in the literature on the thorough literature consolidation of CT. The extensive literature of CT provides a contribution to practitioners and researchers by indicating the advanced issues and implications of CT in order to maximize the educational impact of CT in the modern learning environments.

BACKGROUND

For centuries, philosophy has been considered as an intellectual activity requiring complex cognitive skills and predispositions related to CT (Daniel & Auriac, 2011). Flores et al. (2012) indicated that various definitions of CT are examined to develop a general variable to guide the discussion as CT is linked to constructivism, leadership, and education. CT arises when the learners begin to work at a concrete problem (Dewey, 2004). CT is the main outcome of higher education and a key factor in program accreditation (Staib, 2003). The development of CT skills is high on the agenda for higher education establishments (Kovalik & Kovalik, 2007), and educators' teaching methods need to be adapted to build on the students' current CT skills in order to develop them further.

CT is a highly valued educational outcome throughout the educational spectrum, but particularly so in relation to higher and professional education (Daly, 2001). Success in bringing a substantive concept of CT to college faculty requires the well-planned professional development based on the multiple dimensions of a substantive concept of CT (Paul, 2005). CT is part of higher order thinking (Daniel, 2001). Tractenberg et al. (2013) indicated that higher order thinking is valued across many curricula in higher education. Barak and Dori (2009) stated that students' higher order thinking skills are enhanced in terms of their ability to pose complex questions, present solid opinions, introduce consistent arguments, and demonstrate CT. CT is linked to problem solving (Tyreman, 2000).

The critical analysis of the literature in the areas of CT and reflective practice are described as a background for contemporary work with self-regulated learning theory (Kuiper & Pesut, 2004). CT is foundational to the effective teaching of any subject, and it must be at the essence of any professional development program (Elder, 2005). CT should be expanded beyond its rationalist criteria to focus on the process of inquiry (Bevan, 2009). The process of CT, learning, discussion, and inquiry, involves tentativeness in accepting a belief (Ikuenobe, 2002). Both CT and obstacles to successful CT are commonly found in the activities of everyday life (Alston, 2001). CT strategies (e.g., questioning, debate, role play, and small group activity) are developed and utilized in a professional development program (Simpson & Courtney, 2008).

ADVANCED ISSUES AND IMPLICATIONS OF CRITICAL THINKING

This section emphasizes the overview of CT; the emerging trends of CT in the modern learning environments; CT in nursing education and health care; reflective judgment and CT; constructivist learning and CT; metacognitive strategies, cognitive strategies, and CT; and the promotion of CT in the modern learning environments.

Overview of Critical Thinking

Critical thinking (CT) has drawn a lot of attention both in educational research and practice (Niu, Behar-Horenstein, & Garvan, 2013). Most definitions of CT acknowledge the importance of both cognitive (i.e., skills) and dispositional (i.e., propensity) dimensions in the thinking process (Ku, 2009). The cognitive dimension of critical thinking emphasizes the reasoning and logical thinking skills related to intellectual ability (Sosu, 2013). The important cognitive skills include the interpretation, analysis, and evaluation of arguments (Halpern, 1998). Disposition dimension examines the tendency to do something (Ennis, 1996), or the manner in which individuals organize the tasks (Ku & Ho, 2010).

CT dispositions have been recognized as a pattern of attitudes, intellectual virtues, and habits of mind, thus explaining the way an individual reasons, argues, and make appropriate decisions (Facione, Sanchez, Facione, & Gainen, 1995). Having a disposition to critically think implies having the ability to do so (Sosu, 2013). The disposition to think critically has been linked with improved academic performance, deep learning, good professional practice, professional expertise, anxiety, and ego-resilience (Fahim, Bagherkazemi, & Alemi, 2010). CT dispositions have a significant effect on entrepreneurship levels (Kırbaşlar & Özsoy-Güneş, 2015). Personality characteristics (e.g., openness to experience) are correlated with CT (Clifford, Boufal, & Kurtz, 2004).

CT is conceptualized in terms of processes and skills (Bailin, 2002). There are at least six ways of defining CT (i.e., skill, ability components, pedagogy, ways of being, developmental approach, and integrative approach) (Moon, 2008). Definitions of CT emphasize argumentation, verbal reasoning, hypothesis testing, judging likelihood, uncertainty, and problem solving (Halpern, 2014). Taylor (2006) viewed CT as the rational examination of ideas, inferences, principles, arguments, conclusions, issues, statements, beliefs, and actions. Wilgis and McConnell (2008) defined CT as an intentional, complex phenomenon that involves the analysis, interpretation, and evaluation of information to make judgments and decisions. Cottrell (2011) defined CT as a complex process of deliberation which involves a broad range of skills and attitudes.

CT is a fundamental component of management, decision making, clinical judgment, professional achievement, and effective cooperation in the community (Akyuz & Samsa, 2009). CT requires the reflective thinking toward solving the complex problems (Paul, 2014). The complexity of CT as a multidimensional construct presents many challenges to those seeking to scientifically study its instruction and assessment (Bensley & Murtagh, 2012). The Cornell Critical Thinking Test (CCTT) is one of the many multiple-choice tests with validated questions that have been reported to measure general CT ability (Iwaoka, Li, & Rhee, 2010). The Halpern Critical Thinking Assessment (HCTA) is a reliable measure of CT skills and has been validated with multiple populations and measures of academic success (Butler, 2012).

Emerging Trends of Critical Thinking in the Modern Learning Environments

Thinking and learning are interrelated (Andreou, Papastavrou, & Merkouris, 2014). Many studies have demonstrated significant correlations between CT dispositions and CT ability (Rimiene, 2002). CT ability is related to motivation toward learning (Valenzuela, Nieto, & Saiz, 2011), as well as individual's perceived requirement for the effective thinking processes (Halpern, 2006). Brainstorming strategies have a positive effect on both CT ability and reading comprehension ability (Ghabanchi & Behrooznia, 2014). The constructs of CT dispositions include individual's ability, sensitivity, and inclination to engage in the critical thought (Mathews & Lowe, 2011). CT involves a disposition to a defined process of inquiring, learning, and acquiring knowledge, in terms of rationally evaluating and justifying beliefs (Ikuenobe, 2001). Possessing CT disposition toward thinking refers to the extent to which an individual is inclined to perform a given thinking skill (Valenzuela et al., 2011).

CT is important in contemporary life, where individuals face the large amounts of information, complex problems, and rapid technological changes (Angeli & Valanides, 2009). CT is an approach to organizing and carrying out tasks (Forneris & Peden-McAlpine, 2009). CT connects to the value placed on rationality and agency (Doddington, 2007). CT performance involves, not only various reasoning skills but also CT dispositions (Clifford et al., 2004) and metacognition (Magno, 2010).

Metacognition refers to knowledge, awareness, and control of individual's own cognition (Bensley & Spero, 2014). Psychologists have identified metacognitive monitoring as central to CT (Tarricone, 2011).

The ability to think critically is a valuable asset that benefits students (Yang, 2008), and becomes more important to success in life as the pace of change continues to accelerate (Ben-Chaim, Ron, & Zoller, 2000). Due to what can be considered a rapid increase in the creation of new information every year (Jukes & McCain, 2002), CT skills are needed more than ever in order to assist individuals in becoming more adaptable, flexible, and better able to cope with the rapidly evolving information. Manalo et al. (2013) stated that CT requires making judgment calls about the quality of presented information. Instruction in CT is important because it allows individuals to gain a more complex understanding of information they encounter and promotes effective decision making and problem solving in the real-world applications (Butler et al., 2012).

Reflection, concept mapping, and questioning are some of the best practices identified as effective in developing the CT dimension (Burrell, 2014). Reflection is an active process that involves viewing oneself and the environment to gain discernment related to self and others (Smith, 2011) and is associated with the enhanced thought process, learning, and self-assessment (Burrell, 2014). Teachers may lack the ability to effectively teach CT skills because of the inability to engage in critical reflection of their teaching practices (Choy & Oo, 2012). Concept mapping is an instructional tool that stimulates students to engage in CT (Burrell, 2014). Concept mapping activity involves creating a structured diagram that illustrates the relationship between concepts (Ellermann, Kataoka-Yahior, & Wong, 2006). Applying concept mapping results in a significant improvement in CT skills (Atay & Karabacak, 2012) and creates an opportunity for teaching and learning (Taylor & Wros, 2007). Concept mapping is adopted to evaluate the students' ability to apply conceptual knowledge to their actual cognitive activities (Yu, Lin, & Fan, 2015). Questioning in the classroom is applied to explore the important areas for discussion or with targeted content (Billings & Halstead, 2012) and promotes the refinement of CT in all levels of education (Seker & Komur, 2008).

CT skills are important in the social and interpersonal contexts where good decision making and problem solving are needed on a daily basis (Ku, 2009). Good critical thinkers make better decisions and judgments in complex situations (Gambrill, 2006), engage less in cognitive bias and heuristic thinking (Facione & Facione, 2001) and are more likely to get better grades, become more informed and more active citizens, and are more employable (Barton & McCully, 2007). Critical thinkers have the CT skills to evaluate the credibility of sources, identify reasons, ask suitable questions, synthesize information from a variety of sources, and draw appropriate explanations from the particular type of task (Halpern, 2014).

Critical Thinking in Nursing Education and Health Care

The need for CT in nursing education has been emphasized in response to the rapidly changing health-care environment (Simpson & Courtney, 2002). CT is an essential competency (Wang & Woo, 2010) and has been widely discussed in nursing education (Shin, Lee, Ha, & Kim, 2006). Nursing education continues to work to improve and accommodate the changing need of the nursing student and the health care industry (Paul, 2014). Nursing education improves the abilities of students in solving problems, decision making, and CT in different circumstances (Atay & Karabacak, 2012). Nursing education worldwide is embracing CT as a desirable educational outcome, to the extent that some commentators refer to the importance of CT in nursing as a given (Daly, 2001).

Teaching for CT represents a vital and emerging priority in school health education (Broadbear & Keyser, 2000). The development of CT among nursing graduates is important that education experts refer to it as the main goal of all educational endeavors (Tajvidi, Ghiyasvandian, & Salsali, 2014). Wangensteen et al. (2011) stated that CT should be strengthened in nursing education, and CT is a significant predictor for attitude toward research. Developing a framework of best practices for promoting CT in nursing education can assist nurse educators with integrating CT strategies into nursing curricula (Burrell, 2014). CT drives the acquisition of new knowledge and reassessment of previous assumptions that are the essential entities for nurses (Kaddoura, 2010). CT enhances the nursing students' awareness and decision-making skills (Walthew, 2004).

Hwang et al. (2010) stated that CT is the utilization of self-regulatory judgments to identify the patient's problems and provide the patient care. Chang et al. (2011) stated that CT ability has a positive correlation with nursing competence. Clinical educators and managers are challenged to develop the learning strategies that help students critically think and transfer CT skills into the effective nursing practice (Marchigiano, Eduljee, & Harvey, 2011). CT skills and the inclination to engage in CT are essential for nurse managers to function as transformational leaders capable of influencing staff to effectively align with organizational goals in health care (Zori, Nosek, & Musil, 2010). The development of students' CT abilities is one of the greatest challenges facing contemporary nursing educators (Mun, 2010).

The application of simulations has been considered as opportunities for students to enhance their CT abilities (Shinnick & Woo, 2013). The simulation courseware can achieve substantial CT learning outcomes in nursing education (Shin, Ma, Park, Ji, & Kim, 2015). Contextual learning can be a model of clinical learning in nursing education that develops the contextual, reflective nature of CT (Forneris & Peden-McAlpine, 2007). Knowledge and understanding of nursing students' CT skills and related dispositions are important to nurse educators (Profetto-McGrath,

2003). Reflective clinical reasoning in nursing practice depends on the development of both cognitive and metacognitive skill acquisition (Kuiper & Pesut, 2004).

Reflective Judgement, and Critical Thinking

Like CT, reflective judgment is an important skill for students to acquire and practice, because it facilitates their ongoing acquisition and application of knowledge both inside and outside of school and university (Huffaker & Calvert, 2003). King and Kitchener (1994) indicated that reflective judgment is the individuals' understanding of the nature, limits, and certainty of knowing and how this can affect how they defend their judgments and reasoning in context. Reflective judgment is a metacognitive process that is utilized in the context of CT to judge and make decisions in a reflective manner (Dwyer, Hogan, & Stewart, 2015).

Reflective judgment is considered as a component of CT because reflective judgment allows individual to acknowledge that epistemic assumptions (i.e., assumptions about individual's knowledge) are important to judge a situation in which CT is required (King & Kitchener,1994). Reflective judgment has been equated with metacognition (Hofer, 2004), which informs the understanding of its role in CT as a self-regulatory function of thought (Ku & Ho, 2010). Reflective judgment influences how well an individual applies specific CT skills, for example, influencing the way in which the analysis and evaluation processes are conducted (King, Wood, & Mines, 1990).

Constructivist Learning and Critical Thinking

The goal of constructivist learning is to promote the understanding and utilization of knowledge, self-regulation, and mindful reflection regarding CT skills (Driscoll, 2005). Constructivist learning assumes learners actively construct knowledge for themselves (Schunk, 2012). Exploring the perspective of the constructivist learning environment and the extent to which it is related to students' CT can provide the useful information for teachers and researchers (Kwan & Wong, 2014). Problem-based learning can increase CT and problem-solving skills (Chan, 2013). Problem-based learning is a constructivist perspective related to learning, thus promoting reflective thinking in students (Lim, 2011).

Constructivism identifies two dimensions: the domain of epistemology (a theory about knowledge) and the perspective of pedagogy (McNichols, 1999). The domain of epistemology refers to the acquisition and sources of knowledge and the form of pedagogy indicates types of instructional theories, such as collaborative learning, student-centered learning, and authentic assessment (von Glasersfeld, 1990). Social interaction with other people within their environment is important for the development of CT (Vygotsky, 1978). Constructivist instructional methods are defined as the patterns of learner-centered learning (Ernst & Monroe, 2006) and cooperative learning (Yang, Newby, & Bill, 2005).

Constructivist-oriented program improves the post-test CT of both 9th and 12th grade students (Ernst & Monroe, 2006). Constructivist learning environments making use of the Internet and information technology (IT) have the positive effect on students' CT (Yang et al., 2005). The constructivist learning environment is a factor influencing CT through the existing motivational beliefs and cognitive strategies (Kwan & Wong, 2015). The students' motivational beliefs serve as a mediator in developing their thinking and cognitive engagement (Meece, Blumenfeld, & Hoyle, 1988). Meece et al. (1988) indicated that students in a constructivist-oriented learning environment are reported to have the higher intrinsic motivation, while those in a traditional whole-class learning environment are reported to have less intrinsic motivation. Classroom environment affects the goal motivation of students (Meece et al., 1988).

Metacognitive Strategies, Cognitive Strategies, and Critical Thinking

CT occurs when individuals utilize their metacognitive skills and strategies that increase the probability of a desirable outcome (Magno, 2010). Cognitive strategies are called learning/thinking strategies which are defined as an individual's information processing or thinking tools to learn and remember, transforming incoming information into knowledge, retrieving and applying this knowledge later (Lefrancois, 2000). Cognitive strategies include rehearsal, elaboration, and organization, aiming at recognizing through the method of repeating, associating, and grouping the materials (Duncan & McKeachie, 2005). The ability to metacognitively think about thinking (Ku & Ho, 2010) implies a reflective sensibility and the capacity for reflective judgment (King & Kitchener, 1994). Executive function refers to the metacognitive processes (Gagne, 1985), regarding self-regulate thought, such as attention (Banich, 2009), memory processes (Hofmann, Schmeichel, & Baddeley, 2012), and higher order thinking skills (Miller & Cohen, 2001).

CT is metacognitive in nature as it involves the ability to evaluate the individual's own cognition or the cognition of others and infer the reasonable conclusions from the associated cognitive and metacognitive products of thinking (Dwyer, Hogan, & Stewart, 2012). CT is a metacognitive process that, through the utilization of a number of sub-skills (e.g., analysis, evaluation, and inference), increases the chance of drawing the logical solution (Dwyer, Hogan, & Stewart, 2014). Boekaerts (1988) stated that metacognition involves two key dimensions: an awareness of individual's own cognitive functioning (metacognitive knowledge) and application of individual's cognitive resources for learning and problem solving.

Metacognitive strategies refer to the processes of operating cognitive processes in a self-monitoring manner (Boekaerts, 2010). The metacognitive strategies comprise planning, monitoring, and evaluating (Duncan & McKeachie, 2005). Metacognitive and cognitive strategies are important to CT ability (Kwan & Wong, 2015). Good critical thinkers effectively engage in more metacognitive activities, especially high-level planning and high-level evaluating strategies (Ku & Ho, 2010). Chen (2001) indicated that providing a constructivist-oriented learning environment only is not enough to develop the students' CT. Encouraging students to use the cognitive and metacognitive strategies is more effective (Chen, 2001).

PROMOTION OF CRITICAL THINKING IN THE MODERN LEARNING ENVIRONMENTS

CT has become an educational perspective with most policymakers and educationists calling for the development of students' learning attitudes (Stapleton, 2010). CT is viewed as a skilled process that conceptualizes the information from observation, experience, reflection, inference, and communication in a technical manner (Shin et al., 2006). CT skills are transferable and that they can be applied in practice when appraising, evaluating, and implementing research (Seymour et al., 2003). CT processes, including but not limited to reflection, inference, and synthesizing information, enable individuals to make the reasonable judgments not only in the classroom but in everyday life (Beaumont, 2010).

CT skills are required to meet the learning needs (Lin et al., 2015). Educational activities (e.g., co-curriculum, training, and camp) related to CT and problem-solving skills should be promoted to the students (Rodzalan & Saat, 2015). Promoting CT skills in the classroom is considered as an important educational activity (Seymour, Kinn, & Sutherland, 2003). CT exercises can change the student's focus from remembering to active learning (Simpson & Courtney, 2008). Gaming can help develop students' CT skills (Gerber & Scott, 2011). Asynchronous discussion boards have been applied as the online learning approaches (Pucer, Trobec, & Žvanut, 2014). Asynchronous discussion boards promote active learning behavior, positively affect the student's learning outcomes (Williams & Lahman, 2011), and encourage the utilization of CT (Arend, 2009).

Advances in information and communication technology (ICT) have opened opportunities for online learning by interactivity (Baid & Lambert, 2010). Interactivity is recognized as a driver for CT (Saadé, Morin, & Thomas, 2012), as it offers an educational opportunity to explore the innovative pedagogical solutions that help students develop CT skills in a safe environment with real-life practical problems (Kaddoura, 2010). Learner-centered, interactive approaches, and strategies, such as web-based educational simulations (Salleh, Tasir, & Shukor, 2012) and interactive exercises (Boctor, 2013), allow students to individually learn the modern ways of thinking (Carter & Rukholm, 2008).

Many studies focus on exploring the implicit theories of teachers and students in the teaching field related to the issue of CT in education (Smetanová, Drbalová, & Vitáková, 2015). Acquisition of CT skills is viewed as an important goal in higher education (Heijltjes, van Gog, Leppink, & Paas, 2015). Educational approaches are acknowledged to have the significant effect on the students' CT development (Tiwari, Lai, So, & Yuen, 2006). Although CT is considered as an essential skill for learning and for coping with ever-changing working environments (Halpern, 2014), there is increasing evidence that all students do not improve their CT skills during their studies in higher education (Arum & Roksa, 2011). Higher education has been identified as an extremely adequate context in which to facilitate learning of CT skills (Rapanta, Garcia-Mila, & Gilabert, 2013).

In order to become effective, teachers and students need to develop CT skills (Rahimi & Sajed, 2014). Thinking critically drives a student's learning toward self-directed learning (Lin et al., 2015). Active learning strategies employed in the study are useful to promote student's CT skills (Kim, Sharma, Land, & Furlong, 2013). CT skills are essential in the instructional design process (Baum & Newbill, 2010). CT has become a matter of interest to the assessment offices (Fawkes, O'meara, Weber, & Flage, 2005). Teaching CT in community college classrooms involves helping students overcome the emotional barriers and question (Brookfield, 2005).

The teaching of CT and reasoning in general should strive to bring about the learning attitudes in students (Ritola, 2011). Students who spend more time taking classes and are more fully engaged in the university experience have greater gains in CT (Borg & Stranahan, 2010). Explicit instruction combined with practice is required to improve CT skills (Heijltjes, van Gog, & Paas, 2014). Graduate students themselves have reported that their CT skills are inadequate for their future work (Tynjala, Slotte, Nieminen, Lonka, & Olkinuora, 2007), and employers have expressed similar perspectives regarding students' preparedness in the learning environments (Tynjala, 2008).

Regarding CT, all cognitive processes are governed by information processing, which includes acquiring information, coding information for storage, and retrieving information from memory (Schunk, 2012). There are two assumptions: the memory system is an active, organized processor of information, and prior knowledge plays an important role in learning (Gredler, 2009). The information processing of a human mind is represented by three stages of the human memory system, being responsible for transferring information from one stage to the next and then storing it in the memory (Driscoll, 2005). The level of CT disposition is positively related to the students' learning (Bell & Loon, 2015). CT disposition is recognized as the internal motivation to engage problems and make decisions through the utilization of CT (Facione, Facione, & Giancarlo, 1998) and is a measure of the tendency toward CT (Stedman & Andenoro, 2007). Teacher assistance in group work is helpful in promoting students' CT dispositions in the modern learning environments (Fung, 2014).

FUTURE RESEARCH DIRECTIONS

The classification of the extensive literature in the domains of CT will provide the potential opportunities for future research. CT is a decision-making process used to solve problems. Neuroeconomics determines the practical understanding of the brain's decision-making processes, thus challenging age-old assumptions to support more tailored research, treatment options, and greater insight into how and why judgments are formed (Kasemsap, 2016c). Neuroscience is the study of the brain and the nervous system, drawing from psychology and biology, as well as computer science, chemistry, engineering, medicine, and other fields. Effective problem-solving skills enable employees to analyze problems, identify problem severity, and evaluate the impact of alternative solutions. Talent management is a key component to business success in the current economy as it allows companies to retain top talent while increasing productivity (Kasemsap, 2016d). An examination of linkages among CT, neuroeconomics, neuroscience, problem-solving skills, and talent management would seem to be viable for future research efforts.

CONCLUSION

This chapter highlighted the overview of CT; the emerging trends of CT in the modern learning environments; CT in nursing education and health care; reflective judgment and CT; constructivist learning and CT; metacognitive strategies, cognitive strategies, and CT; and the promotion of CT in the modern learning environments. CT is the intellectually disciplined process of actively and skillfully conceptualizing, applying, analyzing, synthesizing, and/or evaluating information gathered from, or generated by, observation, experience, reflection, reasoning, or communication, as a guide to belief and action. CT is a higher-level cognitive skill that involves thinking about thinking. CT is aimed at achieving the best possible outcomes in any situation.

CT is the practice of methodically gathering, analyzing, and evaluating information. CT is one of the most vital parts of the problem-solving and decision-making process, as it is the act of clearly thinking through options that will lead to a final choice. While decision making is the process that leads to actionable conclusions, CT is the element that defines whether the choice is sound. CT involves reviewing the results of the application of decisions made and implementing change where possible. CT enables individuals to recognize a wide range of subjective analyses of otherwise objective data, and to evaluate how well each analysis might meet individuals' needs.

Robotics and artificial intelligence can facilitate the development of CT skills in the robotic age. The promotion of CT is essential for the learning organizations that seek to serve educators, increase educational performance, strengthen competitiveness, and achieve continuous success in the modern learning environments. Therefore, it is required for the learning organizations to promote their CT and develop a learning plan to regularly check their practical advancements toward satisfying educator requirements. Promoting CT has the potential to enhance organizational performance and reach strategic goals in the modern learning environments.

REFERENCES

Akyuz, H. I., & Samsa, S. (2009). The effects of blended learning environment on the critical thinking skills of students. *Procedia: Social and Behavioral Sciences*, *1*(1), 1744–1748. doi:10.1016/j.sbspro.2009.01.308

Alston, K. (2001). Re/thinking critical thinking: The seductions of everyday life. *Studies in Philosophy and Education*, *20*(1), 27–40. doi:10.1023/A:1005247128053

Andreou, C., Papastavrou, E., & Merkouris, A. (2014). Learning styles and critical thinking relationship in baccalaureate nursing education: A systematic review. *Nurse Education Today*, *34*(3), 362–371. doi:10.1016/j.nedt.2013.06.004 PMID:23830067

Angeli, C., & Valanides, C. (2009). Instructional effects on critical thinking: Performance on ill-defined issues. *Learning and Instruction*, *19*(4), 322–334. doi:10.1016/j.learninstruc.2008.06.010

Arend, B. (2009). Encouraging critical thinking in online threaded discussions. *The Journal of Educators Online*, *6*(1), 1–23.

Arum, R., & Roksa, J. (2011). *Academically adrift. Limited learning on college campuses*. Chicago, IL: The University of Chicago Press.

Atay, S., & Karabacak, Ü. (2012). Care plans using concept maps and their effects on the critical thinking dispositions of nursing students. *International Journal of Nursing Practice, 18*(3), 233–239. doi:10.1111/j.1440-172X.2012.02034.x PMID:22621292

Baid, H., & Lambert, N. (2010). Enjoyable learning: The role of humour, games, and fun activities in nursing and midwifery education. *Nurse Education Today, 30*(6), 548–552. doi:10.1016/j.nedt.2009.11.007 PMID:20044181

Bailin, S. (2002). Critical thinking and science education. *Science & Education, 11*(4), 361–375. doi:10.1023/A:1016042608621

Banich, M. T. (2009). Executive function: The search for an integrated account. *Current Directions in Psychological Science, 18*(2), 89–94. doi:10.1111/j.1467-8721.2009.01615.x

Barak, M., & Dori, Y. J. (2009). Enhancing higher order thinking skills among inservice science teachers via embedded assessment. *Journal of Science Teacher Education, 20*(5), 459–474. doi:10.1007/s10972-009-9141-z

Barton, K., & McCully, A. (2007). Teaching controversial issues where controversial issues really matter. *Teaching History, 127*, 13–19.

Baum, L. M., & Newbill, P. L. (2010). Instructional design as critical and creative thinking: A journey through a Jamestown-era Native American village. *TechTrends, 54*(5), 27–37. doi:10.1007/s11528-010-0434-z

Beaumont, J. (2010). A sequence of critical thinking tasks. *TESOL Journal, 1*(4), 427–448. doi:10.5054/tj.2010.234763

Bell, R., & Loon, M. (2015). The impact of critical thinking disposition on learning using business simulations. *The International Journal of Management Education, 13*(2), 119–127. doi:10.1016/j.ijme.2015.01.002

Ben-Chaim, D., Ron, S., & Zoller, U. (2000). The disposition of eleventh-grade science students toward critical thinking. *Journal of Science Education and Technology, 9*(2), 149–159. doi:10.1023/A:1009474023364

Bensley, D. A., & Murtagh, M. P. (2012). Guidelines for a scientific approach to critical thinking assessment. *Teaching of Psychology, 39*(1), 5–16. doi:10.1177/0098628311430642

Bensley, D. A., & Spero, R. A. (2014). Improving critical thinking skills and metacognitive monitoring through direct infusion. *Thinking Skills and Creativity, 12*, 55–68. doi:10.1016/j.tsc.2014.02.001

Bernstein, D., & Crowley, K. (2008). Searching for signs of intelligent life: An investigation of young children's beliefs about robot intelligence. *Journal of the Learning Sciences*, *17*(2), 225–247. doi:10.1080/10508400801986116

Bevan, R. (2009). Expanding rationality: The relation between epistemic virtue and critical thinking. *Educational Theory*, *59*(2), 167–179. doi:10.1111/j.1741-5446.2009.00312.x

Billings, D., & Halstead, J. (2012). *Teaching in nursing: A guide for faculty*. St. Louis, MO: Elsevier Saunders.

Boctor, L. (2013). Active-learning strategies: The use of a game to reinforce learning in nursing education. A case study. *Nurse Education in Practice*, *13*(2), 96–100. doi:10.1016/j.nepr.2012.07.010 PMID:22910398

Boekaerts, M. (1988). Emotion, motivation and learning. *International Journal of Educational Research*, *12*(3), 229–234. doi:10.1016/0883-0355(88)90001-8

Boekaerts, M. (2010). Motivation and self-regulation: Two close friends. In T. Urdan & S. Karabenick (Eds.), *The decade ahead: Applications and contexts of motivation and achievement* (pp. 69–108). Bingley, United Kingdom: Emerald Group Publishing. doi:10.1108/S0749-7423(2010)000016B006

Borg, M. O., & Stranahan, H. A. (2010). Evidence on the relationship between economics and critical thinking skills. *Contemporary Economic Policy*, *28*(1), 80–93. doi:10.1111/j.1465-7287.2008.00134.x

Broadbear, J. T., & Keyser, B. B. (2000). An approach to teaching for critical thinking in health education. *The Journal of School Health*, *70*(8), 322–326. doi:10.1111/j.1746-1561.2000.tb07266.x PMID:11044963

Brookfield, S. D. (2005). Overcoming impostorship, cultural suicide, and lost innocence: Implications for teaching critical thinking in the community college. *New Directions for Community Colleges*, *2005*(130), 49–57. doi:10.1002/cc.195

Burrell, L. A. (2014). Integrating critical thinking strategies into nursing curricula. *Teaching and Learning in Nursing*, *9*(2), 53–58. doi:10.1016/j.teln.2013.12.005

Butler, H. A. (2012). Halpern critical thinking assessment predicts real-world outcomes of critical thinking. *Applied Cognitive Psychology*, *26*(5), 721–729. doi:10.1002/acp.2851

Butler, H. A., Dwyer, C. P., Hogan, M. J., Franco, A., Rivas, S. F., Saiz, C., & Almeida, L. S. (2012). The Halpern critical thinking assessments and real-world outcomes: Cross-national applications. *Thinking Skills and Creativity*, *7*(2), 112–121. doi:10.1016/j.tsc.2012.04.001

Carlgren, T. (2013). Communication, critical thinking, problem solving: A suggested course for all high school students in the 21st century. *Interchange*, *44*(1), 63–81. doi:10.1007/s10780-013-9197-8

Carmel-Gilfilen, C., & Portillo, M. (2010). Creating mature thinkers in interior design: Pathways of intellectual development. *Journal of Interior Design*, *35*(3), 1–20. doi:10.1111/j.1939-1668.2010.01039.x

Carter, L., & Rukholm, E. (2008). A study of critical thinking, teacher-student interaction, and discipline-specific writing in an online educational setting for registered nurses. *Journal of Continuing Education in Nursing*, *39*(3), 133–138. doi:10.3928/00220124-20080301-03 PMID:18386701

Chan, Z. C. (2013). Exploring creativity and critical thinking in traditional and innovative problem-based learning groups. *Journal of Clinical Nursing*, *22*(15/16), 2298–2307. doi:10.1111/jocn.12186 PMID:23452036

Chang, M. J., Chang, Y. J., Kuo, S. H., Yang, Y. H., & Chou, F. H. (2011). Relationships between critical thinking ability and nursing competence in clinical nurses. *Journal of Clinical Nursing*, *20*(21/22), 3224–3232. doi:10.1111/j.1365-2702.2010.03593.x PMID:21414054

Chen, W. (2001). Description of an expert teacher's constructivist-oriented teaching: Engaging students' critical thinking in learning creative dance. *Research Quarterly for Exercise and Sport*, *72*(4), 366–375. doi:10.1080/02701367.2001.10608973 PMID:11770786

Choy, S., & Oo, P. (2012). Reflective thinking and teaching practices: A precursor for incorporating critical thinking into the classroom. *International Journal of Instruction*, *5*(1), 167–182.

Clifford, J. S., Boufal, M. M., & Kurtz, J. E. (2004). Personality traits and critical thinking skills in college students: Empirical tests of a two-factor theory. *Assessment*, *11*(2), 169–176. doi:10.1177/1073191104263250 PMID:15171465

Cottrell, S. (2011). *Critical thinking skills: Developing effective analysis and argument*. New York, NY: Palgrave MacMillan.

Daly, W. M. (2001). The development of an alternative method in the assessment of critical thinking as an outcome of nursing education. *Journal of Advanced Nursing*, *36*(1), 120–130. doi:10.1046/j.1365-2648.2001.01949.x PMID:11555056

Daniel, M. F. (2001). Philosophical dialogue among peers: A study of manifestations of critical thinking in pre-service teachers. *Advances in Health Sciences Education: Theory and Practice*, *6*(1), 49–67. doi:10.1023/A:1009827102022 PMID:11486139

Daniel, M. F., & Auriac, E. (2011). Philosophy, critical thinking and philosophy for children. *Educational Philosophy and Theory*, *43*(5), 415–435. doi:10.1111/j.1469-5812.2008.00483.x

Dewey, J. (2004). *How we think*. Whitefish, MT: Kessinger Publishing.

Doddington, C. (2007). Critical thinking as a source of respect for persons: A critique. *Educational Philosophy and Theory*, *39*(4), 449–459. doi:10.1111/j.1469-5812.2007.00350.x

Driscoll, M. P. (2005). *Psychology of learning for instruction*. Boston, MA: Pearson Education.

Duncan, T., & McKeachie, W. (2005). The making of the motivated strategies for learning questionnaire. *Educational Psychologist*, *40*(2), 117–128. doi:10.1207/s15326985ep4002_6

Dwyer, C. P., Hogan, M. J., Harney, O. M., & O'Reilly, J. (2014). Using interactive management to facilitate a student-centred conceptualisation of critical thinking: A case study. *Educational Technology Research and Development*, *62*(6), 687–709. doi:10.1007/s11423-014-9360-7

Dwyer, C. P., Hogan, M. J., & Stewart, I. (2012). An evaluation of argument mapping as a method of enhancing critical thinking performance in e-learning environments. *Metacognition and Learning*, *7*(3), 219–244. doi:10.1007/s11409-012-9092-1

Dwyer, C. P., Hogan, M. J., & Stewart, I. (2014). An integrated critical thinking framework for the 21st century. *Thinking Skills and Creativity*, *12*, 43–52. doi:10.1016/j.tsc.2013.12.004

Dwyer, C. P., Hogan, M. J., & Stewart, I. (2015). The effects of argument mapping-infused critical thinking instruction on reflective judgement performance. *Thinking Skills and Creativity*, *16*, 11–26. doi:10.1016/j.tsc.2014.12.002

Eguchi, A. (2016). Educational robotics as a learning tool for promoting rich environments for active learning (REALs). In *Human-computer interaction: Concepts, methodologies, tools, and applications* (pp. 740–767). Hershey, PA: IGI Global. doi:10.4018/978-1-4666-8789-9.ch033

Elder, L. (2005). Critical thinking as the key to the learning college: A professional development model. *New Directions for Community Colleges, 2005*(130), 39–48. doi:10.1002/cc.194

Ellermann, C. R., Kataoka-Yahior, M. R., & Wong, L. (2006). Logic models used to enhance critical thinking. *The Journal of Nursing Education, 45*(6), 220–227. PMID:16780010

Ennis, R. H. (1996). Critical thinking dispositions: Their nature and assessability. *Informal Logic, 18*(2/3), 165–182.

Ernst, J., & Monroe, M. (2006). The effects of environment-based education on students' critical thinking skills and disposition toward critical thinking. *Environmental Education Research, 12*(3/4), 429–443. doi:10.1080/13504620600942998

Facione, P. A., & Facione, N. C. (2001). Analyzing explanations for seemingly irrational choices: Linking argument analysis and cognitive science. *The International Journal of Applied Philosophy, 15*(2), 267–286. doi:10.5840/ijap200115217

Facione, P. A., Facione, N. C., & Giancarlo, C. A. F. (1998). *The California critical thinking disposition inventory*. Millbrae, CA: California Academic Press.

Facione, P. A., Sanchez, C. A., Facione, N. C., & Gainen, J. (1995). The dispositions towards critical thinking. *The Journal of General Education, 44*(1), 1–25.

Fahim, M., Bagherkazemi, M., & Alemi, M. (2010). The relationship between test takers critical thinking ability and their performance on the reading section of TOEFL. *Journal of Language Teaching and Research, 1*(6), 830–837. doi:10.4304/jltr.1.6.830-837

Fawkes, D., O'meara, B., Weber, D., & Flage, D. (2005). Examining the exam: A critical look at the California critical thinking skills test. *Science & Education, 14*(2), 117–135. doi:10.1007/s11191-005-6181-4

Flores, K. L., Matkin, G. S., Burbach, M. E., Quinn, C. E., & Harding, H. (2012). Deficient critical thinking skills among college graduates: Implications for leadership. *Educational Philosophy and Theory, 44*(2), 212–230. doi:10.1111/j.1469-5812.2010.00672.x

Forneris, S. G., & Peden-McAlpine, C. (2007). Evaluation of a reflective learning intervention to improve critical thinking in novice nurses. *Journal of Advanced Nursing*, *57*(4), 410–421. doi:10.1111/j.1365-2648.2007.04120.x PMID:17291205

Forneris, S. G., & Peden-McAlpine, C. (2009). Creating context for critical thinking in practice: The role of the preceptor. *Journal of Advanced Nursing*, *65*(8), 1715–1724. doi:10.1111/j.1365-2648.2009.05031.x PMID:19602012

Fung, D. (2014). Promoting critical thinking through effective group work: A teaching intervention for Hong Kong primary school students. *International Journal of Educational Research*, *66*, 45–62. doi:10.1016/j.ijer.2014.02.002

Gagne, R. M. (1985). *The conditions of learning*. New York, NY: Holt, Rinehart and Winston.

Gambrill, E. (2006). Evidence-based practice and policy: Choices ahead. *Research on Social Work Practice*, *16*(3), 338–357. doi:10.1177/1049731505284205

Gerber, S., & Scott, L. (2011). Gamers and gaming context: Relationships to critical thinking. *British Journal of Educational Technology*, *42*(5), 842–849. doi:10.1111/j.1467-8535.2010.01106.x

Ghabanchi, Z., & Behrooznia, S. (2014). The impact of brainstorming on reading comprehension and critical thinking ability of EFL learners. *Procedia: Social and Behavioral Sciences*, *98*, 513–521. doi:10.1016/j.sbspro.2014.03.447

Gomez, K., Bernstein, D., Zywica, J., Hamner, E., Lee, U., & Cunningham, J. (2016). Robotics community experiences: Leveraging informal design and learning experiences to motivate urban youth in STEM. In Human-computer interaction: Concepts, methodologies, tools, and applications (pp. 120–147). Hershey, PA: IGI Global. doi:10.4018/978-1-4666-8789-9.ch006

Grandgenett, N., Ostler, E., Topp, N., & Goeman, R. (2014). Robotics and problem-based learning in STEM formal educational environments. In *K-12 education: Concepts, methodologies, tools, and applications* (pp. 463–488). Hershey, PA: IGI Global. doi:10.4018/978-1-4666-4502-8.ch028

Gredler, M. E. (2009). *Learning and instruction: Theory into practice*. Upper Saddle River, NJ: Pearson/Merrill Prentice Hall.

Halpern, D. F. (1998). Teaching critical thinking for transfer across domains: Dispositions, skills, structure training, and metacognitive monitoring. *The American Psychologist*, *53*(4), 449–455. doi:10.1037/0003-066X.53.4.449 PMID:9572008

Halpern, D. F. (2006). Is intelligence critical thinking? Why we need a new definition of intelligence. In P. Kyllonen, R. Roberts, & L. Stankov (Eds.), *Extending intelligence: Enhancement and new constructs* (pp. 293–310). New York, NY: Taylor & Francis Group.

Halpern, D. F. (2014). *Thought and knowledge: An introduction to critical thinking*. New York, NY: Psychology Press.

Heijltjes, A., van Gog, T., Leppink, J., & Paas, F. (2015). Unraveling the effects of critical thinking instructions, practice, and self-explanation on students' reasoning performance. *Instructional Science*, *43*(4), 487–506. doi:10.1007/s11251-015-9347-8

Heijltjes, A., van Gog, T., & Paas, F. (2014). Improving students' critical thinking: Empirical support for explicit instructions combined with practice. *Applied Cognitive Psychology*, *28*(4), 518–530. doi:10.1002/acp.3025

Hofer, B. K. (2004). Epistemological understanding as a metacognitive process: Thinking aloud during online searching. *Educational Psychologist*, *39*(1), 43–55. doi:10.1207/s15326985ep3901_5

Hofmann, W., Schmeichel, B. J., & Baddeley, A. D. (2012). Executive functions and self-regulation. *Trends in Cognitive Sciences*, *16*(3), 174–180. doi:10.1016/j.tics.2012.01.006 PMID:22336729

Huffaker, D. A., & Calvert, S. L. (2003). The new science of learning: Active learning, metacognition and transfer of knowledge in e-learning applications. *Journal of Educational Computing Research*, *29*(3), 325–334. doi:10.2190/4T89-30W2-DHTM-RTQ2

Hwang, S. Y., Yen, M., Lee, B. O., Huang, M. C., & Tseng, H. F. (2010). A critical thinking disposition scale for nurses: Short form. *Journal of Clinical Nursing*, *19*(21/22), 3171–3176. doi:10.1111/j.1365-2702.2010.03343.x PMID:21040020

Ikuenobe, P. (2001). Questioning as an epistemic process of critical thinking. *Educational Philosophy and Theory*, *33*(3/4), 325–341. doi:10.1111/j.1469-5812.2001.tb00274.x

Ikuenobe, P. (2002). Epistemic foundation for teaching critical thinking in group discussion. *Interchange*, *33*(4), 371–393. doi:10.1023/A:1021564120979

Iwaoka, W. T., Li, Y., & Rhee, W. Y. (2010). Research in food science education: Measuring gains in critical thinking in food science and human nutrition courses: The Cornell critical thinking test, problem-based learning activities, and student journal entries. *Journal of Food Science Education*, *9*(3), 68–75. doi:10.1111/j.1541-4329.2010.00100.x

Jenkins, D. (2012). Global critical leadership: Educating global leaders with critical leadership competencies. *The Journal of Leadership Studies*, *6*(2), 95–101. doi:10.1002/jls.21241

Jukes, I., & McCain, T. (2002). *Minds in play: Computer game design as a context of children's learning*. Mahwah, NJ: Lawrence Erlbaum Associates.

Kaddoura, M. A. (2010). New graduate nurses' perceptions of the effects of clinical simulation on their critical thinking, learning, and confidence. *Journal of Continuing Education in Nursing*, *41*(11), 506–516. doi:10.3928/00220124-20100701-02 PMID:20672760

Kasemsap, K. (2016a). Analyzing the roles of human capital and competency in global business. In S. Sen, A. Bhattacharya, & R. Sen (Eds.), *International perspectives on socio-economic development in the era of globalization* (pp. 1–29). Hershey, PA: IGI Global. doi:10.4018/978-1-4666-9908-3.ch001

Kasemsap, K. (2016b). Mastering big data in the digital age. In M. Singh & D. G. (Eds.), Effective big data management and opportunities for implementation (pp. 104–129). Hershey, PA: IGI Global. doi:10.4018/978-1-5225-0182-4.ch008

Kasemsap, K. (2016c). The fundamentals of neuroeconomics. In B. Christiansen & E. Lechman (Eds.), *Neuroeconomics and the decision-making process* (pp. 1–32). Hershey, PA: IGI Global. doi:10.4018/978-1-4666-9989-2.ch001

Kasemsap, K. (2016d). Promoting leadership development and talent management in modern organizations. In U. Aung & P. Ordoñez de Pablos (Eds.), *Managerial strategies and practice in the Asian business sector* (pp. 238–266). Hershey, PA: IGI Global. doi:10.4018/978-1-4666-9758-4.ch013

Kim, K., Sharma, P., Land, S. M., & Furlong, K. P. (2013). Effects of active learning on enhancing student critical thinking in an undergraduate general science course. *Innovative Higher Education*, *38*(3), 223–235. doi:10.1007/s10755-012-9236-x

King, P. M., & Kitchener, K. S. (1994). *Developing reflective judgment: Understanding and promoting intellectual growth and critical thinking in adolescents and adults*. San Francisco, CA: Jossey–Bass.

King, P. M., Wood, P. K., & Mines, R. A. (1990). Critical thinking among college and graduate students. *The Review of Higher Education, 13*(2), 167–186.

Kırbaşlar, M., & Özsoy-Güneş, Z. (2015). The effect of critical thinking disposition on entrepreneurship levels: A study on future teachers. *Procedia: Social and Behavioral Sciences, 174*, 199–207. doi:10.1016/j.sbspro.2015.01.647

Kovalik, D. L., & Kovalik, L. M. (2007). Language simulations: The blending space for writing and critical thinking. *Simulation & Gaming, 38*(3), 310–322. doi:10.1177/1046878106298271

Ku, K. Y. L. (2009). Assessing students' critical thinking performance: Urging for measurements using multi-response format. *Thinking Skills and Creativity, 4*(1), 70–76. doi:10.1016/j.tsc.2009.02.001

Ku, K. Y. L., & Ho, I. T. (2010). Metacognitive strategies that enhance critical thinking. *Metacognition and Learning, 5*(3), 251–267. doi:10.1007/s11409-010-9060-6

Ku, K. Y. L., Ho, I. T., Hau, K. T., & Lai, E. C. M. (2014). Integrating direct and inquiry-based instruction in the teaching of critical thinking: An intervention study. *Instructional Science, 42*(2), 251–269. doi:10.1007/s11251-013-9279-0

Kuiper, R. A., & Pesut, D. J. (2004). Promoting cognitive and metacognitive reflective reasoning skills in nursing practice: Self-regulated learning theory. *Journal of Advanced Nursing, 45*(4), 381–391. doi:10.1046/j.1365-2648.2003.02921.x PMID:14756832

Kwan, Y. W., & Wong, A. F. L. (2014). The constructivist classroom learning environment and its associations with critical thinking ability of secondary school students in Liberal Studies. *Learning Environments Research, 17*(2), 191–207. doi:10.1007/s10984-014-9158-x

Kwan, Y. W., & Wong, A. F. L. (2015). Effects of the constructivist learning environment on students' critical thinking ability: Cognitive and motivational variables as mediators. *International Journal of Educational Research, 70*, 68–79. doi:10.1016/j.ijer.2015.02.006

Lefrancois, G. R. (2000). *Psychology for teaching*. Belmont, CA: Wadsworth/Cengage Learning.

Lim, L. A. Y. L. (2011). A comparison of students' reflective thinking across different years in a problem-based learning environment. *Instructional Science, 39*(2), 171–188. doi:10.1007/s11251-009-9123-8

Liu, O. L., Frankel, L., & Roohr, K. C. (2014). Assessing critical thinking in higher education: Current state and directions for next-generation assessment. *ETS Research Report Series, 2014*(1), 1–23.

Liu, X., Guo, C., & Zhang, L. (2014). Scholar metadata and knowledge generation with human and artificial intelligence. *Journal of the Association for Information Science and Technology*, *65*(6), 1187–1201. doi:10.1002/asi.23013

Loes, C. N., Salisbury, M. H., & Pascarella, E. T. (2015). Student perceptions of effective instruction and the development of critical thinking: A replication and extension. *Higher Education*, *69*(5), 823–838. doi:10.1007/s10734-014-9807-0

Magno, C. (2010). The role of metacognitive skills in developing critical thinking. *Metacognition and Learning*, *5*(2), 137–156. doi:10.1007/s11409-010-9054-4

Manalo, E., Kusumi, T., Koyasu, M., Michita, Y., & Tanaka, Y. (2013). To what extent do culture-related factors influence university students' critical thinking use? *Thinking Skills and Creativity*, *10*, 121–132. doi:10.1016/j.tsc.2013.08.003

Marchigiano, G., Eduljee, N., & Harvey, K. (2011). Developing critical thinking skills from clinical assignments: A pilot study on nursing students' self-reported perceptions. *Journal of Nursing Management*, *19*(1), 143–152. doi:10.1111/j.1365-2834.2010.01191.x PMID:21223414

Martincová, J., & Lukešová, M. (2015). Critical thinking as a tool for managing intercultural conflicts. *Procedia: Social and Behavioral Sciences*, *171*, 1255–1264. doi:10.1016/j.sbspro.2015.01.239

Mathews, S. R., & Lowe, K. (2011). Classroom environments that foster a disposition for critical thinking. *Learning Environments Research*, *14*(1), 59–73. doi:10.1007/s10984-011-9082-2

McNichols, T. J. (1999). Deconstructing constructivism: The Kantian connection. *Journal of Philosophy and History of Education*, *49*, 141–146.

Meece, J. L., Blumenfeld, P. C., & Hoyle, R. H. (1988). Students' goal orientations and cognitive engagement in classroom activities. *Journal of Educational Psychology*, *80*(4), 514–523. doi:10.1037/0022-0663.80.4.514

Miller, E. K., & Cohen, J. D. (2001). An integrative theory of prefrontal cortex function. *Annual Review of Neuroscience*, *24*(1), 167–202. doi:10.1146/annurev.neuro.24.1.167 PMID:11283309

Moon, J. (2008). *Critical thinking: An exploration of theory and practice*. Abingdon, United Kingdom: Routledge.

Morlino, G., Gianelli, C., Borghi, A. M., & Nolfi, S. (2015). Learning to manipulate and categorize in human and artificial agents. *Cognitive Science*, *39*(1), 39–64. doi:10.1111/cogs.12130 PMID:25041751

Mun, M. S. (2010). An analysis of narratives to identify critical thinking contexts in psychiatric clinical practice. *International Journal of Nursing Practice*, *16*(1), 75–80. doi:10.1111/j.1440-172X.2009.01803.x PMID:20158552

Niu, L., Behar-Horenstein, L. S., & Garvan, C. W. (2013). Do instructional interventions influence college students' critical thinking skills? A meta-analysis. *Educational Research Review*, *9*, 114–128. doi:10.1016/j.edurev.2012.12.002

Paul, R. (2005). The state of critical thinking today. *New Directions for Community Colleges*, *2005*(130), 27–38. doi:10.1002/cc.193

Paul, S. A. (2014). Assessment of critical thinking: A Delphi study. *Nurse Education Today*, *34*(11), 1357–1360. doi:10.1016/j.nedt.2014.03.008 PMID:24713126

Profetto-McGrath, J. (2003). The relationship of critical thinking skills and critical thinking dispositions of baccalaureate nursing students. *Journal of Advanced Nursing*, *43*(6), 569–577. doi:10.1046/j.1365-2648.2003.02755.x PMID:12950562

Pucer, P., Trobec, I., & Žvanut, B. (2014). An information communication technology based approach for the acquisition of critical thinking skills. *Nurse Education Today*, *34*(6), 964–970. doi:10.1016/j.nedt.2014.01.011 PMID:24581890

Qing, Z., Xiang, W., & Linna, Y. (2007). A preliminary investigation into critical thinking of urban Xi'an high school students. *Frontiers of Education in China*, *2*(3), 447–468. doi:10.1007/s11516-007-0036-5

Rahimi, A., & Sajed, M. A. (2014). The interplay between critical pedagogy and critical thinking: Theoretical ties and practicalities. *Procedia: Social and Behavioral Sciences*, *136*, 41–45. doi:10.1016/j.sbspro.2014.05.284

Rapanta, C., Garcia-Mila, M., & Gilabert, S. (2013). What is meant by argumentative competence? An integrative review of methods of analysis and assessment in education. *Review of Educational Research*, *83*(4), 483–520. doi:10.3102/0034654313487606

Rimiene, V. (2002). Assessing and developing students' critical thinking. *Psychology Learning & Teaching*, *2*(1), 17–22. doi:10.2304/plat.2002.2.1.17

Ritola, J. (2011). Justificationist social epistemology and critical thinking. *Educational Theory*, *61*(5), 565–585. doi:10.1111/j.1741-5446.2011.00421.x

Rodzalan, S. A., & Saat, M. M. (2015). The perception of critical thinking and problem solving skill among Malaysian undergraduate students. *Procedia: Social and Behavioral Sciences*, *172*, 725–732. doi:10.1016/j.sbspro.2015.01.425

Saadé, R. G., Morin, D., & Thomas, J. D. E. (2012). Critical thinking in e-learning environments. *Computers in Human Behavior*, *28*(5), 1608–1617. doi:10.1016/j.chb.2012.03.025

Salleh, S. M., Tasir, Z., & Shukor, N. A. (2012). Web-based simulation learning framework to enhance students' critical thinking skills. *Procedia: Social and Behavioral Sciences*, *64*, 372–381. doi:10.1016/j.sbspro.2012.11.044

Schunk, D. H. (2012). *Learning theories: An educational perspective*. Upper Saddle River, NJ: Pearson/Merrill Prentice Hall.

Seker, H., & Komur, S. (2008). The relationship between critical thinking skills and in-class questioning behaviors of English language teaching students. *European Journal of Teacher Education*, *31*(4), 389–402. doi:10.1080/02619760802420784

Seymour, B., Kinn, S., & Sutherland, N. (2003). Valuing both critical and creative thinking in clinical practice: Narrowing the research–practice gap? *Journal of Advanced Nursing*, *42*(3), 288–296. doi:10.1046/j.1365-2648.2003.02618.x PMID:12680973

Shin, H., Ma, H., Park, J., Ji, E. S., & Kim, D. H. (2015). The effect of simulation courseware on critical thinking in undergraduate nursing students: Multi-site pre-post study. *Nurse Education Today*, *35*(4), 537–542. doi:10.1016/j.nedt.2014.12.004 PMID:25549985

Shin, K. R., Lee, J. H., Ha, J. Y., & Kim, K. H. (2006). Critical thinking dispositions in baccalaureate nursing students. *Journal of Advanced Nursing*, *56*(2), 182–189. doi:10.1111/j.1365-2648.2006.03995.x PMID:17018066

Shinnick, M. A., & Woo, W. A. (2013). The effect of human patient simulation on critical thinking and its predictors in prelicensure nursing students. *Nurse Education Today*, *33*(9), 1062–1067. doi:10.1016/j.nedt.2012.04.004 PMID:22564925

Simpson, E., & Courtney, M. (2002). Critical thinking in nursing education: Literature review. *International Journal of Nursing Practice*, *8*(2), 89–98. doi:10.1046/j.1440-172x.2002.00340.x PMID:11993582

Simpson, E., & Courtney, M. (2008). Implementation and evaluation of critical thinking strategies to enhance critical thinking skills in Middle Eastern nurses. *International Journal of Nursing Practice, 14*(6), 449–454. doi:10.1111/j.1440-172X.2008.00719.x PMID:19126073

Smetanová, V., Drbalová, A., & Vitáková, D. (2015). Implicit theories of critical thinking in teachers and future teachers. *Procedia: Social and Behavioral Sciences, 171*, 724–732. doi:10.1016/j.sbspro.2015.01.184

Smith, E. (2011). Teaching critical reflection. *Teaching in Higher Education, 16*(2), 211–223. doi:10.1080/13562517.2010.515022

Sosu, E. M. (2013). The development and psychometric validation of a critical thinking disposition scale. *Thinking Skills and Creativity, 9*, 107–119. doi:10.1016/j.tsc.2012.09.002

Staib, S. (2003). Teaching and measuring critical thinking. *The Journal of Nursing Education, 42*(11), 498–508. PMID:14626388

Stapleton, P. (2010). A survey of attitudes towards critical thinking among Hong Kong secondary school teachers: Implications for policy change. *Thinking Skills and Creativity, 6*(1), 14–23. doi:10.1016/j.tsc.2010.11.002

Stedman, N. L., & Andenoro, C. (2007). Identification of relationships between emotional intelligence and critical thinking dispositions in undergraduate leadership students. *Journal of Leadership Education, 6*(1), 190–208. doi:10.12806/V6/I1/RF10

Tajvidi, M., Ghiyasvandian, S., & Salsali, M. (2014). Probing concept of critical thinking in nursing education in Iran: A concept analysis. *Asian Nursing Research, 8*(2), 158–164. doi:10.1016/j.anr.2014.02.005 PMID:25030649

Tarricone, P. (2011). *The taxonomy of metacognition*. New York, NY: Psychology Press.

Taylor, B. J. (2006). *Reflective practice: A guide for nurses and midwives*. Maidenhead, United Kingdom: Open University Press.

Taylor, J., & Wros, P. (2007). Concept mapping: A nursing model for care planning. *The Journal of Nursing Education, 46*(5), 211–216. PMID:17547344

Tiwari, A., Lai, P., So, M., & Yuen, K. (2006). A comparison of the effects of problem-based learning and lecturing on the development of students' critical thinking. *Medical Education, 40*(6), 547–554. doi:10.1111/j.1365-2929.2006.02481.x PMID:16700770

Tractenberg, R. E., Gushta, M. W., Mulroney, S. E., & Weissinger, P. A. (2013). Multiple choice questions can be designed or revised to challenge learners' critical thinking. *Advances in Health Sciences Education: Theory and Practice*, *18*(5), 945–961. doi:10.1007/s10459-012-9434-4 PMID:23288470

Tynjala, P. (2008). Perspectives into learning at the workplace. *Educational Research Review*, *3*(2), 130–154. doi:10.1016/j.edurev.2007.12.001

Tynjala, P., Slotte, V., Nieminen, J., Lonka, K., & Olkinuora, E. (2007). From university to working life: Graduates workplace skills in practice. In P. Tynjala, J. Valimaa, & G. Boulton-Lewis (Eds.), *Higher education and working life: Collaborations, confrontations and challenges* (pp. 73–88). Oxford, United Kingdom: Elsevier.

Tyreman, S. (2000). Promoting critical thinking in health care: Phronesis and criticality. *Medicine, Health Care, and Philosophy*, *3*(2), 117–124. doi:10.1023/A:1009973021449 PMID:11079339

Valenzuela, J., Nieto, A. M., & Saiz, C. (2011). Critical thinking motivational scale: A contribution to the study of relationship between critical thinking and motivation. *Electric Journal of Research in Educational Psychology*, *9*(2), 823–848.

von Glasersfeld, E. (1990). An exposition of constructivism: Why some like it radical. In R. Davis, C. Maher, & N. Noddings (Eds.), *Constructivist views on the teaching and learning of mathematics* (pp. 19–29). Reston, VA: National Council of Teachers of Mathematics.

Vygotsky, L. S. (1978). *Mind in society: The development of higher psychological processes*. Cambridge, MA: Harvard University Press.

Walthew, P. J. (2004). Conceptions of critical thinking held by nurse educators. *The Journal of Nursing Education*, *43*(9), 408–411. PMID:15478693

Wang, Q., & Woo, H. L. (2010). Investigating students' critical thinking in weblogs: An exploratory study in a Singapore secondary school. *Asia Pacific Education Review*, *11*(4), 541–551. doi:10.1007/s12564-010-9101-5

Wangensteen, S., Johansson, I. S., Björkström, M. E., & Nordström, G. (2011). Research utilisation and critical thinking among newly graduated nurses: Predictors for research use. A quantitative cross-sectional study. *Journal of Clinical Nursing*, *20*(17/18), 2436–2447. doi:10.1111/j.1365-2702.2010.03629.x PMID:21689180

Weissberg, R. (2013). Critically thinking about critical thinking. *Academic Questions*, *26*(3), 317–328. doi:10.1007/s12129-013-9375-2

Wilgis, M., & McConnell, J. (2008). Concept mapping: An educational strategy to improve graduate nurses' critical thinking skills during a hospital orientation program. *Journal of Continuing Education in Nursing*, *39*(3), 119–126. doi:10.3928/00220124-20080301-12 PMID:18386699

Williams, L., & Lahman, M. (2011). Online discussion, student engagement, and critical thinking. *Journal of Political Science Education*, *7*(2), 143–162. doi:10.1080/15512169.2011.564919

Yang, Y. T. C. (2008). A catalyst for teaching critical thinking in a large university class in Taiwan: Asynchronous online discussions with the facilitation of teaching assistants. *Educational Technology Research and Development*, *56*(3), 241–264. doi:10.1007/s11423-007-9054-5

Yang, Y. T. C., Gamble, J. H., Hung, Y. W., & Lin, T. Y. (2014). An online adaptive learning environment for critical-thinking-infused English literacy instruction. *British Journal of Educational Technology*, *45*(4), 723–747. doi:10.1111/bjet.12080

Yang, Y. T. C., Newby, T. J., & Bill, R. L. (2005). Using Socratic questioning to promote critical thinking skills through asynchronous discussion forums in distance learning environments. *American Journal of Distance Education*, *19*(3), 163–181. doi:10.1207/s15389286ajde1903_4

Yu, K. C., Lin, K. Y., & Fan, S. C. (2015). An exploratory study on the application of conceptual knowledge and critical thinking to technological issues. *International Journal of Technology and Design Education*, *25*(3), 339–361. doi:10.1007/s10798-014-9289-5

Zori, S., Nosek, L. J., & Musil, C. M. (2010). Critical thinking of nurse managers related to staff RNs' perceptions of the practice environment. *Journal of Nursing Scholarship*, *42*(3), 305–313. doi:10.1111/j.1547-5069.2010.01354.x PMID:20738741

ADDITIONAL READING

Ahn, S. H., & Yeom, H. A. (2014). Moral sensitivity and critical thinking disposition of nursing students in Korea. *International Journal of Nursing Practice*, *20*(5), 482–489. doi:10.1111/ijn.12185 PMID:24112319

Alexander, P. A. (2014). Thinking critically and analytically about critical-analytic thinking: An introduction. *Educational Psychology Review*, *26*(4), 469–476. doi:10.1007/s10648-014-9283-1

Angeli, C., Valanides, N., & Papastephanou, M. (2011). The role of the authority of the text on critical thinking. *Interchange*, *42*(3), 307–328. doi:10.1007/s10780-012-9161-z

Chittooran, M. M. (2015). Reading and writing for critical reflective thinking. *New Directions for Teaching and Learning*, *2015*(143), 79–95. doi:10.1002/tl.20137

Colley, B. M., Bilics, A. R., & Lerch, C. M. (2012). Reflection: A key component to thinking critically. *The Canadian Journal for the Scholarship of Teaching and Learning*, *3*(1), 1–19. doi:10.5206/cjsotl-rcacea.2012.1.2

Demir, M., Bacan, H., Tarhan, S., & Dombay, M. A. (2011). Quadruple thinking: Critical thinking. *Procedia: Social and Behavioral Sciences*, *12*, 545–551. doi:10.1016/j.sbspro.2011.02.066

Hatcher, D. L. (2011). Which test? Whose scores? Comparing standardized critical thinking tests. *New Directions for Institutional Research*, *2011*(149), 29–39. doi:10.1002/ir.378

Haw, J. (2011). Improving psychological critical thinking in Australian university students. *Australian Journal of Psychology*, *63*(3), 150–153. doi:10.1111/j.1742-9536.2011.00018.x

Hunter, S., Pitt, V., Croce, N., & Roche, J. (2014). Critical thinking skills of undergraduate nursing students: Description and demographic predictors. *Nurse Education Today*, *34*(5), 809–814. doi:10.1016/j.nedt.2013.08.005 PMID:24018356

Hürsen, Ç., & Kaplan, A. (2014). Determination of critical thinking dispositions of the university students in terms of demographic features. *Procedia: Social and Behavioral Sciences*, *131*, 367–372. doi:10.1016/j.sbspro.2014.04.131

Hyytinen, H., Hyytinen, K., Ursin, J., Toom, A., & Lindblom-Ylanne, S. (2015). Problematising the equivalence of the test results of performance-based critical thinking tests for undergraduate students. *Studies in Educational Evaluation*, *44*, 1–8. doi:10.1016/j.stueduc.2014.11.001

Kim, N. (2015). Critical thinking in wikibook creation with enhanced and minimal scaffolds. *Educational Technology Research and Development*, *63*(1), 5–33. doi:10.1007/s11423-014-9361-6

Leach, B. T., & Good, D. W. (2011). Critical thinking skills as related to university students' gender and academic discipline. *International Journal of Humanities and Social Science*, *1*(21), 100–106.

Lin, C. C., Han, C. Y., Pan, I. J., & Chen, L. C. (2015). The teaching-learning approach and critical thinking development: A qualitative exploration of Taiwanese nursing students. *Journal of Professional Nursing*, *31*(2), 149–157. doi:10.1016/j.profnurs.2014.07.001 PMID:25839955

Mahdyeh, N., & Arefi, M. (2014). A comparison of critical thinking, self-efficacy and academic performance among students of faculty of humanities and engineering. *Indian Journal of Fundamental and Applied Life Sciences*, *4*(1), 153–162.

Minakova, L. Y. (2014). Critical thinking development in foreign language teaching for non-language-majoring students. *Procedia: Social and Behavioral Sciences*, *154*, 324–328. doi:10.1016/j.sbspro.2014.10.157

Mulnix, J. W. (2012). Thinking critically about critical thinking. *Educational Philosophy and Theory*, *44*(5), 464–479. doi:10.1111/j.1469-5812.2010.00673.x

Nair, G. G., & Stamler, L. L. (2013). A conceptual framework for developing a critical thinking self-assessment scale. *The Journal of Nursing Education*, *52*(3), 131–138. doi:10.3928/01484834-20120215-01 PMID:23402245

Pitt, V., Powis, D., Levett-Jones, T., & Hunter, S. (2015). The influence of critical thinking skills on performance and progression in a pre-registration nursing program. *Nurse Education Today*, *35*(1), 125–131. doi:10.1016/j.nedt.2014.08.006 PMID:25194469

Rainbolt, G. W., & Dwyer, S. L. (2012). *Critical thinking: The art of argument*. Stamford, CT: Cengage Learning.

Roberts, M., & Ion, R. (2015). Thinking critically about the occurrence of widespread participation in poor nursing care. *Journal of Advanced Nursing*, *71*(4), 768–776. doi:10.1111/jan.12586 PMID:25472849

Shean, A. (2015). Mindless learning versus critical thinking: Educators must teach how to think. *Journal of Psychological Issues in Organizational Culture*, *6*(2), 101–103. doi:10.1002/jpoc.21176

Shehab, H. M., & Nussbaum, E. M. (2015). Cognitive load of critical thinking strategies. *Learning and Instruction*, *35*, 51–61. doi:10.1016/j.learninstruc.2014.09.004

Sorrell, B. (2015). On crafting introductory remarks: Developing a synthetic conception of critical thinking. *Metaphilosophy*, *46*(1), 127–140. doi:10.1111/meta.12117

Šulavíková, B. (2013). Philosophical counselling based on dialogical critical thinking. *Human Affairs*, *23*(4), 680–688. doi:10.2478/s13374-013-0157-x

Vaidya, A. J. (2013). Epistemic responsibility and critical thinking. *Metaphilosophy*, *44*(4), 533–556. doi:10.1111/meta.12047

Wells, L. A. (2015). Clinical thinking in psychiatry. *Journal of Evaluation in Clinical Practice*, *21*(3), 514–517. doi:10.1111/jep.12324 PMID:25653010

KEY TERMS AND DEFINITIONS

Belief: The assumption that is held to be true, by an individual or a group, regarding concepts, events, people, and things.

Cognition: The mental processes involved in judging, knowing, learning, perceiving, recognizing, remembering, thinking, and understanding that lead to the awareness of the world.

Competency: The important skill that is needed to do a task.

Critical Thinking: The unbiased examination of assumptions underlying current beliefs to assess their correctness and legitimacy, and thus to validate or invalidate the beliefs.

Learning: The measurable change in behavior through experience, instruction, and study.

Process: The sequence of linked procedures which consume one or more resources to convert inputs into outputs.

Skill: The ability and capacity acquired through the deliberate, systematic, and sustained effort to execute the complex activities.

Thinking: The activity of utilizing the individual's mind to consider something.

Chapter 5
The Role of Living Labs in the Process of Creating Innovation

Anna Maria Sabat
Jan Kochanowski University, Poland

Anna Katarzyna Florek-Paszkowska
Jagiellonian University, Poland

ABSTRACT

The paper is based on the research carried out into Living Labs in Canada. The aim of the paper is presenting the essence of Living Labs as a concept facilitating innovation generation in businesses thanks to the cooperation of various actors, e.g. producers with users, inspiring the process of the development of new goods and services. The research questions raised pertain to the clarification how Living Labs create innovation in businesses. The Living Labs functioning in the Ontario region were the subject matter of the research. The described case study is theory-genic in character because of the early development stage of the knowledge. During the research process the multi-directional nature and the impact dynamics of the idea of Living Labs among peer partners of innovative processes have been noted, emphasizing the prosumer idea as well as the possibility of businesses cooperating in Living Labs.

DOI: 10.4018/978-1-5225-1656-9.ch005

INTRODUCTION

Failure is an option here. If things are not failing, you are not innovating enough. (Elon Musk)

Living Labs are a new trend in innovation, which is understood in this paper as creating new. J. A. Schumpeter (1960) introduced it into economic sciences. Innovation is applying new knowledge in the production process (Begg, 1997). Innovation is thus a process which modifies the present possibilities, based on the needs of end users in the nearest future as well as implementing them on the market.

A Living Lab is a modern concept but its roots can be traced back to Knight (1749), who was the first to use the term "living laboratory" (den Ouden, E. et al., 2016). Westerlund and Leminen define living labs as: "physical regions, virtual or relational spaces, where the stakeholders cooperate in a public-private-people partnership of business, public institutions, universities, users and others, everyone involved in the developed product or service, participating in the process of prototyping, validation as well as testing of new technologies, services, products and systems in real life, and not in laboratory conditions" (Leminen, 2013). As such, living labs are used to generate innovations.

There are seven key characteristics of living labs that have been included in the definition presented above (Leminen, 2015):

1. Innovation activities take place in real-life environments (cf. Ballon et al., 2005; Intille et al., 2006).
2. Public-private-people partnerships (4Ps) are formed by participants, who include companies, researchers, authorities, and users (cf. Westerlund & Leminen, 2011).
3. The importance of users, including citizens and customers, is emphasized (Følstad 2008).
4. They are different from testbeds, field trials, and other forms of innovation (cf. Almirall et al., 2012; Ballon et al., 2005; Bergvall-Kåreborn et al., 2009). They feature innovations that are more mature than in-house R&D, where prototyping and field trials are more appropriate, but the innovations are less mature than would be found in pilot projects (Ballon et al., 2005).
5. Multiple stakeholders are employed in living labs (cf. Ballon et al., 2005; Leminen & Westerlund, 2012; Westerlund & Leminen, 2011).
6. Multiple roles are pursued by stakeholders in living labs (Nyström et al., 2014).
7. Collaboration between stakeholders is an essential feature of living labs, which are grounded in the principles of open innovation (cf. Leminen & Westerlund, 2012; Niitamo et al., 2006).

Living labs can be an effective method of implementing innovations and creating open systems, which generate innovations thanks to the cooperation of various entities (e.g. manufacturers) with end-users, who inspire this process and can constitute its driving force. A living lab creates an environment, where innovations come into being with the help of testing processes and experimentation as the result of collaboration of the participants of the production process of new goods and services . In this approach living labs can be perceived as organizations of the future.

Also, the way universities function is changing, as different universities explore how to put their "third role" into practice. In pioneering regions across Europe, universities are becoming active players in their communities, contributing to the quality of life and regional well-being, adding value to regional development processes and anchoring the importance of knowledge in the regional innovation ecosystem. Ideally, this is a co-creation process producing regional services in collaboration with industry, public authorities and citizens (den Ouden, E. et al., 2016). The European Union's Smart Specialization Platform breaks down the active regional contributions by universities into four areas (den Ouden, E. et al., 2016 after Markkula M., Kune H., 2015):

1. Business innovation: closely linked, although not exclusively, to the research function of the university.
2. Human capital development: linked to the teaching function.
3. Community development: linked to the public service role of universities.
4. Institutional capacity of the region: the university contributes through engagement of its management and members in local civil society.

Basis on the information presented above living labs can be perceived as:

- A regional system (Oliveira et al., 2006).
- An innovative system (Ballon et al., 2005; Eriksson et al., 2005).
- An ecosystem (Lievens et al., 2011; Schaffers & Turkama, 2012; Tang & Hämäläinen, 2012).
- A network (Leminen & Westerlund, 2012; Leminen et al., 2015a).
- A complex holistic system (Dutilleul et al., 2010).
- A technology and user system (Bajgier et al., 1991; Intille et al., 2006).
- The context and methodology of innovation creation (Almirall et al., 2012; Bergvall-Kåreborn et al., 2009; Dell´Era & Landoni, 2014; Mulder & Stappers, 2009).
- A method of involving the creator and end-user in such areas as: agricultural innovation (Schaffers & Kulkki, 2007), regional innovation (Juujärvi & Pesso, 2013), intelligent cities (Ballon et al., 2011), prosumer innovations

(Leminen & Westerlund, 2012; Nyström, & Westerlund, 2015a), public-private partnerships (PPP) (Lepik et al., 2010; Niitamo et al., 2006), as well as public-private-people partnerships (the quadruple helix) (Arnkil et al., 2010; Ferrari et al., 2011).
- A system of product development, services and models (Bajgier et al., 1991; Bengtson, 1994; Lasher et al., 1991).
- Running a business and work ethic (Veeckman et al., 2013; Schuurman, & De Marez, 2009).
- An innovation management tool (Edvardsson et al., 2012; Leminen & Westerlund, 2012).

The "Living Lab" term is currently seeking out of the scientific vocabulary stock as it is being colloquially used in business as a fashionable trend for many domains. Observing Living Labs, you can see the broad and comprehensive context of operations conducted within their framework, which do not have a single cohesive method or research algorithm either. Nevertheless, even though there is no single definition, you can still attempt to formulate it and observe common features and types of Living Labs (Table 1).

RESEARCH METHOD APPLIED

It was decided to use a case study, where case refers to a recognizable research object, here both a physical and administrative living lab, functioning in 2015, situated in the Ontario region in Ottawa, Canada. It was researched using three data collection and analysis techniques: source document analysis, observation as well as free-form interviews (Czakon, 2011, p. 46).

Using a case study in the research is justified since Living Labs are presently at an early knowledge development stage and is focused on the current phenomenon in actual conditions (Czakon, 2011, p. 47). The process was started by posing the main research question: How do Living Labs create innovations in businesses?

The variety of data sources has assured credible results since triangulation (Czakon, 2011, p. 155) has been used as a qualitative research method. It has turned out to be the most appropriate one for this type research. It derives from grounded theory methodology and is meant for humanities. Its essence is field research, which are characterized by an intellectual process with regard to the following stages:

- A Description of the Researched Reality: Field stories;
- **The Analytical Process:** A description which includes cause and effect dependencies;

Table 1. Characteristics of different types of living labs

Characteristic	Type of Living Labs			
	Utilizer-Driven	Enabler-Driven	Provider-Driven	User-Driven
Purpose	Strategic R&D activity with preset objectives	Strategy development through action	Operations development through increased knowledge	Problem solving by collaborative accomplishments
Organization	Network forms around an utilizer, who organizes action for rapid knowledge results	Network forms around a region (regional development) or a funded project (e.g., public funding)	Network forms around a provider organization(s)	Network initiated by users lacks formal coordination mechanisms
Action	Utilizer guides information collection from the users and promotes knowledge creation that supports the achievement of preset goals	Information is collected and used together and knowledge is co-created in the network	Information is collected for immediate or postponed use; new knowledge is based on the information that provider gets from the others	Information is not collected formally and builds upon users' interests; knowledge is utilized in the network to help the user community
Outcomes	New knowledge for product and business development	Guided strategy change into a preferred direction	New knowledge supporting operations development	Solutions to users' everyday-life problems
Lifespan	Short	Short/medium/long	Short/medium/long	Long

Source: Leminen et al., 2012

- Applying triangulation; and
- This method researches the actual aspect of interconnections between abstract concepts.

State of the Art

The basis for formulating research questions was a literature analysis of the state of the resources in stock as well as scientific papers pertaining to the functioning of living labs, which was the expected result, apart from observing the process of generating innovations as well as an answer to the research question posed.

Living labs have been present in literature sources since 2004, when the spontaneous formation of living labs was noticed. Despite their functioning for over ten years, there is still no empiric research on how to practically implement the idea of Living Lab (Veeckman et al., 2013).

The knowledge defined in the paper, which is the departure point for case studies (strictly speaking – the level of ignorance and observation of the actual information gap), is an intersubjective state.

The concept of living labs was described by Lasher et al. (1991), when examining information and communication technologies (ICT) businesses with regard to partnership interconnections. Since then the Living Lab term has been permanently connected with ICT and more activities were introduced using only the Living Lab method.

Literature research indicated the existence few publications with regard to the presentation and publication of research on Living Labs (A. Følstad, 2008). The research and discussions based on experience from more than one Living Lab are undertaken by e.g. Ballon et al. (2011) and Niitamo et al. (2006), who presents an overview of 18 case studies, where technological platforms from Finland, Great Britain and the Netherlands were experimented on (Ballon et al., 2005). Amongst the reviewed publications, three can be singled out, which include fairly detailed descriptions of living labs in view of creating innovations and development processes (Kusiak, 2007).

One of notable journals is Technology Innovation Management Review (TiM) from Canada, which systematically publishes issues only on Living Labs. The latest world publications from 2012-2015 have been presented in Table 2. In 2015 TiM Review was awarded as the best journal in the world on living labs.

The distribution used in the Table 2 constitutes a solid foundation for carrying out research and also indicates the early development stage as well as theoretical gap, which is complemented by theoriogenic case studies.

Out of pragmatic considerations, the papers have been divided into three topic areas: defining the structure, typology of a Living Lab, the aspects connected with the management of living labs as well as papers in the area of creating innovation in living labs.

Analyzing the above it is worth noting that there are few literature resources, most significantly in the area of systematizing, standardizing and ordering the concept.

Before carrying out the research in the Ontario region - the place where the idea of living lab was invented, a pilot study was conducted in Świętokrzyskie Voivodeship organizations in Poland. This became the foundation and inspiration of preparing general plan of proper research carried out in Canada and helpful in formulating appropriate research questions.

Table 2. Literature study: Living Lab concept

Publication Date	Authors of Research		
	Structure and Types of Living Labs	Management of Living Labs	Innovation of Living Labs
2012	M. Westerlund, S. Leminen and A. G. Nystrom	B.R. Katzy	E. Almirall, M. Lee, and J. Wareham
		D. Schuurman and L. De Marez	H. Schaffers and P. Turkama
		V.-P. Niitamo, M. Westerlund, and S. Leminen	I. Mulder
2013	C. Veeckman, D. Schuurman, S. Leminen, and M. Westerlund	L. Hakkarainen and S. Hyysalo	S. Juujarvi and K. Pesso
		S. Leminen	D. Schuurman, L. De Marez, and P. Ballon
		P. Femenias and P. Hagbert	A. Stahlbrost
2014			M. Gray, M. Mangyoku, A. Serra, L. Sanchez, F. Aragall
2015	S. Leminen	B.R. Katzy, C. Bucker	C. Veeckman, S. Graaf

Source: own research

CREATING ORGANIZATIONS OF THE FUTURE

Living Labs in Generating Innovations in Świętokrzyskie Voivodeship Organizations

The SWOT analysis of the small and medium-sized enterprises (SMEs) sector in Poland has shown that the main factors to be emphasized and strengthened are service sector prospects, ubiquitous Internet access enabling the ICT development, development of law protected areas (tourism), central geographic location and high adaptation capability and innovation absorption. The threat posed to business development may be the lack of relevant security systems, lack of capital, and unfavorable conditions to run a company (bureaucratic barriers).

The results of pilot project implementation enabled marketing development of business networks participating in it. The involvement in the movement of the new prosumer approach, emphasized by the implementation of the living lab methodology provides extensive applicability in the network of SMEs in Kielce Technology Park. The distribution of achieved solutions by the multiplication of systems and procedures (standards) becomes a key part of the most effective forms of quality improvement in the Świętokrzyskie Voivodeship organizations.

The aim of the Kielce Technology Park is to create and implement an IT system that will not only improve the management over the Kielce Technology Park's know-how and resources, but will also help establish the Park's relationships with lessees and between companies themselves.

Within the CentraLab pilot project in the Kielce Technology Park the opportunity to create business plans and business strategies has emerged. Virtual assistance, content-wise support (analytical, financial and quality) is being verified concerning efficiency and formulating a universal document (naturally with the participation of the companies) enabling the owners and company management to produce a new business plan unassisted. The form of a business plan - designed by experts, allows the user to create a professional plan, together with the calculation of all essential specifications, and then to save and print out the result. The user may take a break from their work and resume it at any time. Besides, this application will be useful not only to people who want to o start their own business, but also to those who are already on the market. A business plan is often required when preparing projects or applying for financial support.

The rapid development of technology, artificial intelligence, and the growth of speed and size of the changes generate the necessity to adapt the undertakings (optimization strategies and future forecasts become less effective than adaptation). The CentraLab network provides the businesspeople with tools ready to enable an instant change and flexible environment adaptation.

Information and communication technology (ICT) has become, within a very short time, one of the basic building blocks of modern society (Anderson, van Weert, 2002). Nowadays, the ICT in Poland have already changed communication both in the "B2B" ("Business to Business") or "B2C" ("Business to Consumer") business models, but also in public administration or health care. We can notice many elements changing the relevance of the ICT technology for the society, however, three of them require emphasis and a short introduction: supply chaining – ICT supports not only business integration but the whole business network: insourcing – ICT leads to the shifting back of external processes to a company's own activity, which results in the simplification, cost limitation and reducing the length of essential processes; the steroids - this term defines useful implementations of ICT such as tablet or barcodes EAN etc.

In the last decade ICT has become an important economic factor especially for individuals and companies, influencing communication and social changes. Therefore, laws for all innovative functions and factors should be appointed to maintain the security and persistency of ICT. According to EU documents, ICT will acquire a 5% share in European GDP, and their market value is estimated at 660 billion

EUR a year (Europe 2020). As the result of constant dynamics of innovations in the ICT sector and their impact on the way other sectors function, ICT contributes to general productivity growth.

The potential of applications of new elements and technological solutions are mentioned in order to understand better the forecasts and phenomena, which may influence the implementation of ICT in businesses. They can change the future of ICT. One of them is a new phenomenon – graphene. Its discovery was followed by the Noble Prize awarded in Physics in 2010. The huge interest in graphene is driven by the considerable and great potential that this material offers in conventional as well as radically new fields of ICT applications (Kinaret et al., 2011). Graphene has been used to produce tunable lasers by exploiting the material's unusual electronic structure that guarantees a wide tunability range. Graphene's properties also imply that the percentage of light absorbed decreases when the incoming light intensity increases, which can be exploited to fabricate very fast pulsed lasers that have great potential in optical communication.

However, despite the great progress in graphene science and technology during the past few years, the field is still very young and many challenges remain. Some of the challenges are fundamental: due to the unique structure of graphene, many of the possibilities it offers are still poorly understood, and their analysis requires highly sophisticated methods; to quote the Nobel Laureate Frank Wilczek from the Nobel Symposium on graphene held in Stockholm in May 2010, "graphene is probably the only system where ideas from quantum field theory can lead to patentable innovations". (Kinaret et al., 2011).

Living Labs as a Tool for Implementing Innovations: The Swietokrzyskie Voivodeship Example

The new framework of innovation policy in Świętokrzyskie Voivodeship should include a new research and innovation strategy based on the concept of "intelligent specialty" (known as Strategies for Smart Specialisations - RIS3). The strategy will foster increasing public support for innovation development, connected with new technologies and new methods of workforce management, production process or marketing organization and in result advance economic development. The Living Lab concept is a key tool in developing "intelligent specialties"; it will enhance development based on innovations and drive socioeconomic transformation of the Voivodeship.

The fact that the idea of living labs may evolve together with competitive position of the Voivodeship is advantageous. Living labs can fit in four industries, which are intelligent specialties in the Świętokrzyskie Voivodeship: metal casting, building, health-oriented tourism, and eco-food. They are supported within three horizontal areas: information and communication technologies (ICT), renewable energy resources technologies and trade fairs.

Living labs are supposed to build competitive advantage by developing strengths in the research and development area and adapting them to business products in order to use favourable market conditions and consequently react to its needs and changes. New economic areas can develop thanks to business identifying process by means of living labs methodology.

There is a demand to elaborate key performance indicators (KPI) for the Świętokrzyskie Voivodeship. Its priority would be creating workplaces and investment attractiveness combining both public and private sector stakeholders. Open living labs is a chance to become the main tool in transforming the Voivodeship both socially and economically. The implementation of living lab methodology aims at strengthening the potential of the business environment together with high quality service.

The EU prospect enables the development of living labs based on partner cooperation between local authorities and businesses involved in effective introduction of industries within the intelligent specialties area. This cooperation is based on a general assumption that financial support from the public sector will be transferred to priority needs of business groups which constitute a strong point of the region in accordance with the concept of "intelligent specialties".

Living labs created by the Business Environment Institutions will encourage businesses to invest in innovations. Nowadays, our enterprises are in the second half of the national rates in reference to innovation activity, innovation investment and participation in new or significantly improved product sales. Industrial enterprises in the Świętokrzyskie Voivodeship are in a better condition comparing to service companies in other regions. Support for innovation development can be precisely directed thanks to the living lab methodology implementation.

In the previous financing period, a large number of companies from the Świętokrzyskie Voivodeship received a grant for innovativeness. It supported 28 per cent of industrial enterprises and almost 28 per cent of service companies active in innovations. Thanks to this it was placed accordingly in fifth and sixth position in Poland. This significantly exceeds the Voivodeship position in terms of business innovation activity.

We notice that funds intended for innovations have not brought visible effects yet, which may suggest, that these days, public intervention should be more precise and implement innovative methods of the living lab development.

To construct a system of innovations, the application of the living lab method fosters synergy at the interface of particular institutions, businesses, processes, expected results, financing, monitoring and evaluation system, implementation plan, policy, tools and mechanisms.

The living labs will strengthen the internal analytical potential of the Świętokrzyskie Voivodeship, in consequence, it will become a research-innovation centre for key industries in the region, not dependent on costly research and analyses. Within the areas of intelligent specialties of the region we can notice the critical mass of businesses, which are open and capable of taking the innovative direction. It will make them more competitive, therefore it is advisable to use the method in those areas.

Major intelligent specialties have large innovative potential and greatly influence the Voivodeship economy. Due to this, further expansion based on knowledge and innovations provides favourable prospects of faster growth and that, in addition, will advance the Voivodeship development and create new workplaces for professionals. Building construction and metal cast industry already have a great impact on the Voivodeship economy and are rooted in its economic landscape.

Areas connected with eco-food and tourism do not influence the Świętokrzyskie economy to such a great extent, however, they possess a vast potential which, thanks to "intelligent" support, will allow them to develop faster. The chosen specialties constitute a combination of well-established strengths of the Voivodeship and perspective thinking mode. Building the living labs net in perspective industries may appear to be an appropriate and safe solution for creating new products and innovativeness, especially because these areas need support and their development will guarantee growth of the region's competitiveness.

SUGGESTED RESEARCH

In the SMEs network created within CentraLab, there is an opportunity to accomplish advanced projects based on research methods of the triangulation model and the methodology of living labs itself. The triangulation method described also as a quality research method seems to be the most suitable here.

Universal research tools should be developed, analysed and verified systematically, which might considerably contribute to the flexible adaptation of strategic decisions to the changing environment. The present information systems used in businesses should be adjusted to new trends and modern management ways, they should include the permanent progress of changes and react to them. This results in the need for flexible industrial information systems, analytical methods, and their representation, which will reflect dynamic future.

It is recommended to analyse conditions of cooperation among network businesses and companies potentially included in the network. Analysing the conditions of synergy generation and opportunities to develop network businesses and the whole cluster.

Figure 1. Positioning living labs amongst innovation paradigms
Source: Schuurman et al., 2013, p. 28.

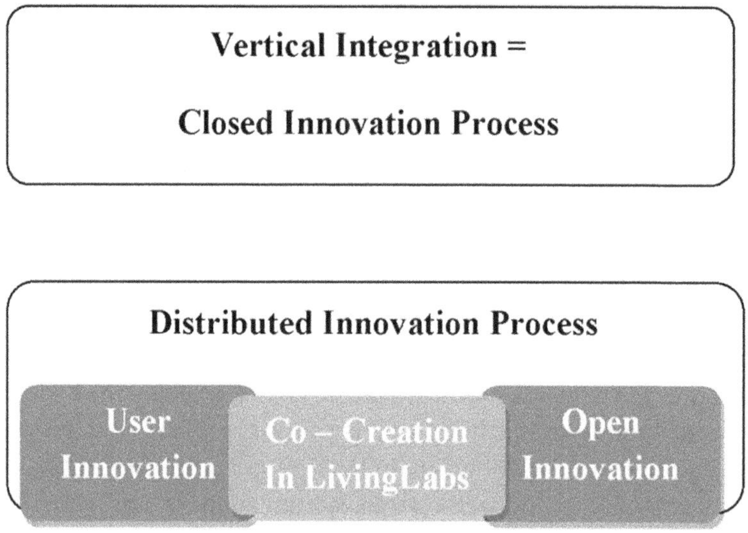

Figure 2 shows the living labs' positioning in the process of innovation emergence and creating living labs. The users – makers' cooperation is based on the vertical integration of innovation system, where the lab is the distributor of knowledge (open and hidden), experience and analyses.

Open innovation can be perceived as a non-linear innovation process with more cooperation between internal R&D departments and the outside the world, and with companies benefiting from the synergies associated with this collaboration. Factors that have favored the shift towards an innovation model that is more open include increased job mobility, recognition of decentralized knowledge and shorter product lifecycles.

Developing of new knowledge management model by means of the living labs involves introducing a system of "gaining" knowledge from the clients who also constitute intellectual capital of a business. This is a manageable thanks to the customer satisfaction assessment methods, however, on this level; it generates information from the past transactions to enable improving the product in the future and operates open knowledge terms. The proposed method of the living labs induces the removal of knowledge sharing barriers rising at the starting point of a product prototype development. This process uses both open and hidden knowledge sources to create innovative solutions.

Figure 2. The "Lead to Win" program in a living laboratory in Carlton
Source: Unpublished material prepared for the study case, Carlton University, Canada, June 2015.

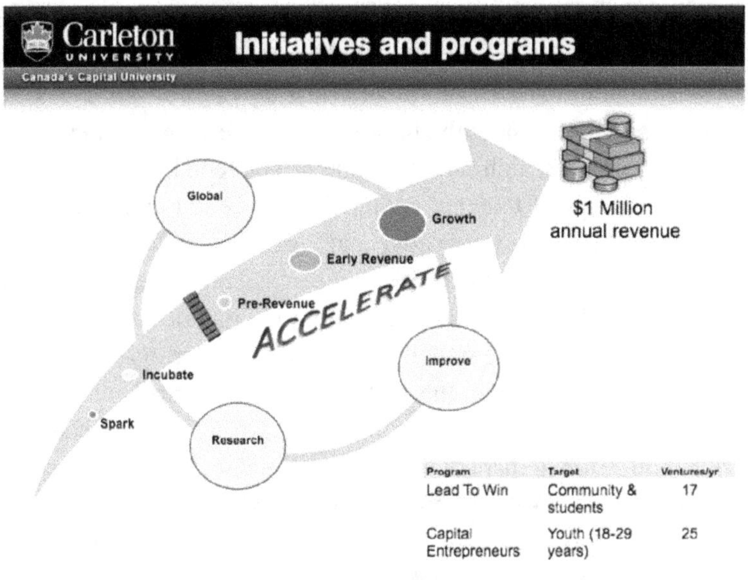

CASE STUDY: CARLTON

In the process of Living Lab selection for the research, The European Network of Living Labs – ENoLL has proved helpful, which is a community, whose aim is to systematically reinforce innovative processes as well as fostering the development and proliferation of the idea open innovation with a large share of users in the whole of Europe and beyond. ENoLL has extensive intercontinental contacts.

According to ENoLL a Living Lab has four main activities (den Ouden, E. et al., 2016):

- **Co-Creation:** Co-design by users and producers.
- **Exploration:** Discovering emerging usages, behaviours and market opportunities.
- **Experimentation:** Implementing live scenarios within communities of users.
- **Evaluation:** Assessment of concepts, products and services according to socio-ergonomic, socio-cognitive and socio-economic criteria.

The examined "1125@Carlton" is a Living Lab, which includes in the process of innovation creation both businesses and scientists, students as well as the local community. It is located in modern buildings of Carlton University, which facilitates the cooperation between Carlton scientists and a broader community in both virtual and real space intended for mutual problem solving.

Many innovation laboratories concentrate on one type of question, e.g. the incubation of new technologies or solving social problems. In the Carlton Living Lab there is broad cooperation based on research and development of ideas from many different domains, which fosters the building of stable communities. For development, innovation, and research an environment is needed, which fosters creative focusing a lot of groups of talented people from different areas can closely work together in order to exchange knowledge and ideas.

Apart from collaboration with a wide range of scientists from all over the world, Carlton cooperates with many business groups, municipalities in order to arrive solutions to complex problems. The stakeholders of 1125@Carlton are amongst others: Freebird Connect, GestureLogic, Integrative Solutions Group Inc., hatch: CUSA's Enterpreneurship Centre, 3ci, C2UExpo, Cisco, CFICE, Construction Resource Initiatives Council, Global Shapers, Just Food, GottaGo, OREC Energy, Play MC2, READ Initiative, timsle, Waterlution, We Are All Musicians, etc.

In Carlton University buildings there is also a Living Laboratory that supports companies organizationally and content-wise, which are starting their activities. However, the enterprise acceleration process takes several months: during this time there following are carried out: research, data analysis, and training for potential entrepreneurs as part of the "Lead-to-win", which provides a variety of free services, and at the same time monitors progress towards generating revenue. Almost from the first day of operation of the company - the Living Lab is testing products and services on international markets and allows rapid globalization. The Living Laboratory controls cash flow that guarantees quick payment from the recipients, helps develop ideas for which customers are willing to pay, improve them and participate in their creation. The Living Laboratory also supports the network of stakeholders, carries out practical and experimental researches in order to increase the probability of success. Thanks to an extensive network, it increases the ability of the company to acquire the first customers, investors, or consult the human resources and knowledge management. Knowledge management is a fundamental element of the functioning of Living Laboratories, which recruit knowledge workers and thanks to its characteristics generates and keeps the greatest talents.

In mid-December 2014, seventeen start-ups received the "green" review score from experts allowing to receive full support in business in order to commercialize innovative ideas and achieve from external stakeholders (recipients) the first million dollars in three years' time from the beginning of the project. As part of ongoing research in Canada, it was possible to observe both the living lab and individual companies: BoardSpace, Canada Micro Markets, Ecelery, Interactive studios, Play

Today, Repair Qoute, Snow, Tapevire, Tarfee, Artifigence, Game it, Abyond, Denilson, Cavara, Social Story, CardAfric and Cloud Base.

CONCLUSION

Comparing the state of the art of living labs presented in the paper and case study, which were conducted using triangulation, it should be pointed out that there are differences, which clearly show the contribution of the case study in the development of knowledge.

Based on the conducted studies, we can conclude that the Living Labs' multi-area character contribute to knowledge diffusion in inner-organizational relationship networks (Olesiński, 2010), and at the same time affect the exchange of the research and development results, feedback submitted by the end user, which leads to the creation of innovation. Their new, unique feature has been found not only in real conditions (not laboratory) new products and services are tested, but also living laboratories can also support organizationally and substantively developing businesses and thanks to the cooperation of all the participants of the process generate synergy.

There is a demand to conduct a vast survey on living labs in Poland (in the world since 2012). The prospects to develop this idea will enable growth in business competitiveness and boost market forces at efficient knowledge management by expanding the area of the process's impact.

Living labs have emerged on the crossroads of open innovation and user innovation frameworks. As open innovation systems, living labs consist of various factors each playing its specific role. The study will provide recommendations and guidelines to further development of the living labs and businesses themselves – which, thanks to the method, will be capable of creating a service network perfectly matched to its needs.

The living labs provide favorable conditions to implement innovations, that is, create open innovations based on different entities' cooperation (for example producers) with users, who inspire the process and may become its driving force. Living labs create an environment, where innovations emerge with the help of testing and experimenting processes, as the effect of mutual cooperation.

It has been concluded; on the basis of the Head Start Program CentraLab, that living labs can involve all the interested parties, especially user communities, at an early stage of the study and innovations to discover scenarios, habits and behaviors and build a new knowledge management model. It is also worth mentioning that this model can be used in advanced technologies, IT, design or renewable resources branches. In this area it is possible to create living labs in Poland for example building autonomic passive houses where potential clients can stay, and their opinions and recommendations serve product improvement.

Living Labs challenge the future. They do not accept the status quo (den Ouden, E. et al., 2016). According to the study, the advantages of living labs as the future organizations are: the ability to expand and process of knowledge - starting sourcing it from customers in real economic conditions; comprehensive solutions to problems, and bringing real value to the life of other people; fostering vertical integration; increasing the dialogue between the stakeholders; sharing experiences; increasing the incubation capabilities of small and medium-sized enterprises; analysis and catalyzing problems; enabling the open cooperation between market actors; increased internetwork cooperation; activity in the central point of generating innovation; involvement of all key actors in the creation of innovation; understanding the process of the generation and commercialization of innovative solutions; acquiring unique knowledge; access to data; knowledge of end users' motivation, increasing the sustainable development opportunities; products localizing; building new business opportunities; change management, defining and developing market niches.

The analysis carried out indicates the need for subsequent case studies and comparative analysis, with which it will be possible to enrich the typology of living labs.

REFERENCES

Almirall, E., Lee, M., & Wareham, J. (2012). Mapping Living Labs in the Landscape of Innovation Methodologies. *Technology Innovation Management Review, 2*(9).

Anderson, J., & van Weert, T. (Eds.). (2002). *ICT in Education*. UNESCO, Division of Higher Education.

Arnkil, R., Järvensivu, A., Koski, P., & Piirainen, T. (2010). *Exploring Quadruple Helix Outlining User-Oriented Innovation Models*. Työraportteja 85/2010 Working Papers. University of Tampere.

Bajgier, S. M., Maragah, H. D., Saccucci, M. S., Verzilli, A., & Prybutok, V. R. (1991). Introducing Students to Community Operations Research by Using a City Neighborhood as a Living Laboratory. *Operations Research, 39*(5), 701–709. doi:10.1287/opre.39.5.701

Ballon, P., Glidden, J., Kranas, P., Menychtas, A., Ruston, S., & Van der Graaf, S. (2011). Is there a Need for a Cloud Platform for European Smart Cities? In P. Cunningham & M. Cunningham (Eds.), *Proceedings of the eChallenges e-2011 Conference*. Dublin: International Information Management Corporation (IIMC).

Ballon, P., Pierson, J., & Delaere, S. (2005). *Test and experimentation platforms for broadbandinnovation: Examining European practice*. The 16th European Regional Conference by the International Telecommunications Society, Porto, Portugal.

Begg, D. (1997). *Microeconomics*. Warsaw: PWE. (in Polish)

Bengtson, P. (1994). Which Comes First, Internal Involvement or External? *Journal for Quality and Participation, 17*(5).

Bergvall-Kåreborn, B., Ihlström Eriksson, C., Ståhlbröst, A., & Svensson, J. (2009). A Milieu for Innovation: Defining Living Labs. In K. R. E. Huizingh, S. Conn, M. Torkkeli, & I. Bitran (Eds.), *Proceedings of the 2nd ISPIM Innovation Symposium: Simulating Recovery – The Role of Innovation Management*.

Czakon, W. (Ed.). (2011). *Methodology bases of research in management sciences*. Warsaw: Wolter Kluwer. (in Polish)

Dell Era, C., & Landoni, P. (2014). Living Lab: A Methodology between User-Centred Design and Participatory Design. *Creativity and Innovation Management, 23*(2).

den Ouden, E., Valkenburg, R., & Blok, S. (2016). *Exploring the future of living labs. Research report - February 2016*. Eindhoven: TU/e LightHouse.

Dutilleul, B., Birrer, F. A. J., & Mensink, W. (2010). Unpacking European Living Labs: Analysing Innovation's Social Dimensions. *Central European Journal of Public Policy, 4*(1).

Edvardsson, B., Kristensson, P., Magnusson, P., & Sundström, E. (2012). Customer Integration within Service Development – A Review of Methods and an Analysis of Insitu and Exsitu Contributions. *Technovation, 32*(7-8), 419–429. doi:10.1016/j.technovation.2011.04.006

Eriksson, M., Niitamo, V.-P., Kulkki, S., & Hribernik, K. A. (2006). *Living Labs as aMulti-Contextual R&D Methodology*. The 12th International Conference on Concurrent Enterprising: Innovative Products and Services through Collaborative Networks, ICE 2006, Milan, Italy.

Europe 2020. (n.d.). *European Strategy for smart, sustainable and inclusive growth*. European Commission.

European Commission Information Society and Media. (2009). *Unit F4 New Infrastructure Paradigms and Experimental Facilities. Living Labs for user-driven open innovation. An overview of the Living Labs methodology, activities and achievements*. Author.

Ferrari, V., Mion, L., & Molinari, R. F. (2011). Innovating ICT Innovation: Trentino as a Lab. In *Proceedings of ICEGOV2011*. doi:10.1145/2072069.2072130

Følstad, A. (2008). Living labs for innovation and development of information and communication technology: a literature review. *Electronic Journal of Virtual Organizations, 10*. Retrieved from http://graphene-flagship.eu/material/graphene/Pages/What-is-graphene.aspx

Innovation Portal of the Polish Agency for Enterprise Development. (n.d.). Retrieved from http://www.pi.gov.pl/parp

Intille, S. S., Larson, K., Munguia Tapia, E., Beaudin, J., Kaushik, P., Nawyn, J., & Rockinson, R. R. (2006). Using a Live-In Laboratory for Ubiquitous Computing Research. In K. P. Fishkin, B. Schiele, P. Nixon, & A. Quigley (Eds.), *Proceedings of PERVASIVE 2006*, (LNCS), (vol. 3968, pp. 349-365). Berlin: Springer-Verlag.

Juujärvi, S., & Pesso, K. (2013). Actor Roles in an Urban Living Lab: What Can We Learn from Suurpelto, Finland? *Technology Innovation Management Review, 3*(12).

Kinaret, J., Ferrari, A., Fal'ko, V., & Kivioja, J. (2011). Graphene-Driven Revolutions in ICT and Beyond. *Procedia Computer Science*, *7*, 30–33. doi:10.1016/j.procs.2011.12.013

Knight, T. (1749). *Reflections upon catholicons, or universal medicines*. Printed for T. Osborne in Gray's-Inn.

Konecki, K. (2000). *Methodology bases of research in management sciences*. Warsaw: PWN. (in Polish)

Kusiak, A. (2007). Innovation: The Living Laboratory Perspective. *Computer-Aided Design and Applications*, *4*(6), 863–876. doi:10.1080/16864360.2007.10738518

Lasher, D. R., Ives, B., & Järvenpää, S. L. (1991). USAA-IBM Partnerships in Information Technology: Managing the Image Project. *Management Information Systems Quarterly*, *15*(4), 551. doi:10.2307/249458

Leminen, S. (2013). Coordination and Participation in Living Lab Networks. *Technology Innovation Management Review, 3*(11).

Leminen, S. (2015). What Are Living Labs? *Technology Innovation Management Review*, *5*(9), 29–35.

Leminen, S., Nyström, A.-G., & Westerlund, M. (2015a). A Typology of Creative Consumers in Living Labs. *Journal of Engineering and Technology Management*, *37*, 6–20. doi:10.1016/j.jengtecman.2015.08.008

Leminen, S., & Westerlund, M. (2012). Towards Innovation in Living Labs Network. *International Journal of Product Development, 17*(1/2), 43. doi:10.1504/IJPD.2012.051161

Lepik, K.-L., Krigul, M., & Terk, E. (2010). Introducing Living Lab's Method as Knowledge Transfer from One Socio-Institutional Context to Another: Evidence from Helsinki-Tallinn Cross-Border Region. *Journal of Universal Computer Science, 16*(8).

Lievens, B., Schaffers, H., Turkama, P., Ståhlbröst, A., & Ballon, P. (2011). Cross Border Living Labs Networks to Support SMEs Accessing New Markets. In P. Cunningham & M. Cunningham (Eds.), *Proceedings of the eChallenges e-2011 Conference: 8*. International Information Management Corporation (IIMC).

Markkula, M., & Kune, H. (2015). Smart Specialization and the Role of Universities in Regional Innovation Ecosystems.Technology Innovation Management Review, 5(10).

Mirijamdotter, A., Ståhlbröst, A., Sällström, A., Niitamo, V.-P., & Kulkki, S. (2006). *The European Network of Living Labs for CWE - user-centric co-creation and innovation*. E-Challenges 2006, Barcelona, Spain.

Molinari, F. (2011). Living Labs as Multi-Stakeholder Platforms for the eGovernance of Innovation. In *Proceedings of the International Conference on Theory and Practice of Electronic Governance (ICEGOV2011)*. doi:10.1145/2072069.2072092

Mulder, I., & Stappers, P. J. (2009). *Co-Creating in Practice: Results and Challenges*. Paper presented at the 15th International Conference on Concurrent Engineering (ICE 2009), Leiden, The Netherlands.

Niitamo, V.-P., Kulkki, S., Eriksson, M., & Hribernik, K. A. (2006). State-of-the-Art and Good Practice in the Field of Living Labs. In *Proceedings of the 12th International Conference on Concurrent Enterprising: Innovative Products and Services through Collaborative Networks* (pp. 349–357). Milan, Italy: Nottingham University Business School.

Nyström, A. G., Leminen, S., Westerlund, M., & Kortelainen, M. (2014). Actor roles and role patterns influencing innovation in living labs. *Industrial Marketing Management, 43*(3), 483–495. doi:10.1016/j.indmarman.2013.12.016

Olesiński, Z. (2010). *Managing Inter-Organizational Relations*. Warsaw: C.H. Beck. (in Polish)

Oliveira, A., Fradinho, E., Caires, R., Oliveira, J., & Barbosa, A. (2006). Successful Regional Information Society Strategy to an Advanced Living Lab in Mobile Technologies and Services System Sciences. In *HICSS '06: Proceedings of the 39th Annual Hawaii International Conference on System Sciences.*

Schaffers, H., & Kulkki, S. (2007). *Living Labs: An Open Innovation Concept Fostering Rural Development. Tech Monitor.*

Schaffers, H., & Turkama, P. (2012). Living Labs for Cross-Border Systemic Innovation. *Technology Innovation Management Review*, *2*(9), 25–30.

Schmid, G. (2003). *Beyond Conventional Service Economics*. Discussion Paper FS I -2003.

Schumpeter, J. A. (1960). *Economic Theory Development*. Warsaw: PWN.

Schuurman, D., & De Marez, L. (2009). User-Centred Innovation: Towards a Conceptual Integration of Lead Users and Living Labs. In *Proceedings of COST298: The Good, the Bad and the Challenging*.

Schuurman, D., De Marez, L., & Ballon, P. (2013). *Open Innovation Processes in Living Lab Innovation Systems: Insights from the LeYLab*. Carleton University.

Tang, T., & Hämäläinen, M. (2012). Living Lab Methods and Tools for Fostering Everyday Life Innovation. In B. Katzy, T. Holzmann, K. Sailer, & K. D. Thoben (Eds.), *Proceedings of the 2012 18th International Conference on Engineering, Technology and Innovation*. Academic Press.

The Nobel Prize in Physics 2010. (2011). Retrieved from http://nobelprize.org/nobelprizes/physics/laureates/2010/

Veeckman, C., Schuurman, D., Leminen, S., & Westerlund, M. (2013). Linking Living Lab Characteristics and Their Outcomes: Towards a Conceptual Framework. *Technology Innovation Management Review, 3*(12).

Veeckman, C., Schuurman, D., Leminen, S., & Westerlund, M. (2013, December). Linking Living Lab Characteristics and Their Outcomes: Towards a Conceptual Framework. Technology Innovation Management Review.

Westerlund, M., & Leminen, S. (2011). Managing the Challenges of Becoming an Open Innovation Company: Experiences from Living Labs. *Technology Innovation Management Review*, *1*(1), 9–25.

Chapter 6
An Imagination of Organizations in the Future:
Rethinking McKinsey's 7S Model

Oya Zincir
Istanbul University, Turkey

Ayşegül Özbebek Tunç
Istanbul University, Turkey

ABSTRACT

Organizations try to survive in a hypercompetitive, changing, and unpredictable environment. The form of this survival continuously changes and requires different tools, solutions, dynamics and drivers according to the actual time. Today organizations face with a big paradigm shift, the industrialization of information age. Organizations should find a new form on the basis of the new age requirements. Some authors have suggested some strategic prescriptions harmonized with the actual term such as McKinsey's 7S Model including strategy, structure, systems, skills, staff, style and shared goals. It is considered that there is a gap related to updating and upgrading these formulas by activating some new concepts such as morphing, organizational ambidexterity and so on. In this study, it is aimed to understand the organizations in the future with internal elements and outer changes which would affect them at the edge of the new era by using this model and offer many propositions.

DOI: 10.4018/978-1-5225-1656-9.ch006

INTRODUCTION

I don't read history. That's in the past. I'm thinking of the future. Henry Ford (Lander, 2014)

Managerial and organizational studies' aspects have changed through the years and so have organizations. According to the well-known study of Alvin Toffler (1980), three types of societies -based on the concept of waves; each wave pushes the older ones aside- have evolved through first, second and third waves as agricultural, industrial and information ages. First wave is the society after Agrarian Revolution which characterized by the enclosure of common land and the introduction of some technological innovations such as the seed drill. The concepts of this age led to the development of job specialization, complex political structures, non-portable possessions, architecture, and the rise of industry and commerce. Second wave is the society during the Industrial Revolution which characterized by mass production, distribution, consumption, education, recreation and destruction that combines those things with standardization, centralization, concentration, and synchronization. The structure of organization was bureaucracy. Third wave is the Post-Industrial Society characterized by actionable knowledge as a primary resource, demassification, diversity, knowledge based productions and non-linear change. The structure of organization is adhocracy. Based on these developments, organizations have evolved into various types through the ages.

With the 21st century, it has been recognized that new social and technical aspects of work systems, organizational structures, managerial practices, competencies and skills are needed. Some authors (Anderson, 2012; Rifkin, 2011, 2014) define this new age as "Third Industrial Revolution" and explain it as "the industrialization of Information Age". It can been said that it is the collaboration age; the modern enterprise is already a cyberfactory with the usage of robotics, internet of things and high-automation (Czarniawska, 2012) and high collaboration is very important for organizations and people who work for them.

Rifkin adresses renewable energy by mentioning its sustainable use and he stresses that this energy will be one of the main power element to change the understanding of world (Rifkin, 2011). Further, Rifkin (online, para.36-37) also explains organizations and workforce of the future by mentioning some important aspects: 'The new high tech workforce of the Third Industrial Revolution will need to be skilled in renewable energy technologies, green construction, IT and embedded computing, nanotechnology, sustainable chemistry, fuel-cell development, digital power grid management, hybrid electric and hydrogen-powered transport and hundreds of other technical fields. Entrepreneurs and managers will need to be educated to take advantage of cutting edge business models, including distributed and collaborative research and development strategies, open source and networked commerce, performance contracting, shared savings agreements, and sustainable low-carbon logistics and supply chain management". Organizations shift from centralized global companies to distributed small and medium size enterprise networks. Klaus Schwab (2016), using the term of Fourth Industrial Revolution, explained future concepts on the topics of mobile supercomputing, self-driving cars, geneting editing and many others by defining them with some megatrends (physical, digital and biological). These developments will affect economies, organizations, governments, societies and individuals. Schwab (2016) also emphasizes that the speed, breadth and depth and the holistic effect of the Fourth Industrial Revolution is a way more different than the other industrial revolutions. It is non-liner; will change many paradigms and recreate them by gathering some of them together and it needs to be considered holistically with countries, companies, sectors and more. He mentions that the new era is the "collaboration era", so that these institutions should collaborate together to do the necessary changes. The developments in advanced robotics, 3D printing, circular economy, internet of things (IoT), sharing economy, artificial intelligence, innovation making, humanistic approaches and many more, will reshape the world. In this manner, we need to understand new organizational concepts and economics systems and they should be renewed according to these developments. Machines will be used even more in doing business and so, not "race against the machine" but "race with the machine" understanding should be complementary. In short, people and machines should be team mates. Data driven developments help goods to be better produced, and this leads to increase asset productivity. Organizations collaborate with different partners and learn from them even more and so forth. Thus, with all these, it can be said that the skill levels and managerial styles might be quite different from now; more agile, sustainable, collaborative, humanistic, flexible, technologic, creative, innovative, etc.

In this study, it is aimed to prescribe organizations of the future by rethinking McKinsey's 7S Model. It is a management model which developed by Robert Waterman and Tom Peters, most often used as an organizational analysis tool to assess and monitor changes in the internal situation of an organization. By using this model, it is aimed to understand organizations in the future with internal elements and outer changes which would affect them at the edge of the new era. On this framework, based on future thinking and new concepts, 'organizations in the future' is prescribed on some topics and furthermore, propositions are presented about some concepts. Some of them are liquid organization model, morphing, organizational ambidexterity, roles (staff who work for not jobs but for multiple roles), cyborganization, holacracy.

LITERATURE REVIEW

Introduced in the June 1980 issue of *Business Horizons*, in an article titled "Structure is not Organization" by Thomas J. Peters, Robert H. Waterman and Julien Phillips, McKinsey's 7S Model maps a constellation of interrelated factors that influence an organization's ability to change on seven elements. It suggests that significant progress in one element of the organization will be difficult without working on the others. The model has been used as an important tool to understand the complexity of organizations and provides a deep understanding of that structure alone doesn't refer to the whole organization.

The 7S are (which can be seen in Figure 1); hard elements that of structure, strategy, systems and soft elements that of skills, style, staff and shared values. As the authors explained (Waterman, Peters & Phillips, 1980), the elements are interconnected, they have effects on each other and do not follow a hierarchy or an importance degree. Strategy is the plan devised to maintain and build competitive advantage over the competition; structure is the way an organization is structured and who reports to whom, and systems is the daily activities and procedures that staff members engage in to 'get the work done'. These are hard elements; easy to define or identify and management can directly influence them. Shared values -called "superordinate goals" when the model was first developed- are the core values of the company that are evidenced in the corporate culture and the general work ethic; style is the style of leadership adopted; staff is the employees and their general capabilities, and skills are the actual skills and competencies of the employees who work for the company. These are soft elements; more difficult to describe, less tangible and more influenced by culture (Waterman, Peters & Phillips, 1980).

Peters (2011, para.15) also explains that now, in the new era, "hard is soft, soft is hard. The plans and the numbers are often 'soft' and the people (staff) and shared values (corporate culture) and skills (core competencies) which are truly 'hard' that is, the bedrock upon which the adaptive and enduring enterprise is built". It illustrates the paradoxical characteristics of the Hard Ss and Soft Ss elements. This perspective grounds on the "people" dimension. People's competences differentiate and change constantly (Dahlgaard-Park, 2013). Hamm (2015, para.3) explains this as that "the traditional hard skills are readily available and they cover a small percentage of what it takes to be 'excellent' as a company. On the other hand, the soft skills are really difficult to master. Finding people with the skills that are highest in demand is difficult". Currently, workforce is changing. Digitalization and other developments brought some serious effects on employees and customers. Employees expect a harmonized work-life balance. Customer expectations are more specialized today. They are the forcing dimension of business (one example is user innovation). In this demand economy, people also choose which organizations that they want to work

Figure 1. McKinsey's 7S Model

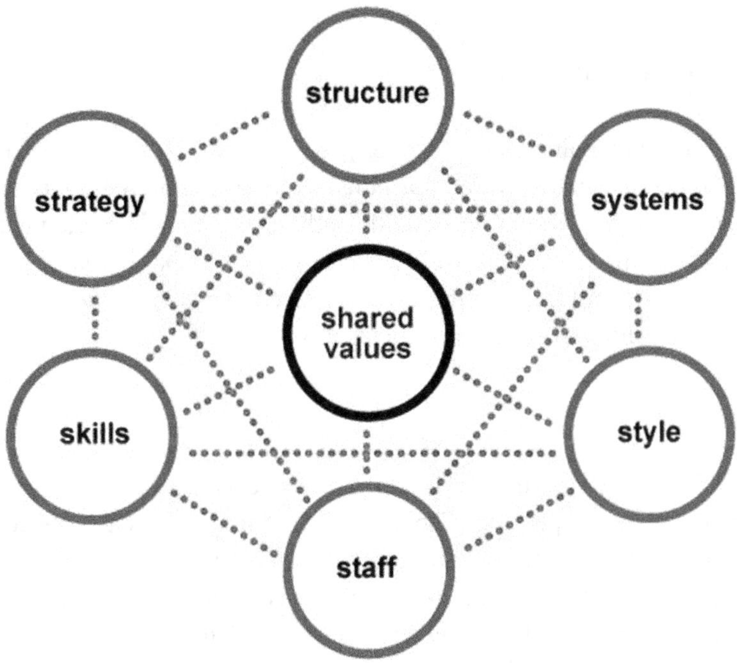

with and there are human resources intermediaries who gather people and organizations. Organizations will have the chance to use "human clouds" to choose its future employees. All these grounds on "people dimension" of the model. It is the main dimension which is effected by the changes in workplace behaviors and working conditions. When considering the model, it can be said that, social aspects of works, leadership/management style and staff' skills and competencies are the pillars of the model now by considering all the elements.

RETHINKING MCKINSEY'S 7S MODEL FOR ORGANIZATIONS IN THE FUTURE

The methodology of this study bases on rethinking McKinsey's 7S Model for the future understanding of organizations. As explained above, the model includes seven elements which interact each other. Mentioning that 'hard is soft, soft is hard' in this new era, in this study, every element are explained on some topics. For example, strategy is defined as a soft element and conceptualized by tactics and maneuvers, emergent strategies, strategic flexibility, transient advantage strategy etc. Staff, another example, is conceptualized by t-shaped skills, roles in organizations, technophilia, etc.

Soft Elements (Soft Ss)

As Peters (2011) explained lately, strategy, structure and systems are the soft elements of the model. In this study, it is considered to explain these elements under this category. When it is first written, the authors explains these elements as hard elements. By the time, business world has changed and so have organizations. First, each element is identified, and than some new concepts are explained and lastly propositions are introduced for each concept.

Strategy

In the model, strategy is defined as the plan devised to maintain and build competitive advantage over the competition. The focus is on 'building sustainable competitive advantage over the competition'. Strategy has been seen as a hard element and business world was (and still is) focused on building sustainable competitive advantage by creating a strong position among rivals and defending it for extended periods of time (McGrath, 2013). Lippitt (2007) underlined the fact that 'execution, not strategy, offers an exclusive competitive advantage'. Strategy execution which aims a competitive advantage is still very important for companies, but companies need to understand new rules of this new era. In today's world, competitors and customers are very unpredictable and amorphous (McGrath, 2013), and high-velocity competitive contexts are fast becoming the norm (Leavy, 2014).

In hyper-changing industries that effected by globalization and the digital revolution, competitive advantage often evaporates in a year. So, if companies want to stay ahead, they need to constantly start new strategic initiatives, building and exploiting many transient competitive advantage at once (McGrath, 2013). For this manner, first, it is essential to define 'transient advantage'. It is a strategy that prioritizes fast innovation to build a pipeline of competitive advantages where business environment requires the capacity to surf through waves of short-lived opportunities.

Proposition 1: In turbulent industries, organizations which follows transient advantage strategy can cope with emergent strategies more easily than the others.

Where being fast and changing fast are important aspects, and strategic flexibility is a vital capability, organizations need to give importance to tactics and maneuvers more. The reason is that tactics are actions that take advantage of opportunities offered by the gaps within a given strategic system. Maneuvers are actions that improve or maintain strategic situation of the company in a competitive environment or avoid a worse situation. 'Strategy is undertaken before the battle; tactics and maneuvers are implemented during the battle'. In this manner, where execution of the strategy is very important, organizations need to prioritize maneuvers and tactics to execute strategies. Strategic flexibility, also another important concept, is defined as the ability to precipitate intentional changes and adopt to environmental changes through the continuous rethinking of current strategies, asset deployment and investment strategies (Roberts & Stockport, 2014).

Proposition 2: In turbulent industries, organizations might prioritize maneuvers and tactics to execute strategies faster.

Proposition 3: Organizations that internalized strategic flexibility can adopt to environmental changes faster and easier.

Waterman et. al (1980), explained the strategy as the way a company aims to improve its position vis-a-vis competition. The focus is now on transient advantage strategies and wining the competition for a short period of time by, for example, creating new strategic initiatives, building organizational dynamics on strategic flexibility, etc. From this perspective:

1. Making innovation an everyday proficiency;
2. Practicing healthy disengagement as a normal, regular activity and continuously reconfiguring resources; and
3. Doing activities to achieve a dynamic balance between stability and agility (Leavy, 2014) are essential for organizations who want to cope with new environment changes and be competitive with the changing dynamics.

    Proposition 4: Transient advantage strategy prioritizes fast innovation practices for organizations to compete among rivals.

    Proposition 5: Transient advantage strategy generates the development of new strategic advantages for the organizations.

Porter (1980) asserted that companies should concentrate for one direction for their strategies to gain competitive advantage. Therefore he recommended to choose either differentiation-oriented or low cost-oriented as strategic positioning. He argues that there are insuperable trade-offs between these strategies, so that organizations have to make explicit choices (Porter, 1996). This assumption was a cornerstone for strategy field and also for companies who practice the findings of related academic studies in 1990s. Porter was not only a management thinker to support one point of view that means making choices that favor one activity over the other. According to this perspective, companies that pursue mixed strategies risk becoming "stuck in the middle," which may lead to poor performance (Raisch & Birkinshaw, 2008). This metaphor obviously presents how the duality is perceived as risky.

This duality is called 'ambidextrous approach' in terms of not only strategy but also structure. Ambidexterity concept is first used by Duncan (1976) who argued that firms need to innovate in order to insure long-term success and should consider a dual structure, one to initiate and one to execute, as a means to achieve these ends. This concept covers the idea that organizations should put in place dual structures, with one unit focusing on exploration and the other focusing on exploitation (Birkinshaw & Gupta, 2013). Exploration includes such things captured by terms such as search, variation, risk taking, experimentation, play, flexibility, discovery, innovation. Exploitation includes such things as refinement, choice, production, efficiency, selection, implementation, execution (March, 1991). It is argued that too much emphasis on exploitation to the exclusion of exploration leads to competency traps, inertia, and ultimately obsolescence. By contrast, too much focus on exploration to the exclusion of exploitation leads to a failure trap in which organizations gains no returns from its knowledge (Simsek et al., 2009). For this reason, some authors support the balance between these contrary ways. Actually today's organizations focus on strategies, in other words, how to formulate the best strategy to survive. On the other hand, future organizations should concentrate on dynamic capabilities, expertise activities, competencies and try to reveal the best mixture to be successful. This mixture provides them flexibility, diversity, harmony capability and holistic paradigm for organizations.

Proposition 6: Ambidexterity strategies might provide flexibility and diversity for organizations.

Proposition 7: Ambidexterity strategies might push the development of dynamic capabilities for organizations.

Structure

Structure is the way an organization is structured and who reports to whom. It divides tasks and then provides coordination, trades off specialization and integration (Waterman, Peters & Phillips, 1980). There are various structure types, such as functional, matrix, adhocratic, etc. A traditional hierarchy is not always the most effective structure; different organizations have different types of structure and these different structures effect the way they perform. Mintzberg (1980) identifies five main successful organizational structure as the entrepreneurial organization, the machine organization (bureaucracy), the professional organization, the divisional (diversified) organization and the innovative organization (adhocracy). Structure effects the organizational practices such as standardization of work processes, unit size, centralization/decentralization etc. It is important to mention that there's no "right" organizational structure, thus, to understand how structure relates to the variety of attributes in an organization is essential.

There is also a new concept which is called 'holacracy'. It is defined as a system of organizational governance in which authority and decision-making are distributed throughout a holarchy of self-organizing teams rather than being vested in a management hierarchy (Rud, 2009). As Robertson (2007) explains, holacracy aligns the explicit structure of an organization with its more organic natural form, replacing artificial hierarchy with a fractal "holarchy" of self-organizing teams ("circles"). A "circle" in holacracy is a self-organizing team. Each circle has an aim (purpose), and the authority to define and assign its own roles and accountabilities. Each circle is a holon -a whole self-organizing entity in its own right-, and a part of a larger circle. Each one connects to other of its sub-circles via a double-link, where a member of each circle is appointed to sit on the other, creating a bidirectional flow of information and rapid feedback loops. Each circle governs itself by uncovering the roles needed to reach the aim of the circle, and assigning circle members to fill them (Robertson, 2007).

Proposition 8: In holacratic organizations, staff might perform innovative work behavior more often.

Proposition 9: In holacratic organizations, staff might be more focused to team and organizational goals.

Proposition 10: In holacratic organizations, staff might get more rapid feedback and information from other members of the organization.

There is another concept that should be mentioned. It is 'morphing' which means continuous ongoing transformation. Morphing is used especially for change management practices. Generally, it mentions on-going changes of the whole organization; for example, the changes of organization's strategies, structure and processes etc. Morphing embraces fluidity and continuous change rather than stability and certainty. It explains the rapid, seamless and more or less total change of the organizations which are mostly in high-velocity industries seek to regenerate their competitive advantage under hyper-competition conditions (Marshak, 2004). As explained above, transient advantage strategy helps organization to win the competition for a short period of time. This means that morphing could be an internal structural change process[1] to fit this strategical movements and flexibility.

Proposition 11: Morphing might help organizations to adapt strategic flexibility more easily.

Proposition 12: Morphing might help organizations to adapt short-term strategic changes more easily.

Liquid organization model is another concept that could be mentioned as a structural practice. Liquefying an organization means disrupting the industrial-age driven assumptions on which rigid structures are designed and move on to make it adaptive, dynamic and anti-fragile. Based on lean management and open collaboration principles, the liquid organization model is flat, meritocratic and value-driven, enabling stigmergic behavior and "organic" effectiveness (Verzera, 2014). Its meritocratic feature defines it as a performance-measured system. Stigmergic behavior means efficient collaboration between members and parts of the organization. As a whole, collaborative working, transparency, lean perspective, concentrated decisional power, agility, etc. are the basic features of the liquid organization. New organizations which could be called as "game changers" ground on these concepts. These entities lead to structural changes and new organizational structures; and new practices that change business world's dynamics.

Proposition 13: Liquid organization model might enhances stigmergic behaviors of staff by enabling collaboration.

Proposition 14: Liquid organization model might helps to improve organic effectiveness of the organization.

Czarniawska (2011) has developed an analytical metaphor of 'cyberfactory' that refers to a news agency where software has a central importance in the control of news production through the professional journalist. This terms brings the affirmation of the interconnectedness of people and machines in perceiving the world and brings another term that is 'cyborganization'. It is a term which describes the concept of a future form of organization, shaped by technology-mediated coordination, social computing and organizational algorithms. It basically describes a hybrid of a human and computer shaped form of organization (Kaminski, 2014). With new arising technologies and collaborative networks of resources and knowledge, one can assume that daily life and work will change dramatically within the next decades which have never seen before. Organizations can basically become structural artifacts that reflects as cyborganizations of the *Second Machine Age* which has been defined as a new era that makes people and software-driven machines substitutes, rather than complements. It is argued that it will bring technology-driven opportunities and challenges which will transform economy in the new era and will require human-beings to learn 'race with the machines' (Brynjolfsson & McAfee, 2014).

Proposition 15: In cyborganizations, staff can gain technological skills and be more innovative.

Proposition 16: Cyborganizational structure can act as a catalyst between staff and technological assets of an organization.

In terms of organizational paradigm, chaos theory refers to a subset of chaos incorporates principle of quantum mechanics and presents them in a complex system environment. On the basis of this theory, organizations face with the layout in chaos as an inevitable part of the environment (Ozbebek Tunc & Turan, 2013). In such an environment possibly in the future, fractal organization structure can help companies to survive. Because this type of organization offers open, self-organized, reconfigurable, nonlinear, self-contained and acquired structure. It discusses on the meaning of authority and hierarchy. From this aspect, it shows similarity with holacratic organization. This structure is self-similar, too irregular, goal-oriented, dynamically restructured and suitable for a turbulent environment (Ryu & Jung, 2004).

Proposition 17: Fractal organizations can help to link with internal regularity and external irregularity.

Systems

Systems are formal and informal daily activities and procedures that staff members engage in to 'get the work done'. Financial systems, promotion and reward systems, recruitment systems, and information systems are some examples of internal systems. Through these organizational systems, all the processes, information flows and key activities are carried out (Dahlgaard-Park, 2013). The way systems mirror the state of an organization is very important to enhance organizational effectiveness.

Getting Things Done (GTD) approach focuses on workflow and time management. It rests on the idea of moving planned tasks and projects out of the mind by recording them externally and then breaking them into actionable work items which allows one to focus attention on taking action on tasks (Allen, 2001). GTD approach can help staff and organizations simplify tasks, arrange them and engage in. This can increase efficiency and reduce time waste. These are really important for all types of organizations.

Proposition 18: GTD approach can help organizations to increase efficiency.

Proposition 19: GTD approach can help organizations to be more effective when doing business-related activities.

Integrative Decision Making (IDM) is defined as a decision-making process in which an individual balances tensions between opposing variables. It follows a four-step process. The first step (salience) seeks to define the relevant aspects of a problem. The second step (causality) seeks to determine the relationships between related and seemingly unrelated parts of the problem. The third step (architecture) involves the creation of a model that outlines the relationships defined in the previous two steps. The fourth step (resolution) outlines the decision and how it was reached (Business Dictionary).

Proposition 20: IDM can help staff to gain effectiveness in decision making process.
Proposition 21: IDM can help organizations to be more faster in strategy implementation process by making decision making process more effective.

All groups of individuals or organizations get things done via a 'crowd of wisdom'. Collective intelligence (CI) is defined as groups of individuals doing things collectively that seem intelligent. Wikipedia is an example of collective intelligence. CI refers to anyone in a large group who chooses to do an activity, without being assigned by someone in a position of authority (Malone et. al., 2009). Collaboration platforms such as Wikipedia allows organizations to connect and engage people and information, anywhere, anytime, and on any device. More data means more potential information that can be gathered about anything. This can lead to better decision making, being able to filter the information and make sense of it is crucial (Morgan, 2014). Thus, it can be said that knowledge and information gathering is essential for people and organizations so much more, especially for making new innovations, and CI helps to gather people for these purposes.

Proposition 22: CI can help staff to harness knowledge from other members of an organization, interest people and other organizations.
Proposition 23: CI can help staff to collaborate of other members of the organization, interest people and other organizations by displaying collective behavior.
Proposition 24: CI can help staff to learn new ways of doing business.

A meme is an idea, behavior, style or usage that spreads from person to person within a culture. It acts as a unit for carrying cultural ideas, symbols, or practices that can be transmitted from one mind to another through writing, speech, gestures, rituals, or other imitable phenomena with a mimicked theme (Dawkins, 1990). Techno-memes (temes) are memes which live in technological artifacts instead of the human mind. Communication systems spread memes over longer distances, and broadcasting increases the number of copies sent out; sound and image recording systems raise longevity by storing previously ephemeral information; and new technologies raise fidelity enormously by making the shift from analogue to digital recording. The systems that do these are increasingly copying, storing and propagating information without human help (Blackmore, 2009). The Internet of Things is helping to create the vast amounts of big data. Devices will be able to "talk" to each other and to people on their own without human intervention or activation. This has the potential to make lives and workplaces easier and more streamlined as well as help people to better understand themselves about how they work, and how they live (Morgan, 2014). The current developments for the Internet of Things (IoT) can be a way of exchange temes between machines and this can reshape most industries by changing the way of doing the formal and informal daily activities and procedures.

Proposition 25: Temes can reflect as a factor that increase efficiency of an organization.

Proposition 26: Temes can reflect as a factor that make work-related activities more easily.

Artificial intelligence (AI) is a field of science and engineering concerned with the computational understanding of what is called 'intelligent behavior' and with the creation of artifacts that exhibit such behavior (Shapiro, 1992). It is defined as the study and design of "intelligent agents (machines)" which refers to the system that perceives its environment and takes actions that maximize its chances of success. AI promises to replace some human jobs, freeing people for other pursuits by automating manufacturing and transportation, self-modifying, self-writing, and learning software which relieves programmers of the burdensome task of specifying the whole of a program's functionality. AI can see relationships in enormous or diverse bodies of data that a human could not (Rajani, 2011).

Proposition 27: AI can help to increase staff's innovative work behavior by freeing them on thinking more about work-related issues.

Another concept that is interested in this chapter is biomimicry. It was first used by Benyus (1997) and defined as 'an approach to innovation that seeks sustainable solutions to human challenges by emulating nature's time-tested patterns and strategies' by Biomimicry Insitute (online). In short, it is about inspring from the nature. Companies inspire from the nature related to innovation, structure, strategies, business models, etc. It can be assessed that this inspiration is a systemic challenge because of its holistic effect to the whole. If a company design all activities, roles, competencies, goods and services on the basis of biomimicry view, it can be said that the organization is systematically biomimetic and so adapt to the environment as an eco-system. They can come up with organizational problems by considering nature's solution systems and ways. For example, they can model a kind of bird on producing new airplane or an ant subsystem on organizing team work.

Proposition 28: Organizations adopt biomimicric thinking can change their systems to the nature and so adapt easily to the future environment.

Hard Elements (Hard Ss)

As Peters (2011) explained lately, skills, staff, styles and shared goals are the hard elements of the model. Related new concepts are categorized and explained theoritically under every element. After that they are linked with future organization abilities and some propositions have been offered.

Skills

Skills are the actual skills and competencies of the employees who work for the company. They are the distinctive capabilities possessed by individuals, groups, and an organization as a whole. The skills variable can be referred to as the core competencies of an organization. It is a strong component of competitive advantages of organizations (Dahlgaard-Park, 2013).

Intrapreneurial behavior (IB) refers to the voluntary employee behavior and is a "bottom-top" process related to the spontaneous behaviors of individual employees which is aimed at the perception of opportunities, generation of ideas, creation of new products, as well as the development of new business lines. IB is an extra-role behavior which is self-initiated and reflects in individuals' risk-taking, innovation, and proactiveness (Antoncic & Zorn 2004; De Jong et al. 2011).

Proposition 29: By the time, the mission of being change agent has shifted from corporations to individuals.

Shirky (2010) argues that people are learning how to use free time more constructively for creative acts rather than consumptive ones, particularly with the advent of online tools that allow new forms of collaboration. People use online tools such as social media platforms and tools and create something collectively. Cognitive surplus (CS) is defined as the free time of the world's educated citizenry as an aggregate (Shirky, 2010). Cognitive surplus frames the abundance of free time as a collective resource that society may be able to use more effectively (Tomlinson & Silberman, 2012).

Proposition 30: Future employees can simultaneously and collaboratively create, contribute and use information via online tools.

Proposition 31: Future employees can spend their free time to learn and create something.

T-shaped skills as Hansen (2010) explains, is a metaphor used in job recruitment to describe the abilities of people in the workforce. It represents as a method for assessing résumés to build interdisciplinary work teams for creative processes. T-shaped people have two kinds of characteristics: The vertical stroke of the "T" is a depth of skill that allows them to contribute to the creative process that can be from any number of different fields such as an industrial designing, architecture, social science, etc. The horizontal stroke of the "T" is the disposition for collaboration across disciplines which is composed of two things:

1. Empathy that allows people to imagine the problem from another perspective; and
2. Enthusiasm about other people's disciplines, to the point that these people may actually start to practice the disciplines (Hansen, 2010).

   Proposition 32: T-shaped skills can make workforce more collaborative.

   Proposition 33: T-shaped people can have deeper understanding about theirselves and the others.

Staff

Staff is the employees of an organiation and their general capabilities. Staff is the crucial asset of an organization, it is a "sine qua non", a must. An organization should consider its staff and their general capabilities. The workforce is changing. New employee behaviors entering organizations are challenging the conventional idea of how employees work and what they expect from an organization. A new generational workplace means new behaviors, approaches, attitudes, and expectations about work and the workplace. Mobility allows employees to stay connected and working even when they are on the go. This helps make employee location independent. Organizations, large and small, can be comprised of international teams. Organizations can also develop a presence in any part of the world and work without boundaries (Morgan, 2014).

First, Y Generation entered the business life; in the near future, Millennials will enter. And they are quite different from other generations; they are shaped by technology (can be called as techno-born). They value community and creativity in their work. They have invested in human capital more than previous generations. They will contend with the effects of starting their careers during a historic downturn for years to come (Millennials Report, 2014). In short, the future employee wants to work for an organization that believes and invests in sustainability and community development, corporate social responsibility, health and wellness, employee happiness, ethical choices, and in making the world a better place (Morgan, 2014).

'Role' in an organization means one specific activity that the organization is counting on. For example, creating and maintaining a project release plan is a Project Manager's role. Whenever a new role is defined, it is then assigned to a member of the organization to fill and execute. This formally gives that member control to do what is needed to enact the accountabilities of the role, within any defined limits. It also gives others in the organization the reasonable expectation that they can ask the individual to account for any of the role's explicit accountabilities. In holacracy (explained above), the title of a role becomes secondary, the role is the list of explicit accountabilities. With the detailed accountabilities used in holacracy, any given role may have dozens of accountabilities, and any given individual may fill multiple roles (Robertson, 2007).

Proposition 34: Role-shaped organizations can help staff to gain new talents and competences.

Technology decisions are now in the hands of anyone and everyone within an organization, not just IT department or management. Cloud-based technologies can decrease the time to deploy, make upgrades and deployments easier and faster, and allow organizations to put together their own "stacks" made up of components from various vendors (Morgan, 2014). Technophilia is a term that refers to the enthusiasm toward technology with its rewarded and knowledgeable adoption (Ronit, 2011). It can reflect as techno-creativity of staff and, with the staff-driven activities, organizations can become techno-creative and change some industries' parameters and dynamics.

Proposition 35: Staff that reflects technophilic behaviors can help organizations to become techno-creative organizations.

Green-collar workers can be defined as workers who have interest, expertise, and knowledge in environmental issues and practices within organizations and are employed primarily for these reasons. A green-collar worker is an environmentalist, someone whose functions at the workplace include protecting the environment, encouraging sustainable workplace practices, and raising environmental awareness among co-workers. They advocate for the environment within their organizations and are cognizant of whether the organization acts in a socially responsible manner with regard to environmental issues (Harvey et. al., 2010).

Proposition 36: Green-collar workers can help an organization to give more importance to sustainability-related activities.

Proposition 37: Green-collar workers can help co-workers to be aware of environmental issues and social responsibilities of an organization.

Style

Style is the style of leadership adopted. It refers to the issues of how key managers behave in achieving organizational goals. It is also considered to encompass the cultural style of an organization. All organizations have their own specific culture and management style. The style/culture includes the dominant values, beliefs, norms, and traditions that are relatively enduring features that permeate organizational life (Dahlgaard-Park, 2013). Morgan (2014:91) explains that the future managers needs to "be a leader, follow from the front, understand technology. They need to lead by example, embrace vulnerability, believe in sharing and collective intelligence, be a fire starter, practice real-time recognition and feedback, be conscious of personal boundaries and adapt to the future employee". To be a fire-starter means to challenge conventional ideas about management and work not just take things at face value. A fire-starter must be curious ditch the template and the management formula. Lead by example means to be a role-model for others in every actions.

Proposition 38: When management/leadership style become more people-centric, staff can be more committed to the organization.

Proposition 39: When manager/leader is a 'fire starter', staff can be more innovative.

Some organizations can be 'managerless'. They are managed by employee-appointed committees. But how could they be managed? They can be guided by coaches and mentors. They can be called as 'employee-managed' organizations. Transparency, freedom with accountability, mentorship and support, communication and long-term approach are very important for these organizations. They must reflect values, hire the right people and understand that employees have their own lives (Morgan, 2014). 'Self-management' concept can reflect for these organizations.

Proposition 40: Self-management can help staff to be harmonized by collaborated and cooperated activities.

Proposition 41: Self-management can help staff to make decisions faster.

Proposition 42: In employee-managed organizations, staff can be more committed to the organization.

Shared Goals

Shared Values -called "superordinate goals" when the model was first developed-are the core values of a company that are evidenced in the corporate culture and the general work ethic. This element is considered to be the interconnecting center of all other variables and refers to the guiding concepts and meaning or the purpose of organizations' existence that are shared among all organizational members (Dahlgaard-Park, 2013). Morgan (2014) explains that future organizations will focus on creating a place of "want" instead of a place of "need", adapt to change faster, innovate anywhere and all the time, shift from profit to prosperity, and be more flatter. It can be said that future organizations will be more people-centric, humanistic, lean, innovative, technologic and flexible.

Proposition 43: When organizations become more people-centric, staff can have more flexibility with their roles, functions and work styles.

Proposition 44: When organizations become more flexible, staff can adapt to change faster.

Proposition 45: When organizations focus on creating prosperity, staff can be more people-centric and sustainable.

DISCUSSION AND CONCLUDING REMARKS

In this chapter, it is tried to relate between business concepts and futuristic trends for organizations. It is mentioned about the paradigm shift over years and described the dynamics of today. On the basis of today, it is assumed that some propositions for the future. The major purpose of this study is to prescribe the organizations of the future by rethinking McKinsey's 7S Model by mentioning some concepts such as morphing, holacracy, biomimcry etc. All these concepts are newly discussed theoretically and practically. Even though they are new concepts for today, they will be pillars and corner stones of future. They do shape not only companies but also whole business environment. Market, companies and even employees will have to change theirselves to this aura.

Furthermore, in this study, some propositions are offered to understand the organizations in the future with internal elements and outer changes which would affect them at the edge of the new era which is called as Second Machine Age or Third/Fourth Industrial Revolution by considering different dynamics and concepts. This revolution forces companies to turn traditional management roles upside down and inside out via digital mind on the basis of digitalization. This big change create a paradigm shift about business manners. No doubt that today's organizations should not pass over the symptoms of the new era to be a good example of 'future organization'. For this, they will have to be highly sensitive to all variables of environment and fast-changing as well as the world. This chapter aimed to spotlight on the question 'what is the future?' by offering some futuristic predictions which will be probably realized.

On the other part of this study, McKinsey's 7S Model maps a constellation of interrelated factors that influence an organization's ability to change on seven elements which have been called as hard and soft elements. Hard elements have been conceived as strategy, structure and systems. Soft elements have been conceived as staff, skills, shared values and style. Being hard means a 'hardly changeable, easy to define' structure, and being soft lies upon an 'uneasy to define' concept according to the model. But conversely, in this new era, which grounds on chaotic, hyper-changing and competitive environment, it can be said that 'hard is soft, soft is hard'. This reverse view let companies think all variables and parameters as deeply embedded parts of the system and people as its center. It is because that the model basically grounds on the 'people' dimension now. Human-beings are developing more and more each day and gain new skills and competences. Furthermore, the growth rate of technologic developments is incrementally rising and human-machine collaboration occurs more often. In the near future, it is expected to see many new technologies and developments that grounds on it. This may lead new line of businesses and challenges in the current ones. Organizations may face with a big paradigm shift. The radical changes may push companies to be more innovative but unstructured, holistic but niche-based activities. Organization may look for new ways to build enduring but flexible structures and evolve from self-managed to unmanaged. They may have indulgence to democracy of ideas coming from employees, partners, customers, and even society. In other words, organizations may have to find a new form on the basis of the new age.

All in all, this chapter let anyone to rethink actual strategic models and imagine a future organization. It provides to find an answer to the basic question 'How may be a future organization?'. Future organizations may be structured with no hierarchy, holacratic, liquid, open to morphing, cyborganization, role-shaped, people-centric and fractal formation. They may adopt the systems including getting things done approach, integrative decision making process and temes. They may let to improve collective intelligence and artificial intelligence as a whole. They may prefer to formulate emergent tactics and maneuvers rather than long-term strategies, so they have internalized strategic flexibility. They also may plan to attain transient advantage which is contrary to today's dynamics. They may focus on contrary competencies as being ambidextrous and having duality. They may let their employees to be more entrepreneur, collaborative, innovative, t-shaped and learn to learn at their free time. They may prefer to work with green-collar workers and encourage them to be self-managed people. In conclusion, companies which do the best practice for these concepts in a sustainable manner may be able to survive with destructive competition. No doubt this survival story of companies may start with the exploration of their managerial DNA and then mapping out a route. For companies, the future requires to define uncertainties and make them certainties as soon as their rivals.

REFERENCES

Anderson, C. (2012). *Makers: the new industrial revolution*. New York: Crown Business.

Antoncic, B., & Zorn, O. (2004). The mediating role of corporate entrepreneurship in the organizational support performance relationship: An empirical examination. *Managing Global Transitions*, *2*(1), 5–14.

Benyus, J. M. (n.d.). *Biomimicry: Innovation Inspired by Nature*. Retrieved from https://books.google.com.tr/books?id=mDHKVQyJ94gC&printsec=frontcover&dq=biomimicry&hl=tr&sa=X&ei=WOSvVKW7JYX3UKOFg6gH&redir_esc=y#v=snippet&q=3.8&f=false

Biomimicry Enstitute Web Site. (n.d.). Retrieved 15 October 2015 from http://biomimicry.org/what-is-biomimicry/

Birkinshaw, J., & Gupta, K. (2013). Clarifying The Distincitve Contribution of Ambidexterity to the Field of Organization Studies. *The Academy of Management Perspectives*, *27*(4), 287–298. doi:10.5465/amp.2012.0167

Blackmore, S. (2009). Dangerous Memes; or, What the Pandorans let loose. In S. Dick & M. Lupisella (Eds.), *Cosmos and Culture: Cultural Evolution in a Cosmic Context* (pp. 297–318). NASA.

Brynjolfsson, E., & McAfee, A. (2014). *The Second Machine Age: Work, Progress and Prosperity in a Time of Brilliant Technologies*. New York: W.W. Norton & Company.

Dictionary, B. (2015), *Integrative Decision Making*. Retrieved from http://www.businessdictionary.com/definition/integrative-thinking.html

Czarniawska, B. (2011). *Cyberfactories: How News Agencies Produce News*. Edgar Publishing. doi:10.4337/9780857939135

Dahlgaard-Park, S. M. (2013). Seven-S Framework. In E. H. Kessler (Ed.), *Encyclopedia of Management Theory*. Thousand Oaks, CA: Sage Pub.,Inc. doi:10.4135/9781452276090.n217

Dawkins, R. (1990). *The Selfish Gene*. New York: Oxford University Press, Inc.

De Jong, J. P. J., Parker, S. K., Wennekers, S., & Wu, C. (2011). Corporate entrepreneurship at the individual level: measurement and determinants. In *EIM research reports*. Zoetermeer: EIM.

Duncan, R. (1976). The Ambidextrous Organization: Designing Dual Structures for Innovation. In R. H. Killman, L. R. Pondy, & D. Sleven (Eds.), *The Management of Organization, C.1* (pp. 167–188). New York: North Holland.

Hamm, D. (2015). *Investing in soft skills pays off for workers, employers: create a work environment that values communication, decision making*. Retrieved 1 October 2015 from http://www.insightonmfg.com/ArticleText.aspx?articleId=3854&magzId=2

Hansen, M. T. (2010). *T-Shaped Stars: The Backbone of IDEO's Collaborative Culture: An Interview with IDEO CEO Tim Brown*. Retrieved 5 October from http://chiefexecutive.net/ideo-ceo-tim-brown-t-shaped-stars-the-backbone-of-ideoae%E2%84%A2s-collaborative-culture/

Harvey, D. M., Bosco, S. M., & Emanuele, G. (2010). *The impact of "green-collar workers" on Integrative Decision Making*. Retrieved 5 October 2015 from http://www.businessdictionary.com/ definition/integrative-thinking.html #ixzz3si4xw9uY

Kaminski, J. (2014). *Cyborganization*. Retrieved 15 October 2015 from http://jermainkaminski.com/ cyborganization/

Lander, S. (2014). *Conversations with History*. Hay House, Inc.

Leavy, B. (2014). Strategy, organization and leadership in a new transient-advantage world. *Strategy and Leadership*, *42*(4), 3–13. doi:10.1108/SL-05-2014-0038

Lippitt, M. (2007). Fix the disconnect between strategy and execution. *T&D, 61*(8), 54-57.

Malone, T. W., Laubacher, R., & Chrysanthos, D. (2010). The collective intelligence genome. *MIT Sloan Management Review*, *51*(3), 21–22.

March, J. G. (1991). Exploration and Exploitation in Organizational Learning. *Organization Science*, *2*(1), 71–87. doi:10.1287/orsc.2.1.71

Marshak, R. J. (2004). Morphing: The leading edge of organizational change in the 21st century. *Organization Development Journal*, *22*(3), 8–21.

McGrath, R. M. (2013, June). Transient advantage. *Harvard Business Review*, 64–70.

Mintzberg, H. (1980). Structure in 5's: A synthesis of the research on organizational design. *Management Science*, *26*(3), 322–341. doi:10.1287/mnsc.26.3.322

Morgan, J. (2014). *The Future of Work: Attract New Talent, Build Better Leaders and Create a Competitive Organization*. Wiley.

Peters, T. (2011). *A Brief History of 7S (McKinsey's 7S) Model*. Retrieved 15 August 2015 from http://tompeters.com/docs/7SHistory.pdf

Porter, M. E. (1980). *Competitive strategy: techniques for analyzing industries and competitors*. New York: Free Press.

Porter, M. E. (1996). What is strategy? *Harvard Business Review*, *74*(6), 61–81. PMID:10158474

Raisch, S., & Birkinshaw, J. (2008). Organizational Ambidexterity: Antecedents, Outcomes, and Moderators. *Journal of Management*, *34*(3), 375–409. doi:10.1177/0149206308316058

Rajani, S. (2011). Artificial intelligence: Man or machıne? *International Journal of Information Technology and Knowledge Management*, *4*(1), 173–176.

Report, M. (2014). *15 economic facts about millennials*. Executive Office of the President of the United States: The Council of Economic Advisers. Retrieved from 10 October 2015 https://www.whitehouse.gov/sites/default/files/docs/millennials_report.pdf

Rifkin, J. (2011). *Third industrial revolution: how lateral power is transforming energy, the economy, and the world*. London: Palgrave Macmillan.

Rifkin, J. (2014). *The zero marginal cost society: the internet of things, the collaborative commons, and the eclipse of capitalism*. London: Palgrave Macmillan.

Rifkin, J. (n.d.). *Book Summary: Third industrial revolution: how lateral power is transforming energy, the economy, and the world*. Retrieved from http://www.wermutham.com/pdf/The%20Third%20Industrial %20Revolution.pdf

Roberts, N., & Stockport, G. J. (2014). Defining Strategic Flexibility. In The Flexible Enterprise, Flexible Systems Management (vol. 1, pp. 37-45). Springer Publishing. doi:10.1007/978-81-322-1560-8_3

Robertson, B. J. (2007). *Organization at the Leading Edge: Introducing Holacracy*. Retrieved 15 October 2015 from http://www.integralesleben.org/fileadmin/user_upload/images/DIA/Flyer/ Organization_at_the_Leading_Edge_2007-06_01.pdf

Ronit, R. (2011). Technophilia: a new model for technology adoption. *UK Academy for Information Systems Conference Proceedings*. Retrieved from http://aisel.aisnet.org/cgi/ukais2011/41

Rud, O. P. (2009). *Business intelligence success factors: tools for aligning your business in the global economy*. John Wiley & Sons, Inc.

Ryu, K., & Jung, M. (2004). Fractal Approach to Managing Intelligent Enterprises. In Creating Knowledge Based Organizations. Idea Group. Retrieved from http://books.google.com.tr/books?id=7aHM3hjWsjUC&lpg=PA319&dq=fractal%20organization&hl=tr&pg=PA348#v=onepage&q=fractal%20organization&f=false

Schwab, K. (2016). *The fourth industrial revolution*. World Economic Forum.

Shapiro, S. C. (1992). Artificial Intelligence. In S. C. Shapiro (Ed.), *Encyclopedia of Artificial Intelligence* (2nd ed.; pp. 54–57). New York: John Wiley & Sons, Inc.

Shirky, C. (2010). *Cognitive surplus: how technology makes consumers into collaborators*. New York: Penguin Press.

Simsek, Z., Heavey, C., Veiga, J. F., & Souder, D. (2009). A Typology for Aligning Organizational Ambidexterity's Conceptualizations, Antecedents, and Outcomes. *Journal of Management Studies*, *46*(5), 864–894. doi:10.1111/j.1467-6486.2009.00841.x

Toffler, A. (1980). *Third wave*. New York: William Morrow and Co.,Inc.

Tomlinson, B., & Silberman, M. S. (2012). The cognitive surplus is made of fossil fuels. *First Monday*, *17*(11). doi:10.5210/fm.v17i11.4120

Tunc, O., & Aysegul, A. T. (2013). Businesses as Fractal Organisms in Chaotic Aura. *International Journal of Business and Management Studies*, 10-20.

Verzera, S. (2014). *Liquid organizations: building the next evolutionary stage of anti-fragility*. Academic Press.

Waterman, R. Jr, Peters, T., & Phillips, J. R. (1980). Structure is not Organization. *Business Horizons*, *23*(3), 14–26. doi:10.1016/0007-6813(80)90027-0

KEY TERMS AND DEFINITIONS

Artificial Intelligence: Artificial Intelligence refers to computational understanding of intelligent behavior which works as a system that perceives its environment and takes actions that maximize its chances of success.

Cyborganization: Cyborganization refers to a hybrid of a human and computer shaped form of organization that shaped by technology-mediated coordination, social computing and organizational algorithms.

Holacracy: Holacracy refers to an organizational form which grounds on a peer-to-peer workplace that of flexible structure, self-organized holons, unique decision making process, etc.

Morphing: Morphing means continuous ongoing transformation of an organization that of structural change, strategical change etc.

Second Machine Age: The Second Machine Age refers to the new age that grounds on software-driven technological developments and human-machine collaboration.

Third Industrial Age: The Third Industrial Age refers to the industrialization of information age that grounds on digitalization and renewable, distributable and smart energy for every continent of the world.

Transient Competitive Advantage: Transient Competitive Advantage is a strategy that grounds on short-lived opportunities and focuses to gain short-term strategic advantage of competition among rivals.

ENDNOTES

1 It is not only a 'structural' practice; in this section, it is used for explaining possible future practices.

Chapter 7
The Organization of the Future and the Marketing Function:
Marketers' Competencies in the Era of Information Technology

Mario Gonzalez-Fuentes
Trinity University, USA

ABSTRACT

The past two decades—and the technology advancements experienced throughout them—have left marketers with a new context that has provided new business opportunities. This new context has prompted a change in the focus of the marketing function and demanded a shift in marketing imperatives and competencies. This chapter provides a comprehensive review of the technological changes experienced by the marketing function in a company, as documented by both scholars and practitioners. It also provides a thorough discussion of the ongoing academic debate regarding the new set of technical skills that have defined employability in the marketing circles for the past couple of decades and the challenges ahead for future professionals and executives.

DOI: 10.4018/978-1-5225-1656-9.ch007

INTRODUCTION

The advent of new technologies like Internet and powerful computing capabilities in hardware and software has opened opportunities to expand the marketer's portfolio of tools. For example, the commercialization of Internet provided companies with a permanently open window to showcase their offerings to customers around the world. Similarly, being able to track online shopping behavior allowed for the emergence of machine learning algorithms to refine recommendations by online retailers. Improved storage capabilities have fostered the appearance of Costumer Relationship Management (CRM) software and solutions to help companies manage their interactions with clients at every single point of contact.

New opportunities for companies have emerged as a result of these new technological capabilities; particularly in the way they interact with customers and how they use the information resulting from these interactions. This has prompted a shift in marketing focus and strategy. Away from a product-based approach to a customer-centric one, from a transactional perspective to one where relationships are nurtured with clients. A dominant paradigm has emerged as a result of these changes that sees the marketing function as a continuous social and economic process in which intangible resources are paramount (Vargo & Lusch, 2004). However, these opportunities brought together new challenges for marketing departments across all industries. At all ranks of positions a gradual and continuous, but also substantial, revision of competencies and skills has occurred.

This chapter addresses the most important advances in information technology and artificial intelligence as they have impacted the practice of marketing among companies. More particularly, this chapter explores the way these changes have shifted the strategic vision of marketing departments and the needs for new skills and competencies that were spawned by it. Thus, this chapter has two objectives. The first is to provide a comprehensive review of the technological changes experienced by the marketing function in a company, as documented by both scholars and practitioners. This will lay the basis for readers to understand the technological framework in which modern organizations base their marketing strategies. The second goal is to provide a rich and thorough discussion of the ongoing debate in the literature regarding the new set of technical skills that have defined employability in the marketing circles for the past couple of decades and the challenges ahead for future professionals and executives.

The next three sections address the first of this chapter's objectives and discuss the recent technological developments in three waves: the commercialization of Internet, the surge of social media and the era of big data and artificial intelligence. For each of these three waves, an analysis of the change in marketing's strategic focus is offered. A fourth section addresses the second objective by providing an examination of the evolving notion of interactivity under the new technological paradigm, connecting it with the new imperatives in skills and competencies required in marketing positions.

THE INTERNET OF THINGS: THE RISE OF TRADIGITAL MARKETING

During the first half of the 1990s, marketing professionals faced the challenge to learn "the Internet of things" in order to adapt their strategies to this new media form. The commercialization of Internet, through the World Wide Web, provided the public with instant access to a vast array of information about a myriad of topics and interests (Roberts & Zahay, 2012). Concepts and terms such as e-mail, browsers and webpages became common currency in conversations all over the world. If you had a computer and a telephone, you had access to a wealth of information about a wide array of topics.

One of the first opportunities that the new technology provided for companies and marketers was another outlet where people could learn about their existence and the products or services they were offering. A company's website became the virtual brochure of a firm's activities and whereabouts. In other words, the emergence of websites and webpages provided a virtual space for companies to provide information readily available every day at every hour to consumers without virtually any physical and geographical boundaries. This significantly influenced the way consumers experience their buying-decision process, which entails the steps and activities a person goes through when confronted with an unsatisfied or poorly satisfied need (Frambach et al., 2007). One of the first steps is the search for information and the evaluation of the alternatives in the market. This implied that marketing tactics and campaigns had to be ready to be changed more frequently than in the past. The result of this was a surge in online advertising and a need for creative talent that was able to adapt marketing strategies to online formats (Evans, 2009).

However, it can be argued that the Internet changed the way people communicated with one another. More particularly, it made it very easy for people to connect and interact with others. No more long distances or having to physically displace to where our relatives or friends were in order to establish an exchange of ideas or a conversation. We were able to, almost instantaneously, contact somebody or, at least, be sure they will receive our note or message next time they were online.

Moreover, the idea of leveraging consumers' digital footprint became relevant in light of the new possibilities of interaction that the new technology facilitated. Online ads, such as banners, interstitials or pop-ups were able to track the number of clicks received and whether or not the advertisement effort resulted in a purchase-related action. Internet Protocol (IP) addresses permitted websites to track down the places where most visits landed from, the different pages explored per visit, and the time spent in each of them. In addition, web browsers facilitated companies with information regarding the words used by people when looking for information about products or services and that ultimately served as clues to find their websites (Evans, 2008).

In this sense, Internet facilitated what companies such as AC Nielsen started with their TV ratings[1] but in a larger and richer scale (Silk et al., 2001). Before the popularity of the Internet, households volunteered to have a device attached to their television sets that would track time spent watching television at different times of the day. These devices recorded channels and specific shows watched and behaviors such as switching programs during advertisements, etc. However, as households' consumption of television entertainment evolved, the challenges to track households' behaviors piled up and extended beyond the prevalent technology limits. For instance, there was no way to know whether a family that did not switch channels when advertisements were shown was actually paying attention to these ads. They could alternatively take the advantage to read a book or a magazine, cook, and visit the restroom or talk, among other options. In addition, and as the number of television sets per household grew up in society, the assumption that by tracking the usage of one television one could learn about the household patterns of media entertainment consumption was no longer valid or, in other words, hard to believe. Some part of the family may have been watching other program –and thus, exhibiting other type of behavior- in a different television set that was not being monitored. This also meant not being sure of how many people were exposed to a specific campaign or advertising effort. With Internet initial usage in personal computers, media analysts know one thing for sure: the person's eyes are on the screen in front of them.

The Internet was so disruptive not only in the real world but also in the academic circles. Some theoretical concepts and ideas learned in the area of marketing had to accommodate changes, such as the introduction of the category "Interactive Marketing" to the list of tools of Non-personal (Mass) Communications to distinguish it from Direct Marketing (Deighton, 1996). Among practitioners the use of the term *Tradigital Marketing* started getting traction as a way to distinguish this approach from the traditional paradigm (Tuten & Solomon, 2015).

GOING VIRAL: THE POWER OF SOCIAL MEDIA

In 2004, right after the launch of Facebook, it was evident that another challenge had arrived. Its name was social media. The widespread use of Internet, as well as the larger volume of information available through the World Wide Web, facilitated the formation of online communities. More particularly, the ability to find people online with the same interests, preferences, and opinions gave rise to the organic and voluntary grouping of users in different platforms such as message boards and forums (van Dijck, 2013). The emergence of social media highlighted one phenomenon that redefined marketer's digital agendas permanently: the horizontal revolution (Tuten & Solomon, 2015).

As it was discussed in the previous section, companies initially adopted Internet as a new channel through which they could expand their communication tactics while leveraging some of its interactive features. However, these tactics were still based in the traditional top-down model, that is, information flowed vertically from organizations to people (Mangold & Faulds, 2009). As people started forming communities online, they also started exchanging and sharing information among themselves, in a horizontal fashion. This represented a challenge for companies, namely they were not in full control of the message transmitted to their audiences (Berthon, Pitt, & Campbell, 2008). Moreover, it appeared that customers were increasingly interested not only in participating in the exchange of information that was taken place but also in the purchase-decision process of other customers through their opinions and advice.

The phenomenon of the horizontal revolution tipped off in part thanks to technological advancements in the area of connectivity (Bruce & Solomon, 2013). Wireless modems became the norm and the idea of needing a phone connection to use Internet quickly turned out to be obsolete. The introduction of new devices, such as smartphones and tablets, allowed people to be online and connected to other people more frequently without sacrificing mobility.

The formation of organic online communities quickly evolved into the concept of social networks, that is, virtual spaces where we could stay in touch with our social circles. Social platforms such as Facebook and Twitter successfully emerged as leaders after outlets like MySpace failed to gain traction (Boyd and Ellison, 2007). The network effects enjoyed by these platforms significantly determined their rapid penetration across the world (van Dijck, 2013). In other words, everybody wanted to belong to these social networks because everybody was part of them.

The surge of the Internet as a popular source of information and entertainment, and the emergence of social networks combined to produce the phenomenon of the *second screen*. People started interacting with other devices while watching television which was, until then, the predominant channel used by people to satisfy their entertainment needs at home (Giglietto & Selva, 2014). Empirical studies have identified the widespread use of social media as a catalyst for the emergence of mobile devices, in particular, as second screens for televisions (Lochrie & Coulton, 2011). No longer did viewers have to wait until the next morning to share with friends, colleagues, or classmates and discuss the last episode of their favorite show.

They could now do it instantly, in real time, as events were unfolding, with as many people as they wanted.

The implication for marketers was that their efforts had to be more responsive and engaging. A significant number of studies appeared to indicate that the old marketing adage "the customer is king" was truer than ever and more importantly, that practitioners had only scratch the surface of this adage in the past. Customers wanted to be connected with other customers (Hannah et al., 2011); be engaged by brands and products (Wallace et al., 2014); and, their opinions to be taken into account (Cambria et al., 2013).

As a result of these trends, content became a key element in companies' online marketing strategies (Rowley, 2008). Content was the material used to attract people to form ever-growing groups with interests in common, or communities. In this sense, the *push* strategy that companies had been using traditionally, even with the outset of Internet and the popularity of online advertising, gradually shifted focus to tactics promoting an attraction orientation, also referred to as *pull* strategies. Good and engaging content attracts inbound traffic; attention is then gained via interactivity with content and users which, in turn, promotes sharing and participation (Kilgour et al. 2015; Campbell et al., 2011). Social networks became one of the most important communication channels for companies and the marketing function embraced concepts like content management and creation, blogging and micro blogging, interactive multimedia, and viral marketing, among others (Ho & Dempsey, 2010).

These new opportunities came also with new challenges. One of them was the fragmentation of audiences that came with the multiplicity of devices (Webster & Ksiazek, 2012). The image of the whole family gathered around the only television set owned by the household is light years apart from the current situation in modern households. Parents may be watching television and interacting with their second screens while children are being entertained surfing the web or on any of the multiple social media channels available. This significantly complicated the task of accurately tracking media consumption in the household since different people could use the same device at different times. For companies this meant not being sure if their messages were being delivered to the right member of the household ("Counting Couch Potatoes", 2013). However, the next wave of technology proved to offer an opportunity to overcome part of this challenge.

CUSTOMIZATION: THE ERA OF BIG DATA AND ARTIFICIAL INTELLIGENCE

The previous two decades –and the technology advancements experienced throughout them- left marketers with a new context that demanded a shift in marketing imperatives and competencies. On the one hand, Internet facilitated a permanently open global marketspace with a multiplication of sales channels and customer interfaces that fueled the fragmentation of markets. On the other hand, social media, with its openness and multiplicity of interactive platforms, transformed customers' access to and engagement with companies and their brands, increasing the volume of information substantially.

One important consequence of both, the penetration of Internet and the advent of social media, was the possibility for firms to track customers' interactions with their brands online (Lemon & Verhoef, 2016). Whether it was a customer visiting a company's website, shopping one of their products online or reacting to some of their social media content or updates, these interactions became valuable pieces of information to target their efforts to the most effective segments and markets. The difference this time is that the information compiled was no longer restricted to the structured type, such as demographics and other easily measurable data. Marketers were now also able to collect unstructured data, the type that needs interpretation to extract meaning and classify it. Along with the information that website visits and search queries were providing, social media platforms offered detailed information about their customers' social life as well as their reactions or changes in attitude towards their brands. This has given rise to the extensive use of the term "Big Data" when referring to this bulk of information available (Fan et al., 2015).

This detailed information allowed companies to manage personalized relations with clients and to customize their offers and promotions. The focus on customization has been the driving force behind most marketing efforts in the past couple of years (Kumar, 2015) since studies estimate that the cost of making a sale to a new customer is five to seven times as much as the cost of retaining a customer (Reichheld, 2001). Companies that embraced this approach quickly started adopting the term "customer-centric" to describe their business values and focus. Under this approach, a company's marketing efforts seek to satisfy the needs and wants of each individual customer (Simon et al., 2016; Sheth et al., 2000). Subsequently, customer-centric companies started amassing large volumes of individualized data about their audiences. For this, they invested heavily in Customer Relationship Management (CRM) software to store, manipulate and access this data to support an integrated strategy applied across all of a company's departments and divisions.

This new technological context appears to have fostered another intersection such as the one identified and discussed previously between the marketing and communications fields. This time it was marketing and computer science, and within the latter the subfields of machine learning and artificial intelligence. One of the main applications of these two subfields has been the development of algorithms that learn to adapt themselves to new inputs or data, also called self-learning systems

(Van Otterlo, 2013). These systems are at the core of the development of robots and other human surrogates.

One of the first and most popular contributions of self-learning systems to marketing is the development of algorithms to conduct sentiment analysis. Liu (2012) defines it as "the field of study that analyzes people's opinions, sentiments, evaluations, attitudes and emotions from written language". Sentiment analysis was quickly adopted by marketing professionals responsible for a company's social media strategy to take the pulse on the impact of different campaigns or events on the image of their brands.

For instance, customers increasingly use tags, also referred as user-defined content, to classify media content created by brands, such as articles, videos, or podcasts (Taneja et al., 2012). These tags are then shared by users via social media and, in turn, associated with similar or different tags used by other users, creating a map or cloud of tags. Tag clouds ultimately offer marketers a way to gauge on the sentiment towards a brand (Nam & Kannan, 2014) through the development of algorithms capable of classifying comments by tags and tags by sentiments. Marketers use this information to understand the emotional and psychological connections that customers may have with their brands and use them to establish bonding mechanisms through carefully crafted communications and campaigns.

As a result of this, industry and academic circles have advocated for a change in focus, away from "products" and closer to the concept of "experiences". This has given birth to the subject of experiential marketing (Schmitt & Zarantonello, 2013).

Self-learning systems have also been used to enhance customers' experiences through recommendation engines and solutions (Forrest & Hoanca, 2015). According to Dawar (2013) companies can build a strong and sustainable competitive advantage by focusing on downstream activities, that is, those in charge of delivering a product or service to consumers. Consider how the examples of eBay and Amazon support this idea. Their recommendation engines focus on one aspect of these downstream processes: reducing customers' perceived costs and risks implicit in the purchase process. They provide user-friendly information regarding the evaluation of different alternatives and other customers' assessments, facilitating customers' decisions.

More recently, cases such as the launch of Lenovo's customizable logo reflect the integration of these two trends, customization and experiential branding. The new logo displays a static typeface with a dynamic color and background. The adaptations span through a variety of media outlets, from online channels to traditional ones, such as in-store signage (Bulik, 2015).

However, with greater access to more data and information, more challenges arise. Since marketers have more diverse ways to measure and track consumers reactions, the pressure of quantifying the impact of every single marketing effort over a company's business intensifies. A number of studies present evidence that suggest a greater concern among companies to hold marketing executives accountable for the return on their investments and actions (Giamanco & Gregoire, 2012; Hoffman & Fodor, 2010). Whether it is an online advertising campaign or a social media strategy, the value of every marketing tactic can now be easily quantified and associated with a number of financial performance metrics.

The changing nature of the marketing function discussed in this section, in the context of the three waves of recent advancements in information technology, suggests important implications for the job skills and competencies needed in marketing professionals. The next section argues that the changes in the marketer's skillset can be traced by the fluid notion of interactivity between customers and companies.

THE EVOLVING NOTION OF INTERACTIVITY AND THE MARKETERS' SKILLSET

Scholars have documented the change on the marketer's skillset due to the aforementioned technological advancements. Schlee and Harich (2010) provide evidence of the increase in technical skills needed in current marketing jobs at all levels as compared to the ones required in the past. These changes in the skillset of marketing professionals, however, have not traveled smoothly and efficiently into the job market. Gibbs et al. (2011) document the mismatch between employers and graduates with respect to their views regarding the computing skills needed to succeed in the job market. In their study, employers seemed to have very particular expectations about specific troubleshooting computing skills whereas graduates felt confident employers would provide training if specific skills were needed. As a corollary, pedagogical research in the field of marketing has recently started to emphasize the need to redesign the marketing curriculum to incorporate the new digital realities of the profession (Wymbs, 2011).

As it was previously argued, Internet represented a change in the way people communicated with each other and, in turn, with organizations. A number of studies argue that information technologies have shifted the marketing function focus from the "marketplace" to the "marketspace", where marketers need to engage in a bidirectional exchange of goods, services and information with customers (Foster, 2015; Rayport & Sviokla, 1994; Vandermerwe, 2014). The key term used extensively during the nineties to describe some of the responsibilities of marketing positions was *interactivity* (Lordan, 2006).

We can distinguish two ways of understanding the concept of *interactivity* during this initial wave of technological advancements. On the one hand, *interactivity* meant that companies were able to establish communications with customers that resembled conversations, as opposed to monologues or one-way exchanges of information, as it has been under the traditional model. E-mail and chat rooms facilitated timely and expedient interaction between companies and customers. Marketing professionals had to learn how to handle these conversations in real-time and in the most effective way. This implied marketers needed to show good oral and written communication as well as proficiency with the new information technology tools. In their comprehensive study, Casner-Lotto and Barrington (2006) support this claim. Incidentally, this set of skills came to be known as Internet Literacy among companies, practitioners and academicians (Livingstone, 2008).

On the other hand, *interactivity* also meant that customers were able to modify or alter the environments in which communications take place (Brodie et al., 2103; Hollebeek et al., 2014; Yadav & Pavlou, 2014). In the case of computer-mediated environments, such as the Internet, interactivity meant the ability of customers to tinker or adjust the form and content of communications. In response to this, marketing communications through the digital marketspace quickly adopted a wide variety of elements that prompted a reaction or response from users. Elements such as the call-to-action became an essential element in every digital communication (Bampo et al., 2008; Basheer & Ibrahim, 2010). Examples of these include the use of hypertext in emails and the design of dynamic content in banners or pop-up advertising. For marketing positions, the latter translated into an increasing need for graphic visual and textual design skills. Website design and management became one of the top priorities in every marketer's list of daily responsibilities during the decade that followed the introduction of the Internet to the public. Marketing job descriptions around the world gradually incorporated terms such as web design, newsletter, pay-per-click, web traffic, email lists, search engine optimization, among others (Wymbs, 2011).

With the outset of social media and its importance as a new digital communication channel the understanding of *interactivity*, at least from a marketing point of view, changed once again. This time, interactivity meant participation and engagement. Incidentally, a number of scholars have spotted this strategic shift. For example, Lovejoy and Saxton (2012), and Sashi (2012) argue that social media created opportunities for customer dialogue and engagement that are qualitatively different to those offered by websites. Vargo and Lusch (2004) suggested companies' promotional efforts needed to be understood as a dialogue with consumers, one in which questions were asked and answers had to be provided. In other words, promotion had to be addressed as a communications process. Consumers did not respond passively to this change in approach. They took the leading role initiating these dialogues and managing the resulting conversations to indicate organizations what were their most important concerns and topics of interest (Briones et al., 2011; Prahalad & Ramaswamy, 2000).

Under this new paradigm, marketing and communications become partners in the creation of value, as it was previously argued. In this respect, Ballantyne and Varey (2006) highlight that many marketing functions and responsibilities paralleled those of communication positions, with a major emphasis in listening to customers, finessing their messages and finding the right audiences to receive them. Communication became one of the most important channels through which value-added was delivered by companies. However, this time there was a significant difference: customers were seen more as partners, or co-creators, than just merely on the receiving end of the communication process. Companies understood that the secret to create sustainable value was to incorporate customers in the co-creation of their marketing offerings because that will, in turn, mobilize more customers (Ramaswamy & Gouillart, 2010). It was time to expand the notion that the purpose of a business is to create a customer that Peter Drucker introduced in the 1950s (Drucker, 1954) to assert that its job was to create customers who create other customers (Singh, 2010).

The intersection at which marketing and communication theories arrived when social media became popular was quickly reflected in the job market. In marketing departments across the board new positions were created, such as social media managers and community managers. Some other popular names used for these positions were: Social Media Editor, Social Media Marketing Manager, Social Media Communications Manager, Project Social Media Manager, and Social Media Strategist, among others. Some of the responsibilities included in the job descriptions for these positions were: create, tag and title content; deliver new dynamic content; leverage third-party content; analyze trends in social media tools; analyze discussions to develop innovative ideas, programs and appropriate messaging; manage strategic messages; editorial management and development; report on performance metrics; monitor and respond to the fan community; nurture relationships; optimize the fan experience; help educate and counsel other colleagues and departments on social media; grow relationships with key influencers, bloggers, highly followed personalities, and reviewers (Tuten & Solomon, 2015).

From the list of responsibilities displayed above, it becomes apparent that creative writing skills and the ability to generate engaging content were paramount. Good writers and storytellers were among the talents most demanded. Proficiency in the skillful crafting of messages designed to appeal to niche audiences represented a valuable competency for employers. Professionals that understood how to identify and influence opinion leaders within these audiences became highly valued assets for companies (Pulizzi, 2012; Royle and Laing, 2014).

The newly adopted skills and competencies bore fruits. People were successfully *pulled* to companies' marketing resources, attracted by the carefully crafted content and engaged with it in a wide variety of ways. As it was discussed in the previous section, for marketers the result was an even richer and more complete picture of the interactions that a customer was having with their companies than before. This wealth of information represented an opportunity to understand their clients better and to deliver more value through the customization of their offerings.

Oliver, Rust and Varki (1998) predicted that marketers would face a model in which interactivity meant co-designing continuously changing offerings to meet customers' particular preferences and tastes. These authors referred to this as "real-time" marketing and characterized it as an integration of mass customization and relational marketing. More recently, this paradigm is referred to as the service-dominant logic (Rust and Huang, 2014; Vargo & Lusch, 2004)

In light of this, it can be argued that the notion of *interactivity* changed one more time to incorporate two new features. One of them is the notion that customers want to extend the co-creational aspects of content and messages to the design of products and services. The second feature reflects a shift towards a more holistic view of the way customers perceive the value generated by a company's offerings. They no longer limit the delivery of value to the consumption of a company's offering. Customers derive value from the whole *experience* provided by the interaction with a company, from the initial contact to every single touch point in which they have the opportunity to express themselves and learn more about a brand.

For marketers' skillsets this change of focus suggests a need to learn how to cultivate successful customer relationships (Melaia et al., 2008). As a consequence, marketing executives have increasingly incorporated software tools to store and analyze a wide array of data as part of their every-day work. For instance, studies have found that the specific skills needed at entry and lower-level marketing positions are more extensive than before (Schlee & Harich, 2010). Among these new skills, employers highlight those that refer to the analysis of information in databases, such as command of Structure Query Language (SQL) and Extensive Markup Language (XML), and experienced used of business management solutions, such as SAP and SharePoint.

The introduction of self-learning systems into the marketer's tool kit brought out the need for more advanced analytical skills (Webber, 2013). In particular, advanced training in multivariate statistics, mathematical programming and visual data techniques (Watson, 2012) are found within the desired skills and knowledge included in recently published job descriptions are analyzed. This trend has also promoted the emergence of low to middle-level positions with descriptors such as CRM managers; data scientists; business or market intelligence analysts; and consumer or marketing insight analyst (Stone and Woodcock, 2014). Some of these positions are being filled with professionals with mathematical or computer science background given the acute shortages of analytical skills perceived and reported by some employers. However, these same employers complain about the lack of appropriate knowledge of business context or skills to communicate results to decision-makers that these more technical profiles display (Webber, 2013). As the flow of information is not only expected to increase in diversity and complexity, it is anticipated the overwhelming needs for analytical skills to grow. The recent proliferation of knowledge surrogates such as smart homes and appliances represent one way in which this trend is materializing ("Ordinary Home Appliances Are About to Get Really Sexy", 2016).

With respect to the experiential aspect of the new paradigm of interactivity, Atwal and Williams (2009) call attention to the importance for marketers to gather resources to focus on the intangible aspects of a business, such as brand-related experiences. Similarly, Webster et al. (2005) report brand equity building as one of the most pressing challenges faced by modern marketing managers in both industries, business-to-consumer and business-to-business. Vargo and Lusch (2004) describe how the orientation towards the more intangible elements of a company implies the development of competencies geared towards the provision of services or experiences as opposed to goods.

Finally, several studies that reflect the practitioner's perspective (Finch et al., 2012; Royle & Laing, 2014) highlight the need for marketing professionals equipped with skills oriented to develop and incorporate return-on-investment metrics to the marketing function and tactics. Baker and Holt (2004) warn us about the dangers of failing to address the planning-measuring gap in marketing strategies. They conclude that under those limitations, marketers will be perceived as unaccountable by the rest of the organization and as unable to demonstrate a return on investment in the strategies they have control over.

Thus, the third wave of technological advancements implied a redefinition of marketers' skills in which a balance between the analytic and synthetic competencies was demanded. The new times call for marketing professionals that are both sensitive to other people's emotions and their expressions while at the same time aware of the need to quantify the progress of their strategies.

CONCLUSION

This chapter was conceived with two goals. The first one is to provide the reader with a review of the most important technological changes that impacted the marketing function in modern companies. The second is to present a thorough discussion of the changes marketing job positions experienced over the past two decades as a result of technology changes and new strategic imperatives.

The first of the two objectives was presented dividing the review of major technological advancements in three waves. The first wave represented the commercialization of Internet through the introduction of the World Wide Web. What the Internet provided for companies was a new way to communicate with customers. It offered the possibility to showcase their offerings at all times worldwide. The Internet also introduced the first definition of audiences' digital footprints and allowed companies to keep track of them. The second wave was embodied by the surge of social media and the challenges this phenomenon brought with it: customers' engagement through content creation, the emergence of the second screen and the fragmentation of audiences. The third and most recent wave is marked by the era of big data and the development of information surrogates, such as artificial intelligence applications for sentiment analysis and recommendation engines.

The second objective is tackled with a discussion of how these three waves of technological change accompanied an evolving notion of interactivity from the customers' point of view and an adaptation of marketers' competencies and skills. From interactivity as the ability to establish conversations and adjust the environment where communications take place to one where companies are seen as providers of highly customizable experiences. From marketing responsibilities that relied on graphic design skills to ones in which competencies to ensure a sustainable and engaging relationship with customers were paramount.

REFERENCES

Achrol, R., & Kotler, P. (1999). Marketing in the Network Economy. *Journal of Marketing*, *63*, 146–163.

Atwal, G., & Williams, A. (2009). Luxury Brand Marketing – The Experience is Everything! *Journal of Brand Management*, *16*(5), 338–346.

Baker, S., & Holt, S. (2004). Making Marketers Accountable: A Failure of Marketing Education? *Marketing Intelligence & Planning*, *22*(5), 557–567.

Ballantyne, D., & Varey, R. J. (2006). Creating Value-in-Use through Marketing Interaction: The Exchange Logic of Relating, Communicating and Knowing. *Marketing Theory*, *6*(3), 335–348.

Bampo, M., Ewing, M., Mather, D., Stewart, D., & Wallace, M. (2008). The Effects of the Social Structure of Digital Networks of Digital Networks on Viral Marketing Performance. *Information Systems Research*, *19*(3), 273–290.

Basheer, A.M. & Ibrahim, A. (2010). Mobile Marketing: Examining the Impact of Trust, Privacy Concern and Consumers' Attitudes on Intention to Purchase. *International Journal of Business and Management*, *5*(3), 28–41.

Berthon, P., Leyland, P., & Campbell, C. (2008). Ad Lib: When Consumers Create the Ad. *California Management Review*, *50*(4), 6–30.

Boyd, D. M., & Ellison, N. B. (2007). Social Network Sites: Definition, History, and Scholarship. *Journal of Computer-Mediated Communication*, *13*(1), 210–230.

Briones, R., Kuch, B., Liu, B. F., & Jin, Y. (2011). Keeping up with the digital age: How the American Red Cross Uses Social Media to Build Relationships. *Public Relations Review*, *37*(1), 37–43.

Brodie, R., Ilic, A., Juric, B., & Hollebeek, L. (2013). Consumer Engagement in a Virtual Brand Community: An Exploratory Analysis. *Journal of Business Research*, *66*(1), 105–114.

Bruce, M., & Solomon, M. R. (2013). Managing for Media Anarchy: A Corporate Marketing Perspective. *Journal of Marketing Theory and Practice*, *21*(3), 307–318.

Bulik, B. S. (2015). Worldwide PC Leader Lenovo Rebrands For Post-PC World. *Advertising Age*. Available at http://adage.com/article/digital/lenovo-rebrands-logo-renewed-post-pc-push/298773/

Cambria, E., Schuller, B., Xia, Y., & Havasi, C. (2013). New Avenues in Opinion Ming and Sentiment Analysis. *IEEE Intelligent Systems*, *28*(2), 15–21.

Campbell, C., Pitt, L. F., Parent, M., & Berthon, P. R. (2011). Understanding Consumer Conversations around Ads in a Web 2.0 World. *Journal of Advertising*, *40*(1), 87–102.

Casner-Lotto, J., & Barrington, L. (2006), *Are they really ready to work?Employers' Perspectives on the Knowledge and Applied Skills of New Entrants to the 21st Century U.S. Workforce*. The Conference Board, Corporate Voices for Working Families, Partnership for the 21st Century Skills and Society for Human Resource Management.

Dawar, N. (2013). When Marketing is Strategy. *Harvard Business Review*, *91*(12), 101–108.

Deighton, J. (1996). The Future of Interactive Marketing. *Harvard Business Review*, *74*(6), 151–160.

Drucker, P. (1954). *The Practice of Management*. New York: Harper & Row.

Evans, D. S. (2008). The Economics of the Online Advertising Industry. *Review of Network Economics*, *7*(3), 359–391.

Evans, D. S. (2009). The Online Advertising Industry: Economics, Evolution, and Privacy. *The Journal of Economic Perspectives*, *23*(3), 37–60.

Fan, S., Lau, R. Y. K., & Leon Zhao, J. (2015). Demystifying Big Data Analytics for Business Intelligence through the Lens of Marketing Mix. *Big Data Research*, *2*(1), 28–32.

Finch, D., Nadeau, J., & O'Reilly, N. (2013). The Future of Marketing Education: A Practitioner's Perspective. *Journal of Marketing Education*, *35*(1), 54–67.

Forrest, E., & Hoanca, B. (2015). Artificial Intelligence: Marketing's game changer. In T. Tsiakis (Ed.), Trends and Innovations in Marketing Information Systems, (pp. 45-64). Hersey, PA: IGI Global.

Foster, T. (2015). Industrial Marketing Communication: A (R)evolutionary Journey from Marketplace to Marketspace. In H. Spotts (Ed.), Revolution in Marketing: Market Driving Changes, (pp. 111-118). Springer.

Frambach, R., Roest, H., & Krishnan, T. (2007). The Impact of Conusmer Internet Experience on Channel Preference and Usage Intentions Across the Different Stages of the Buying Process. *Journal of Interactive Marketing*, *21*(2), 26–41.

Giamanco, B., & Gregoire, K. (2012). Tweet Me, Friend Me, Make Me Buy. *Harvard Business Review*, *90*(7), 88–93. PMID:22299507

Gibbs, S., Steel, G., & Kuiper, A. (2011). Expectations of Competency: The Mismatch Between Employers' and Graduates' Views of End-User Computing Skills Requirements in the Workplace. *Journal of Information Technology Education*, *10*, 371–382.

Giglietto, F., & Selva, D. (2014). Second Screen and Participation: A Content Analysis on a Full Season Dataset of Tweets. *Journal of Communication*, *64*(2), 260–277.

Hannah, R., Rohm, A., & Crittenden, V. L. (2011). We're all connected: The Power of the Social Media Ecosystem. *Business Horizons*, *54*(3), 265–273.

Ho, J. Y. C., & Dempsey, M. (2010). Viral Marketing: Motivations to Forward Online Content. *Journal of Business Research*, *63*(9-10), 1000–1006.

Hoffman, D., & Fodor, M. (2010). Can You Measure the ROI of your Social Media Marketing? *MIT Sloan Management Review*, *52*(1), 41–49.

Hollebeek, L., Glynn, M., & Brodie, R. (2014). Consumer Brand Engagement in Social Media: Conceptualization, Scale Development and Validation. *Journal of Interactive Marketing*, *28*(2), 149–165.

Kilgour, M., Sasser, S. L., & Larke, R. (2015). The Social Media Transformation Process: Curating Content into Strategy. *Corporate Communications: An International Journal*, *20*(3), 326–343.

Kumar, V. (2015). Evolution of Marketing as a Discipline: What Has Happened and What to Look Out For. *Journal of Marketing*, *79*(1), 1–9.

Lemon, & Verhoef. (2016). Understanding Customer Experience throughout the Costumer Journey. *Journal of Marketing*.

Liu, B. (2012). Sentiment Analysis and Opinion Mining. *Synthesis Lectures on Human Language Technologies*, *5*(1), 1–167.

Livingstone, S. (2008). Internet Literacy: Young People's Negotiation of New Online Opportunities. In T. McPherson (Ed.), Unexpected Outcomes and Innovative Uses of Digital Media by Youth, (pp. 101-121). Cambridge, MA: The MIT Press.

Lochrie, M., & Coulton, P. (2011). Mobile Phones as Second Screen for TV, Enabling Inter-audience Interaction. In *Proceedings of the 8th International Conference on Advances in Computer Entertainment Technology*.

Lordan, E. J. (2006). Interactivity: The Latest Trend in Effective Communications. *Public Relations Quarterly*, *51*(3), 27–29.

Lovejoy, K., & Saxton, G. D. (2012). Information, Community, and Action: How Nonprofit Organizations Use Social Media. *Journal of Computer-Mediated Communication*, *17*(3), 337–353.

Mangold, W. G., & Faulds, D. J. (2009). Social Media: The New Hybrid Element of the Promotion Mix. *Business Horizons*, *52*(4), 357–365.

McWilliam, G. (2000). Building Stronger Brands through Online Communities. *Sloan Management Review*, *41*(3), 43–54.

Melaia, S., Abratt, R., & Bick, G. (2008). Competencies of Marketing Managers in South Africa. *Journal of Marketing Theory and Practice*, *16*(3), 233–246.

Nam, H., & Kannan, P. K. (2014). The Informational Value of Social Tagging Networks. *Journal of Marketing*, *78*(4), 21–40.

Oliver, R., Rust, R., & Varki, S. (1998). Real-Time Marketing. *Marketing Management*, *7*(Fall), 28–37.

Ordinary Home Appliances Are About to Get Really Sexy. (2016, January 6). *Fortune*. Available at http://fortune.com/2016/01/06/home-appliances-ces-2016/

Potatoes, C. C. (2013, February 9). *The Economist*. Available at http://www.economist.com/news/business/21571471-measuring-tv-audiences-keeps-getting-harder-counting-couch-potatoes

Prahalad, C.K. & Ramaswamy, V. (2000). Co-opting Customer Competence. *Harvard Business Review*, *78*(January-February), 79–87.

Pulizzi, J. (2012). The Rise of Storytelling as the New Marketing. *Publishing Research Quarterly*, *28*(2), 116–123.

Ramaswamy, V., & Gouillart, F. (2010). *The Power of Co-creation. Build It With Them to Boost Growth, Productivity, and Profits*. New York, NY: Free Press.

Rayport, J. F., & Sviokla, J. J. (1994). Managing in the Marketspace. *Harvard Business Review*, *72*, 141–150.

Reichheld, F. (2001). *Loyalty rules! How today's leaders build lasting relationships*. Boston: Harvard Business School Press.

Roberts, M. L., & Zahay, D. (2012). *Internet Marketing: Integrating Online and Offline Strategies*. Stamford, CT: Cengage Learning.

Rowley, J. (2008). Understanding Digital Content Marketing. *Journal of Marketing Management*, *24*(5-6), 517–540.

Royle, J., & Laing, A. (2014). The Digital Marketing Skills Gap: Developing a Digital Marketer Model for the Communication Industries. *International Journal of Information Management*, *34*(2), 65–73.

Rust, R. T., & Huang, M.-H. (2014). The Service Revolution and the Transformation of Marketing Science. *Marketing Science*, *33*(2), 206–221.

Sashi, C. M. (2012). Customer Engagement, Buyer-Seller Relationships, and Social Media. *Management Decision*, *50*(2), 253–272.

Schlee, R. P., & Harich, K. R. (2010). Knowledge and Skill Requirements for Marketing Jobs in the 21st Century. *Journal of Marketing Education*, *32*(3), 341–352.

Schmitt, B., & Zarantonello, L. (2013). Consumer Experience and Experiential Marketing: A Critical Review. *Review of Marketing Research*, *10*, 25–61.

Sheth, J. N., Sisodia, R., & Sharma, A. (2000). The Antecedents and Consequences of Customer-Centric Marketing. *Journal of the Academy of Marketing Science*, *28*(1), 55–66.

Silk, A. J., Klein, L. R., & Berndt, E. R. (2001). The Emerging Position of the Internet as an Advertising Medium. *NETNOMICS: Economic Research and Electronic Networking*, *3*(2), 129–148.

Simon, M., van den Driest, F., & Wilms, T. (2016). Driving Customer-centric Growth: A Practical Roadmap. *Journal of Advertising Research*, *56*(2), 159–168.

Singh, S. (2010). *Social Influence Marketing Trends for 2010*. Available at http://www.slideshare.net/shivsingh/social-influence-marketing-trends-2967561

Stone, M., & Woodcock, N. (2014). Interactive, Direct and Digital Marketing. *Journal of Research in Interactive Marketing*, *8*(1), 4–17.

Taneja, H., Webster, J., Malthouse, E. C., & Ksiazek, T. B. (2012). Media consumption across platforms: Identifying user-defined repertoires. *New Media & Society*, *14*(6), 951–968.

Taylor, K. (2003). Marketing Yourself in the Competitive Job Market: An Innovative Course Preparing Undergraduates for Marketing Careers. *Journal of Marketing Education*, *25*(2), 97–107.

Tuten, T. L., & Solomon, M. R. (2015). *Social Media Marketing*. London, UK: SAGE Publications.

Van Dijck, J. (2013). *The Culture of Connectivity: A Critical History of Social Media*. New York, NY: Oxford University Press.

Van Otterlo, M. (2013). A Machine Learning View on Profiling. In M. Hildebrandt & K. de Vries (Eds.), *Privacy, Due Process and the Computational Turn. Philosophers of Law Meet Philosophers of Technology*. New York: Routledge.

Vandermerwe, S. (2014). *Breaking Through*. Palgrave Macmillan.

Vargo, S., & Lusch, R. (2004). Evolving to a New Dominant Logic for Marketing. *Journal of Marketing*, *68*(1), 1–17.

Wallace, E., Buil, I., & de Chernatony, L. (2014). Consumer Engagement with Self-expressive Brands: Brand Love and WOM outcomes. *Journal of Product and Brand Management*, *23*(1), 33–42.

Watson, H. (2012). The Necessary Skills for Advanced Analytics. *Business Intelligence Journal*, *17*(4), 4–7.

Webber, R. (2013). The Evolution of Direct, Data and Digital Marketing. *Journal of Direct, Data and Digital Marketing Practice*, *14*(4), 291–309.

Webster, F., Malter, A., & Ganesan, S. (2005). The Decline and Dispersion of Marketing Competence. *MIT Sloan Management Review*, *46*(4), 35–43.

Webster, J. G., & Ksiazek, T. B. (2012). The Dynamics of Audience Fragmentation: Public Attention in an Age of Digital Media. *Journal of Communication*, *62*(1), 39–56.

Wellman, N. (2010). The Employability Attributes Required of New Marketing Graduates. *Marketing Intelligence & Planning*, *28*(7), 908–930.

Wymbs, C. (2011). Digital Marketing: The Time for a New "Academic Major" Has Arrived. *Journal of Marketing Education*, *33*(1), 93–106.

Yadav, M. S., & Pavlou, P. A. (2014). Marketing in Computer-Mediated Environments: Research Synthesis and New Directions. *Journal of Marketing*, *78*(1), 20–40.

ENDNOTE

1 http://www.nielsen.com/us/en/solutions/measurement/television.html

Chapter 8
Algorithm as Demiurge:
A Complex Myth of New Media

Jan Kreft
Jagiellonian University, Poland

ABSTRACT

Plato's Demiurge is the quintessence of perfection and power. "Whatever comes from me is indestructible unless I, myself, wish it to be destroyed" - says the creator of the gods, speaking to them in Timaeus. The gods and Demiurge are believed to collaborate on the creation of people in accordance with the standard of excellent ideas; soon the world, as we know, will be created. Plato's Demiurge is also the good, and the platonic gods are righteous. Without Demiurge the world is a chaos, an environment of irrational chances. Nowadays, the myth of Demiurge can be related to the perfection of creativity. Demiurge becomes equal to the anticipated, all-powerful driving force. Omnipotent, yet tamed and friendly. Demiurge is also the leader in the tradition of social research, the "divine" constructor of the economy, the originator of development. In the new media environment, Demiurge is a convenient metaphor for the presentation of the algorithm: mysterious, error-free, resistant to influence, free from human weaknesses. A transcendent being. The aim of this publication is to present new concept, the core myth of new media organizations - the myth of Demiurge associated with the operation of algorithms and critical analysis of myths created around it, which accompany the social, political and business role of algorithms. Considering the aspect connected with the interpretation of digital media operation and their social and business role, algorithms have not been so far analysed in the context of the presence of myths in organisational functioning. The author believes, however, that the common factor in the perception of algorithms in new media is their mythical aureole and mythical thinking associated with them.

DOI: 10.4018/978-1-5225-1656-9.ch008

INTRODUCTION

Considering the discussed question from the perspective of managing a media organisation, myths are present in various aspects. Organisations themselves, their particular parts as well as their perceived and created characteristics are mythologized. Mythical tales refer to the representatives (heroes) of organisations, mainly: their leaders (owners, managers). The myth may also refer to the employees and clients/users of an organisation. The attributes of an organisation may be mythologized too – the pioneer character of organisational activities at the first place.

On one hand, the myth is connected with intra-organisational management, and on the other hand, it is related to the external image of an organisation: creating images of the whole organisation and its particular, important elements.

In the current activities of an organisation, also at the strategic level, myths define the operational criteria to be applied by an organisation, and they legitimize their results. A function performed by myths which is particularly desired is providing a cover for the real objectives (Boje, Fedor, Rowland, 1982). It may refer, first of all, to the fundamental economic objectives of a media organisation: profits, welfare of its shareholders, and its market share – the objectives which are not transparent for the receivers and users of media organisations (especially of social media).

The presence of myths (mythical thinking) in digital media may be analysed from various points of view, and it may be done during the research on their mutual relations.

BACKGROUND

There is a great number of publications on the development of digital media, and the research on algorithms in digital media mainly refers to the reception and placement of advertisements (Mager, 2012). From the market perspective, algorithms are often discussed because they are to decide about the business success or failure, they come as the basis of new business models and business strategies. Their functioning contributes to the higher value of an organisation more than the value of the provided contents.

At the same time, considering the deluge of applications and analyses on IT science, the research on their social and political consequences appear to be very scarce (Lohr, 2012).

For example, during the research done on algorithms in the social context, a new paradigm of hegemony has been formulated (Lash, 2007). The significance of the Wall Street algorithms for education, sport organisations, etc. has been also analysed (Steiner, 2012).

Among various analytical approaches, the reflections on the concept of "structuration" presented by A Giddens are often referred to; people's actions and discourses influence social systems, however, they are in turn affected by social systems. Algorithms as elements of technology are perceived equally as a part of a system, and as a part of social activity, which maintain their mutual, dynamic relation. This concept is referred to by Wiebe E. Bijker (1997) who writes about a technological frame, Wanda Orlikowski and Debra Gash (1994) who write about the concept of a technological frame of reference. Considering the aspect of the institutional theory, algorithms are analysed in an abstract way as institutions understood as procedures, standards or guidelines (Scott, 2013).

As an invisible part of communication, algorithms are analysed as a part of the universe of a data flow that is realised among the machines (Hayles, 2006). They are a part of "a cognisphere," a dynamic flow of symbols and instances among people, animals and machines. They also come as a part of the world in which human contribution to communicating information is not indispensable.

Algorithms are also analysed in the aspect of digital exclusion – as an element of invisible processes which are connected with the access to information and knowledge. They define the value and qualifications of people, the levels of access to the whole range of significant spaces and services (Graham, 2004). Algorithms also perform an important role in organising opportunities, as David Lyon defines it: "social sorting" (Manovich, 1999).

In the research on digital media, algorithms are treated, first of all, as useful or indispensable, because having them we do not need to sort out irrelevant results by ourselves (Spring, 2011). Their use constitutes "a new era of algorithms," "a new age," and finally "a new society." From the users' perspective, their connection to databases makes them useful and comfortable to be applied in one device. As L. Manovich describes it vividly, data and algorithms come as two halves of the ontology pertaining to the world of computers (Manovich, 1999).

The research on algorithms focuses on their usefulness and on the fact that they simplify our moving around the more and more complex and fragmented environment. Searching and content recommendation systems are frequently analysed.

Algorithms in connection with enormous databases are supposed to contribute to our better understanding of the nature of society, the operation of organisations at the micro- and macro-levels – this belief has appeared to be particularly tempting in the consideration of the scientific aspect. Researchers representing various scientific fields, especially those of social and natural sciences and humanities, have been interested in calculating and computing techniques, hoping that they can provide new answers to some fundamental questions.

Considering the aspect connected with the interpretation of digital media operation and their social and business role, algorithms have not been so far analysed in the context of the presence of myths in organisational functioning. The author believes, however, that the common factor in the perception of algorithms in new media is their mythical aureole and mythical thinking associated with them.

Such mythical thinking about algorithms often results from the fact that they are surrounded by business secrecy, they are difficult to understand and created by programmers in a specific "language," and there have been numerous myths created around algorithms which are applied by some leading organisations such as Google, Facebook, Twitter, etc.

Mythical thinking about digital media refers to the role they play in connecting us with other people/suggesting "friends," customizing our offer of information, books, drama performances – all that is done by mythical algorithms which also create articles and information; they "construct the reality," modify our interactions, and manage our access to media contents.

In accordance with the definition assumed in the chapter, myths stem from tales which have been detached from their context, and they have become separated from "the roots of their epoch," the ones which have turned to be universal (Kostera, 2008). These are stories which refer, among others, to the extremes (things we have not experienced, a parallel reality, desired behaviour and common values) (Armstrong, 2004).

A myth is characterised by objective falseness and subjective belief that it is true. It depends on the changeable demand for its existence. It comes as a concept, or a set of beliefs concerning the reality which has been created and presented by media.

A myth is true because it is efficient, not just because it provides some specific information. If it does not provide any new insight into a more profound sense of life – it fails. If a myth provides that, then it also provides a new hope. It becomes a guide, it tells us what to do to enrich our life (Armstrong, 2004). It is a pattern to be followed in all our operations, a model of behaviour – this is its main role.

Myths provide us with explanations, they bring reconciliation of contradictory concepts and they help us to solve dilemmas (Bolman & Deal, 1987). They also perform other important functions: they explain, they express, they maintain solidarity and cohesion, they legitimize, they communicate our unrecognized needs and conflicts, they mediate in contradictions and provide narration to anchor our present in the past (Cohen, 1969).

Comparing myths in primal societies to current myths, the latter ones are characterised by some obscurity and "blurredness." As there are formal difficulties to determine phenomena referred to as myths, the notion of "mythical thinking" is often applied.

In the chapter, it has been assumed that mythical thinking is not focused on explanation but on presentation. It is unreflectively accepted because of ritual communication, participation in common activities and common events. It is difficult, however, to determine some specific boundaries of mythical and scientific thinking, as they overlap and complement each other. At the same time, mythical thinking is highly efficient in epistemological explanation – however, it requires the acknowledgement of some assumptions. It is full of understatement, and it is logical – however it is the logic of holistic visions and reconciling contradictions, joining cultural, empiric and non-empiric aspects.

Mythical thinking aims at blending human experience; it creates an image from various types of experience. The world is understood through the assimilation of impressions taken from things which are real and imagined. The boundaries between the reality of media and its perception have been disappearing. The way of perceiving the reality and the reality itself have become one and the same thing.

In the author's opinion, the relation of myths and media is to be traced mainly in such phenomena as:

- Granting a mythical status to media (in general). This is the status of indispensable entities – their lack is supposed to result in the sense of losing our freedom and social integrity. The "natural" presence of media does not allow us to reflect upon their temporary status.
- Granting the power to create our reality to media (the main, sometimes the only one source of the knowledge about the world).
- Granting the status of a factor having a non-logical influence realised by the common experience to media (for example, in social media).
- Exercising media's own authority in the attempts of falsifying their own role of media organisations.

MAIN FOCUS OF THE CHAPTER

The subject of the following analysis is the role of the myth in the context of media functioning, with particular consideration of mythical thinking about an organisation or its part, and the significance of such thinking from the perspective of managing media organisations.

The suggested concept of a complex myth - that is namely: the myth of Demiurge for new media organisations - facilitates creating and supporting myths by media organisations. It results from the unique possibilities of media to present and interpret the reality.

The subject of the analysis is the relation between mythical thinking and algorithms with their features, such as operation that is difficult to interpret for the Internet users, a mystery connected with their functioning.

Issues, Controversies, Problems

In the mythology of Ancient Greece, Plato's Demiurge comes as the essence of perfection. "θεοὶ θεῶν, ὧν ἐγὼ δημιουργὸς πατήρ τε ἔργων, δι' ἐμοῦ γενόμενα ἄλυτα ἐμοῦ γε μὴ ἐθέλοντος. τὸ μὲν οὖν δὴ [41β] δεθὲν πᾶν λυτόν, τό γε μὴν καλῶς ἁρμοσθὲν καὶ ἔχον εὖ λύειν ἐθέλειν κακοῦ: δι' ἃ καὶ ἐπείπερ γεγένησθε, ἀθάνατοι μὲν οὐκ ἐστὲ οὐδ' ἄλυτοι τὸ πάμπαν, οὔτι μὲν δὴ λυθήσεσθέ γε οὐδὲ τεύξεσθε θανάτου μοίρας, τῆς ἐμῆς βουλήσεως μείζονος ἔτι δεσμοῦ καὶ κυριωτέρου λαχόντες ἐκείνων οἷς ὅτ' ἐγίγνεσθε συνεδεῖσθε" (Plato, 1903).

The translation has it as "Whatever comes from me is indestructible unless I, myself, wish it to be destroyed" – says the creator of gods speaking to them in *Timaeus*. Gods are to cooperate with Demiurge on the creation of people, following the pattern of perfect ideas and soon the world we know shall be created. Demiurge creates the world from the matter, in accordance with ideas, however, he is not a creator – he is a constructor.

Plato's Demiurge is pure goodness, and Plato's gods are just. People, however, may be deprived of such justice as they are not perfect morally, although they are exceptional beings and their goodness comes as a noticeable value. It should be also mentioned that in Plato's philosophy the dialogue is a good "god," and the concept of "goodness" is placed above Demiurge. The dialogue and the parties who are involved in it, together with Demiurge, form transcendence.

Plato defines his tale as a myth, however he treats myths seriously: "You may think it is a myth – an old nurse's fairytale – and you look down on it. It would not be strange to despise such things if we could ever find anything better and truer while looking for it everywhere." This is an answer which Sokrates gives to Callicles in *Gorgias*.

Without Demiurge, the world would be a chaos, a universe of coincidences. Nowadays the myth of Demiurge refers to creative perfection. A demiurge – with no capital letter – becomes the "driving force." Following the mythical original – the omnipotent force which has been tamed and tailored to the human size, and which is now friendly to people.

Considering the aspect of management, a strong leader of an organisation may be a demiurge: an entrepreneur – a „divine" constructor of economy, an originator of development and a guarantor of the bright future and immortality of an organisation. But also a non-human being, the result of human labour whose unusual capabilities/features become mythical. In digital media such a demiurge is the algorithm with its reliability, faultlessness and resistance to any influence. Soulless everyday perfection. The algorithm which comes as a transcendent and perfect being, also as a model of journalist exemplar. It is objective, resistant to any pressure and just. A hero deserving the title of a superman – a hero of unknown origin, long-time awaited and promising to quench our thirst for fair, fully democratic media.

MYTHICAL THINKING IN MEDIATIZED SOCIETY

In new media, which in principle claim to have the right to the dominating sphere of culture, the mediatized present is to confer significance and to provide gratification in the form of answers obtained to various chaotic questions. Constant creation of media contents and their maintenance by algorithms is to result in the immediate satisfaction of newly emerging needs, and in the community of media experience the subsequent patterns of behaviour appear. However, they can be defined as short-term patterns, they come with the promises of a constant inflow of new contacts, with their volatile emotions and faint mutual obligations.

In such environment the stronger the sense of change –which is one of the important component factors of post-modernity - becomes, the more significant role of myths appears to be. Facing constant experience of changes occurring to the environment created and presented by media, some footholds for the social order must be found: temporary, spontaneous and planned, under and beyond our control – all of them often supported by mythical thinking. Such thinking is desired when we consider our volatile material environment, changeable patterns of behaviour modified by efforts of marketing, volatile membership in a group, aspirations that change dynamically and promises to satisfy them – the environment in which a social pattern is exemplified by *the nowist culture* and *hurried culture* (Bertman, 1998).

Mystical Environment of Algorithms: The Internet

In media environment myths refer to the functioning of organisations, although they are rarely aware of these myths (Morgan, Frost, Pondy, 1983). If the functioning of an organisation is far from the ideal, the vision of a perfect organisation and myths explain the anomalies and contradictions. It also refers to the ambiguity of myths, however they can distract our attention from the essence of the problem, and to distort the perspectives. Yet, they are indispensable, as Lee Bolman and Terry Deal metaphorically suggest, they let us stay sane and rational (Bolman, Deal, 1987).

In new media myths accompany users and organisations every day. New media are supposed to become the common space of freedom and unrestricted expression – the sphere which is over/beyond the citizen-market relations, immersed in the net ethos of mythical "founding fathers" of the Internet, a group of outstanding characters who, like Demiurge, have created a new universe out of the wasteland. This virtual world is to be outside/beyond any national borders; these borders are supposed to have some organic character only, they are supposed to be just a reflection of the diversity presented by the Internet users, who are the citizens of particular countries (Goldsmith, Wu, 2006).

In such a created world the Web 2.0 develops as a predecessor of the Web 3.0, the next stage of constant development, the universe where intentions of the Internet users will be recognised based on the analysis of their behaviour and the context of the data they provide (Haendler, 2009).

The mythical status of the Internet is founded first of all on:

1. Emphasizing the fact that there are no particular authors and owners of the Internet,
2. Treating the Internet as a common piece of work which states the beginning of "the digital era,"
3. Interpreting the Internet as a universe of common wisdom, the treasury of all the human knowledge (of unknown location),
4. Multifunctional presence in all the fields of social, economic and political life,
5. The factor which overcomes the boundaries of time and space,
6. Using the Internet as a platform of "real" social relationships,
7. A factor of disintegrating social and cultural barriers, and
8. The space of unlimited expression of human creativity.

Such a status makes the Internet an autonomous being: indispensable and desired, an element of elevated media which acknowledge and allow us to cultivate our individuality, and –thanks to new media - to participate in a better social life.

Algorithm as a Demiurge

In such a mythical Internet environment the algorithm with its reliability, faultlessness, resistance to influence and soulless perfection, comes as a media ideal. It is a hero of unknown origin, long-time expected and promising to satisfy our yearning for fair, fully democratic media which arrange our knowledge in order, etc.

Such a demiurgic algorithm may be interpreted at several levels:

- As a factor which facilitates the dialogue between people, a creator of a common dialogue,
- As a factor which promises to overcome human limits with the use of some mythical creative power that can increase human capabilities (knowledge),

- As a perfect element of media which can replace imperfect people (e.g. journalists), and
- As an element which shapes the perfection of an organisation and confirms its omnipotence.

On one hand, algorithms are treated as powerful entities sorting, regulating, classifying or putting the world in order in some other ways. On the other hand, they are volatile, mysterious and virtually unstudiable. Hidden behind trade secrecy, they are inaccessible and therefore valuable and precious.

Such mythical algorithms:

- Take advantage of „the participating ethos" of the Internet, they encourage us to participate in common voluntary service, to share information about ourselves, to experience our power to make things happen,
- Release us from thoughtful reflections,
- Offer time saving, and
- "Guarantee" satisfaction from the participation in "elevated, sublime time" which is detached from the historical time.

Attributing algorithms with mythical meaning is even more natural as it fits in the interpretation of the reality understood as a mathematical construction with its logic and rules which result from, among others, algorithms that are gradually discovered by the science. Nowadays, this mathematical vision is expressed in media technologies, it comes as a foundation of modernity, with digital media, applications, languages and interfaces. In the past, these were theological visions of the world presented as mathematical constructions (St. Augustine) or mathematical visions of the cosmos (Kepler) based on the mathematical understanding of the world presented by Cartesius. At present, the belief and possibilities of designing and managing the social infrastructure come as important features of the societies of developed countries; Wendy Hui Kyong Chun refers to this state as to the sense of "programmability" (Chun, 2011). The society is interpreted as a network of actors and technological and institutional infrastructure of various types. This network is managed by a specific logic of activities and relations.

Algorithm as Ariadne's Thread

Generally, algorithms help us, first of all, to move around our more and more complicated and fragmented environment. The crucial factors in this process are searching and content recommendation systems, because facing the rapidly growing offer, previous search engines proved to be insufficient.

Algorithm as Demiurge

Algorithms are most frequently discussed when they suggest "friends," information, books, or drama performances. Also, when they create articles and information, when they "construct the reality," manage our activities, modify our interactions, emphasize the selected contents and "exclude" some other contents.

Algorithms decide about the functioning of content aggregators and browsers, they affect the selection of items in the recommendation systems, the selection of products and services in on-line shops; they also select the information displayed in social networking. They come as "the fundamental logic" regulating the flow of information, they allow and attribute information with the meaning (Langlois, 2013).

In this regard, they fit into the mythical, Platonic interpretation of a dialogue from the transcendent perspective, as they create extremely desired values.

In Greek mythology, Ariadne offered Theseus a clue of thread to help him get out of Minotaur's labyrinth. The algorithm comes as a mysterious gift offered by a media organisation (Ariadne) to help us out of the informational chaos (the labyrinth). A gift indeed, as Google or Yahoo services are offered free of charge.

For the users the thread (the algorithm) comes as an exceptional value which allows them to avoid irrelevant information (false traces in the labyrinth). It leads to the meta-information, the knowledge of matching other pieces of information. This ability constitutes the basis for the development of media content aggregators.

The algorithm which leads us out of the informational chaos turns out even more valuable when the Internet user/Theseus willingly interprets the surrounding digital space as the informational chaos. Then the chaos becomes a natural/objective state, so reaching out for the thread becomes obvious.

This belief is supported by another myth – about the common creation of media contents. It shades a true picture of media users' activities. In fact, active writers who provide articles, fandom participants, filmmakers, or users who display their creativity in any other possible way come as a relatively small but resourceful avant-garde group that precedes passive masses of users who follow the path to the Web 2.0.

Thus, freedom of creation in new media is exercised only by a relatively small number of users, and massive creativity is nothing else but traditional consumptive behaviour with a different name (Van Dijk, Nieborg, 2009). The behaviour in the environment full of disparities in the access to the Internet and to the use of the opportunities it offers, which is connected with another myth.

Algorithm as a Demiurge of Democracy and Interactivity

Modern so called new media, the Internet in particular, function in the aureole of revival referring to a number of social attitudes and new phenomena which were supposed to be constrained by unequal relations between broadcasters and passive receivers, characteristic for old mass media.

The interpretation of the role performed by new media includes statements about democratization which is realised with the use of social media platforms that base their operations on algorithms. They are supposed to contribute to "the end" of the politics in its current form.

This myth accompanies glorification of the strongest existing myth of the Internet journalism – that is namely: the desired ideal state in which the audience members become the crucial element of the websites allowing the cultural transformation of journalism to take place (DEuze, 1999). This interpretation presents the Internet as an element of "non-controllable communication." The users establish communication within these frames "omitting the classical role of journalism."

Lev Manovich writes about a myth of interactivity as of an exceptional feature characteristic for new media. He arguments that, for example, traditional art is, in fact, interactive, because it requires cooperation, whereas digital versions require receivers; they only come as different forms of sound, visual or typographical versions. Indeed, each interpretation becomes an interaction, as an interaction is a psychological process (Manovich, 2001).

While providing an analysis which refers to the role of algorithms, we should - first of all - emphasize the potential of the opportunities connected with their use in the management of user accessibility. The dominant position is taken by the organisations which have access to data and which have such tools as algorithms which can be efficiently used to run campaigns applying technologies of "managing" the electorate. Those who provide data about themselves gradually accept the role of users of mechanisms which select and suggest information, opinions and recommendations. The place of media organisations selecting and interpreting information which employ journalists is taken by algorithms.

The Myth of Common Access to Resources

Unrecognizing sets is an important feature of search engine algorithms, and databases (for example the archives) come as a set of data that are possible to be found, not as an obligation to find them.

The functioning of media organisations which use algorithms is not usually accompanied by any objections referring to the scope of information penetrating done by algorithms. Quite the opposite: they suggest facilitation of the access to some "forgotten"/niche contents. Meanwhile, the part of the Internet which is indexed by the standard browsers includes only the resources which are recognised by search engines (HTLM usually). The remaining resources, hardly accessible for search engines, are defined as the Invisible Web or Deep Web. The Invisible Web contains mainly databases, catalogues (often accessible in the net, also as HTLM websites) and websites partially or totally created in Flash technology. Search engines perform poorly with graphic, sound and video files, for example Google search engines which cannot handle text in graphics – the change of the situation has only just started. Another "inaccessible" part includes websites that are not linked with other websites.

The Myth of Objectivity

The on-line universe is considered to be deeply rooted in everyday social practices, however the functioning of algorithms is accompanied by mythical thinking which separates them from human intervention and human imperfectness. Algorithms are followed by the promise of faultless "knowledge production" which not only emphasizes human limitations but also proves human greatness and the fact that we have purposefully reached beyond human cognition.

The idea of entrusting computers (algorithms) with the task of creating and selecting media contents, as their actions would be "socially unadulterated," refers to the myth of perfection: a faultless entity free from prejudice, weakness, vicious intentions. Such an entity, even if it meets expectations only in a partial way, has already overtaken our imagination.

There are still some unanswered questions about "the neutral character" and "justice" of algorithms. Their "objectiveness" is also difficult to define, even if we assume such criteria of algorithm quality assessment as their reliability, efficiency and precision.

The subject of our consideration is also the influence of algorithm operations on the limitation of rationality, or its increase and expectation. This discourse opens a field for considerations pertaining to morality of algorithms, philosophical dilemmas of moral autonomy and fault of machines, the influence of the Internet archival resources and behavioural patterns of users that create "determinant paths" of the obtained results based on the "accumulated fault" of algorithms (Gandy Jr., 2010).

Considering search engines, defining the results of the search with the use of algorithms as "best results" or "favourite results" suggests not only what the algorithm intended but also what should be assumed as intended. It is supposed to balance the standards referring to the objectiveness of journalism – indicating things which are significant and remarkable, and issues that should not be the concern of a public discourse. In opposition to these standards, algorithms are supposed to be free from the imperfections of journalism.

The objectiveness of media created by people, in which the depositary of trust is a journalist and a journalist organisation, comes as a specific form of practice and an attitude toward the collection of information, its processing and dissemination. This is a neutral attitude toward the subject of reference, the lack of bias – that means refraining from taking sides in disputes and the lack of favouritism. It is the attachment to genuineness (however, objectivity is not identical with the concept of truth) and the lack of latent motives (McQuail, 2010). In a word, the ideal standard of journalist practice which refers to freedom and independence.

Although defining objectivity is problematic, it should be emphasized that one of the media objectivity criteria is "adequacy" referring to the process of selection, a requirement for its rules to be clear and cohesive with regulations which define what is important for the users and for the society. The promise of algorithm objectivity, which is based on much less institutionalized standards, lacks such clarity (transparency).

It should be also added that objectivity comes as the ideal state, which is therefore impossible to be achieved, however much desired. Following the rules of objectivity may lead to a new bias. It can be defined as inconsistent with a general concept of media freedom (which does not differentiate false and true answers). Additionally, some receivers do not expect any objective or complete information.

The Page Rank algorithm comes as an example of successful pursuit for ideal objectivity. Its description was initially defined as "a technical document," but then it has been mythologized with time as a central element which distinguishes Google from its competitors, and it has been presented as a fully "democratic" calculation logic offered by the organisation to "provide goodness to the whole world." The Page Rank algorithm is supposed to be free from assessing judgement and personal prejudice of journalists, free from pressure of editorial boards, managers and owners. It is to be consistent with the Google manifesto "Ten Things We Know to Be True," in which the users find the following assertion: "Our users trust our objectivity and no short-term gain could ever justify breaching that trust."

Such mythologisation of the Page Rank facilitates Google to be perceived as a new organisation, not a corporation. Its academic roots legitimize the fact that Google is presented as an academic entrepreneurship which works on artificial intelligence, scientific information, linguistics, and which, at the same time, by mere coincidence, earns money.

The exceptional status is also emphasized by the mission of Google, that is namely: "to organise the world's information and make it universally accessible and useful." Such a mission is to encourage the users and reviewers of Google activities to feel grateful because of the fact that such classification comes as extremely expensive enterprise. In the same way, the business matters of Google should be interpreted as convergent with social issues.

At the same time, objectivity and clearness of the algorithm are shaded by a legend which legitimizes its social significance and which, by assumption, focuses social and business interest. Its strongly emphasized objectivity increases its value, although algorithms are at the same time promoted as excellent tools for advertisers, therefore algorithm construction should also consider their needs and expectations.

In a broader sense, legitimization of algorithms based on their „objectivity" gives us grounds to acknowledge them as the right intermediaries in the process of providing information and knowledge. In this situation, such companies as Google can (at least in a declarative way) put their business interest at stake – an example can be withdrawing business from China in 2010. It was done in the glory of an organisation which cherishes its values more than its business interests, however a compelling factor of this decision was the pressure exercised by the Chinese competitors. Additionally, this decision strengthened the position of all information providers who use algorithms, and Google allowed itself to become a guardian of the "holy fire" of neutrality. Since then, the corporation has been able to emphasize the fact that it does not manipulate the results of the algorithm operations, however it has been burdened with "spiritual responsibility" for assuming the role of the most important guardian of information in the world (Morozow, 2011.

The Myth of Diversity

Another important problem that puts the multi-dimensional diversity of digital media in question is the similarity of results obtained by various browsers (Hindman, 2008). The similarity refers to multiplication of information sources in traditional and new media. Considering this aspect, P. Napoli indicates the influence of algorithms on institutional isomorphism, on the consequences of the fact that media contents producers use algorithms which analyse historical data and consequently, on the fact that media contents are provided on the basis of the same sources, for example, while making decisions about film production (Napoli, 2013).

SOLUTIONS AND RECOMMENDATIONS

The metaphorical presentation of on organization is accompanied by a selection of symbolic and interpretative analyses and metaphors describing the organization. In accordance with this research approach, an organization is a complex, joint product of culture, technology, social and physical structure "immersed" in the environment (Hatch & Cunliffe, 1997). It is a network of meanings created and supported by people who share common values, tradition and customs.

Such a research approach applies to new media organizations as well, that is namely: to search engines, social networking sites, user-generated services – all with algorithms.

The factor which distinguishes media organizations is a particular, unique opportunity to support existing myths and to create new ones, as well as the ability to displace living myths out of the social awareness - all these topics are related to the social impact of the media. Therefore, the creation of myths serves as self-creation of an organization and, in practice, it becomes a process of symbol manipulation.

Not only do media continue to function as centers creating and "distributing" myths in primal societies, but they also continue the tradition of poets and storytellers. They are Homers of the digital era, creating myths about themselves as well (Rothenbuhler & Coman, 2005).

On the one hand, myths are becoming a part of organization management, on the other hand, they are an element of the image of an organization, creating its representations. Myths and mythical thinking are becoming a frame which systematizes the knowledge of an organization.

In their current operations, but also at a strategic level, myths define the criteria for organizational activities, and they also legitimize the effects of those activities. Hiding the real objectives of an organization may be a particularly desired function of myths (Boje, Fedor, & Rowland, 1982).

Mythical thinking commonly feeds on metaphors which come as useful means of identifying and understanding the essence of a given phenomenon (Hatch & Cunliffe, 1997).

As the author proposes, the demiurge metaphor is a fundamental metaphor common to new media organizations which use algorithms. A metaphor that, even metaphorically, allows us to describe media organizations in their various social roles.

The factor which unites algorithms of social networking sites and search engines is their common mythical matrix (halo): not only are algorithms supposed to "rule the world" of media – that by itself metaphorically meets all the expectations connected with governance - but they also allow users to define themselves in the network and to exercise control over such self-definition. They create new opportunities, but they also have control. Their use should constitute "a new era of algorithms," "a new age," "a new society."

With its reliability, correctness and resistance to influence, the algorithm emerges as an ideal media universe. It is a mythical hero of unknown origin and incomprehensible performance. It promises to fulfil our longing for the righteous and fully democratic media.

On the one hand, algorithms are regarded as mythical, powerful entities sorting, governing, classifying or disciplining the world in various ways. On the other hand, they are mysterious and virtually unstudiable. Hidden behind a trade secret, they are extremely valuable.

Most often, they are a subject of reflection when they suggest "friends," information, books, performances, and when they formed articles, information. Also, when they "construct the reality," guide our actions, control interactions and expose some contents while "excluding" some other contents and when they manage our attention. We use algorithms to determine what is significant, what you need to know, as if we appealed to the knowledge of experts, scientific research, common sense, or even to the words of God (Gillespie, 2014).

From the market perspective, they decide about the success and failure whenever they become the basis for new business models and strategies.

The myth is true because it is effective, not because it provides specific information. It is a pattern for all human activities, and it determines the model behavior – that is its main role. The following new media myths aspire to this role:

- **Demiurge of Democracy:** Computing technologies have given a rise to the combination of extensive databases and computational practices that allow users to receive power. Now they are supposed to have a decisive impact on social campaigns and persuasive techniques in the political and commercial spheres.
- **Hidden Demiurge:** Media users are convinced that they remain in touch with other users. Meanwhile, they are in touch with the algorithm. The power of the algorithm is based on the continuous encouragement to be active within the principles of "participation" and under the threat of "invisibility" for other users.
- **Demiurge of Universal Visibility:** The combination of algorithms and databases makes the algorithms useful and convenient for the users to be taken advantage of. In this case, their power comes down to "ignoring" particular datasets and the phenomenon of Invisible Web. The algorithm acts as a gatekeeper.
- **Objective Demiurge:** This myth is the core concept of the power of algorithms. The algorithm should be free from any divisions related to gender, sexuality, race, ethnicity, or social class. It has to be "socially unadulterated," "fair."
- **Demiurge of Common Contacts:** The design of algorithms should provide contacts with all network users.

The myth of Demiurge and mythical thinking about algorithms are the desired sources of knowledge, free from imperfections of positivistic knowledge; they are the consequences of a disintegrated vision of constant, linear development; they emphasize faults and ambiguities of scientific and technological interpretation of the reality. In accordance with interpretations, in times when each piece of information is questioned and turns into another hypothesis, the technical, economic, political and cultural spheres are joined only by the market, the individualism develops, "dressed in" virtual communities, and economy, technology, politics and culture become more and more separated. The belief in the relative character of axioms and assumptions increases (Vattimo, 1982). The knowledge stemming from myths, adequate for our understanding, faces such conditions, and the opposition between myths and rationality disappears.

FUTURE RESEARCH DIRECTIONS

Discuss future and emerging trends. Provide insight about the future of the book's theme from the perspective of the chapter focus. Viability of a paradigm, model, implementation issues of proposed programs, etc., may be included in this section. If appropriate, suggest future research opportunities within the domain of the topic.

The concept of a complex myth of Demiurge in the aspect of operations performed by new media organisations may be considered not only in the context of mythical thinking about algorithms, but also in the context of relationship between this myth and other myths which, first of all, refer to leadership of media organisation, the position of such an organisation on the labour market and contact with providers.

Further research may also refer to the role of myths at the strategic level, the way in which they define operational criteria of an organisation and, or maybe first of all, legitimization of the results of operations performed by an organisation.

The myth of Demiurge may be analysed in the context of social networking operations, websites created by the users, blogs, VoIP communicators or net services – that is namely: organisations understood as innovative versions of solutions that go beyond traditional organisational structures that reach for the environment, organisations the boundaries of which have been disintegrated. It refers to the concept of "management 2.0." that is namely: to organisations the functioning of which is supposed to fit in the times of "turbulence and chaos" in a better way (Kotler & Caslione, 2009). Such organisations are more flexible, operationally perfect, they predict, they make decisions quickly and they process information. They are also administered in a more efficient way by "employees 2.0." preferably from the Y generation (Caseau, 2011). They are focused on exercising their influence, oriented toward the improvement of IT and communication systems.

CONCLUSION

Mythologisation of the operations performed by a media organisation comes as a frame which organises the knowledge about the organisation. Mythical thinking about media organisations refers to organisations themselves, their parts and the way they are perceived. It may also refer to organisational attributes, such as their power and indispensability. It is also possible to make references to the necessity and obviousness of market behaviour (for example, considering search engines, it may refer to the way of using personal data and their aggregation).

The suggested concept of the complex myth, that is namely: the myth of Demiurge for new media organisations, allows media organisations to create and to support myths in a more convenient way. It results from the unique possibilities of media thanks to which they can present and interpret the reality.

Plato's Demiurge comes as the essence of perfection – nowadays the myth of Demiurge may be referred to creative perfection. Demiurge becomes equal to the almighty and expected driving force. It is an omnipotent force, however it is also tamed and friendly. In the environment of new media, the algorithm becomes a mythical demiurge: mysterious, faultless, resistant to influence, free from human imperfections. A transcendental entity which creates the universe of Google, Facebook and a number of other organisations.

The myth of Demiurge allows us to approach an organisation from a broader perspective, it becomes a basis which allows us to define particular behaviour of a media organisation. Such a complex myth favours the formation and supporting of other myths which accompany the functioning of algorithms: the myth of democratization, visibility, accessibility.

A particularly desired advantage of the complex myth of Demiurge may be the fact that it provides a matrix of (mysterious) business objectives of media organisations.

REFERENCES

Armstrong, K. (2004). *A short history of myth* (Vol. 1). Canongate Books.

Bertman, S. (1998). *Hyperculture: The human cost of speed*. Westport, CT: Praeger Publishers.

Bijker, W. E. (1997). *Of bicycles, bakelites, and bulbs: Toward a theory of socio-technical change*. Cambridge, MA: MIT Press.

Boje, D. M., Fedor, D. B., & Rowland, K. M. (1982). Myth making: A qualitative step in OD interventions. *The Journal of Applied Behavioral Science*, *18*(1), 17–28. doi:10.1177/002188638201800104

Bolman, L. G., & Deal, T. E. (1987). *Modern approaches to understanding and managing organizations*. San Francisco: Jossey-Bass.

Caseau, Y. (2011). *Processus et Entreprise 2.0*. Paris: Dunod.

Chun, W. H. K. (2011). *Programmed visions: Software and memory*. Cambridge, MA: MIT Press. doi:10.7551/mitpress/9780262015424.001.0001

Deuze, M. (1999). Journalism and the Web An Analysis of Skills and Standards in an Online Environment. *International Communication Gazette*, *61*(5), 373–390. doi:10.1177/0016549299061005002

Gandy, O. H. Jr. (2010). Engaging rational discrimination: Exploring reasons for placing regulatory constraints on decision support systems. *Ethics and Information Technology*, *12*(1), 29–42. doi:10.1007/s10676-009-9198-6

Gillespie, T. (2014). 9 The Relevance of Algorithms. *Media technologies: Essays on communication, materiality, and society*, 167.

Goldsmith, J. L., & Wu, T. (2006). *Who controls the Internet?: illusions of a borderless world* (Vol. 89). New York: Oxford University Press.

Graham, S. (2004). The software-sorted city: Rethinking the 'digital divide'. *The Cybercities Reader*, 324-332.

Hatch, M. J., & Cunliffe, A. L. (1997). *Organization Theory: Modern. Symbolic-Interpretive and Postmodern Perspectives*. Oxford, UK: Oxford University Press.

Hayles, N. K. (2006). Unfinished work from cyborg to cognisphere. *Theory, Culture & Society*, *23*(7-8), 159–166. doi:10.1177/0263276406069229

Hendler, J. (2009). Web 3.0 Emerging. *Computer*, *42*(1), 111–113. doi:10.1109/MC.2009.30

Hindman, M. (2008). *The myth of digital democracy*. Princeton University Press.

Kostera, M., Aggestam, M., & Eriksson-Zetterquist, U. (2008). *Introduction to the Trilogy: Mythologies of Organizational Everyday Life. In Organizational Olympians: Heroes, heroines and villains of organizational myths*. London: Palgrave Macmillan.

Kotler, P., & Caslione, J. A. (2009). *Chaotics: the business of managing and marketing in the age of turbulence*. AMACOM Div American Mgmt Assn.

Langlois, G. (2013). Participatory Culture and the New Governance of Communication The Paradox of Participatory Media. *Television & New Media*, *14*(2), 91–105. doi:10.1177/1527476411433519

Lash, S. (2007). Power after Hegemony Cultural Studies in Mutation? *Theory, Culture & Society*, *24*(3), 55–78. doi:10.1177/0263276407075956

Lohr, S. (2012). Big data's impact in the world. *The New York Times*.

Lyon, D. (2013). *The Electronic Eye: The Rise of Surveillance Society-Computers and Social Control in Context*. Minneapolis: John Wiley & Sons.

Mager, A. (2012). Algorithmic ideology. *Information Communication and Society*, *15*(5), 269–787. doi:10.1080/1369118X.2012.676056

Manovich, L. (1999). Database as symbolic form. *Convergence (London)*, *5*(2), 80–99. doi:10.1177/135485659900500206

Manovich, L. (2001). *The language of new media*. Cambridge, MA: MIT Press.

McQuail, D. (2010). *McQuail's mass communication theory*. Sage Publications.

Morgan, G., Frost, P. J., & Pondy, L. R. (1983). Organizational symbolism. *Organizational Symbolism, 3*, 35.

Morozov, E. (2011). Don't be evil. *The New Republic*, 13-07.

Napoli, P. M. (2013). *The algorithm as institution: Toward a theoretical framework for automated media production and consumption*. Fordham University Schools of Business Research Paper.

Orlikowski, W. J., & Gash, D. C. (1994). Technological frames: Making sense of information technology in organizations. *ACM Transactions on Information Systems*, *12*(2), 174–207. doi:10.1145/196734.196745

Plato. (1903), Platonis Opera. Oxford University Press.

Rothenbuhler, E. W., & Coman, M. (Eds.). (2005). *Media anthropology*. Thousand Oaks, CA: Sage Publications.

Scott, W. R. (2013). *Institutions and organizations: Ideas, interests, and identities*. Los Angeles, CA: Sage Publications.

Spring, T. (2011). *How Google, Facebook and Amazon run the internet*. PC Advisor.

Steiner, C. (2012). *Automate This: How Algorithms Took Over Our Markets, Our Jobs, and the World*. New York: Penguin.

Van Dijck, J., & Nieborg, D. (2009). Wikinomics and its discontents: A critical analysis of Web 2.0 business manifestos. *New Media & Society*, *11*(5), 855–874. doi:10.1177/1461444809105356

Vattimo, G. (1992). *The transparent society*. Cambridge, MA: Polity Press.

KEY TERMS AND DEFINITIONS

Organization: A complex, joint product of culture, technology, social structure, culture and physical structure "immersed" in the environment; network of meanings created and supported by people who share common values, traditions and customs.

Chapter 9
Decision-Making Models of the Human-Operator as an Element of the Socio-Technical Systems

Nina Rizun
Gdansk University of Technology, Poland

Tatyana Shmelova
National Aviation University, Ukraine

ABSTRACT

The authors of the chapter proved that the fundamental intellectual processes, which lie on the basis of decision-making behavior of the human-operator, could be identified on the bases on the analogies with the devices (elements). The basic intellectual processes of the Rational decision-making models can be adequately identified by the transient processes of the PID-controller; the intellectual processes of the Bounded Rationality and Intuitive decision-making models can be identified by the transient processes of the nonlinear elements. Taxonomy of the most typical automatic control theory elements and their compliance with certain decision-making models with a point of view of decision-making processes specificity and on human-operator behavior in the context of the socio-technical concept was obtained Authors of the chapter also accept the suggestion that the instruments of collective multi-criteria decision-making and social-network analysis theory have similar mathematical and methodological bases.

DOI: 10.4018/978-1-5225-1656-9.ch009

INTRODUCTION

The appearance of the term Socio-Technical Systems (STS) is the result of a rapidly developing interaction between the social and technical factors that create new, sometimes unpredictable, a synergistic effect on the performance of the contemporary company. Large-scale, high- technology systems such as nuclear power generation and aviation have been called STS because they require complex interactions between their human and technological components

STS theory is theory about cooperation between the social aspects of people and society and technical aspects of machines and technology. Sociotechnical theory refers to the interrelatedness of "social" and "technical" aspects of an organization. Sociotechnical theory, as distinct from STS, proposes a number of different ways of achieving joint optimization.

The STS also tend to have two common features: high technologies and high risk activities. As such, they require much less direct operation due to the fact the technology replaces the human-operator (H-O). On the other hand, require much more remote operator's supervision due to the modern tendency to supervise the technology by distance. The systems' work is not transparent due to increased difficulty to know exactly what the technology is being used. The systems are also highly hazardous and of high-risk, and have greater potential for catastrophic.

The relationship between social and technical factors of those systems may be of two types:

- **Linear:** A predictable, manageable. Is associated with expected conditions, the solving of routine, repetitive and well-structured problems, based on pre-established rules, regulations, as well as operator's experience and technical knowledge; and
- **Nonlinear:** Unpredictable, illogical. Is associated with the occurrence of uncertainty. They are characterized by semi-structured and unstructured problems. Solutions, which have received in the non-programmed, unexpected situations, contain the risk. These solutions require specific personal and professional qualities and creativity.

An addition, STS as a system with *feedback*, on the one hand, is a powerful tool for changing the socio element as part of the modern enterprise. The dynamism of the science and technology development, the rate of occurrence and processing of information, the increasing of the work's pace and tension of and more modifies the behavioural characteristics of a human, the specifics of his/her thinking, the models of the DM. On the other hand, H-O creates new requirements for its environmental technology, which should satisfy both his/her professional and behavioural characteristics. The specificity of contemporary cyber-organizations as STS necessitates the formation of modern system of the computational methods and models H-O's DM behaviour.

That's why one of the urgent concerns of the contemporary management – to *find the new approaches to the formation of a socio-technical concept of human operators decision-making models, algorithms and techniques which should be based on modern technical, technological and social categories, phenomena and methodologies.*

BACKGROUND

On the basis of the socio-technical concept of H-Or's decision-making models (DMM), suggested by the authors, the following major heuristics are suggested:

Heuristic 1: The development of modern automated control systems, the complication of hardware components' systems and technology, the extensive usage of computer technology in the organization and management of the production process lead to an increased human's role in the management of and restructuring of its activities (Timochko, 2007; Pavlenko, 2007).

Thus on the one hand, the following typical for STS human-operator's activities:

- **Efficient Type of Thinking:** Acquiring extended number of specific characteristics (indicators), which describe the particular *conditions* for the implementation of the thinking process;
- **Methods:** That are involved in the process of thinking and interpretation of operator's *choice* (the *decision*), the main of which the following: the close, actually transforming into unity, connection between perception and comprehension of the rapidly changing information and conditions of the situation; *decision* often merges with the process of its *implementation*;
- **Decision-Making Process:** İncreasingly fraught with a *stressful* situation, characterized by *unstructured, incomplete* and at the same time *congestion* contradictory initial information coming; and
- **Decision-Making Process:** İncreasingly associated with rules of strictly *limited time*.

On the other hand, the primary modes of operation by the STS human operators are still the following (Strelkov, 2001)

- **Normal Conditions:** (The operator only monitors the working of the technological process, without interfering in it);
- **Emergency Conditions:** (Operator works in semiautomatic or mechanized mode, much depends on the accuracy of its sensorimotor actions and ability to assess the situation);
- **Boundary Conditions:** The technological process still goes on within the prescribed limits, but is already close to its borders (the problem statement – keep the process within the required parameters of technology); and
- **"Free" Conditions:** The operator constructs new mode of working (problem – the empowerment of operating system, saving of the material part, energy and own forces).

Balance and order in the above mentioned provisions can make a modern DM theory, which divides the situations that require decisions into two categories: programmed and unprogrammed:

- **Expected Situations:** Associated with the solving in certainty, routine, repetitive and well-structured problems, based on pre-established rules, regulations, as well as operator's experience and technical knowledge; and
- **Unexpected Situation:** Associated with the occurrence of uncertainty and risk. They are characterized by semi-structured and unstructured problems. Solutions, which have received in the non-programmed situations, contain the risk. These solutions require specific personal and professional qualities and creativity.

Heuristic 2: Representation of the systems in the form of a STS first of all makes possible to take into account the influence of social, cultural environment of people who make decisions. Culture surrounds people and affects their values, convictions and behaviour, which they share along with other members of different social groups. Culture serves to bind us together as members of groups and to provide clues as to how to behave in both normal and unusual situations. The psychologist Hofstede suggests that culture is a "collective programming of the mind" (Makarov, 2000). Thus fatal mistakes can be committed by normal, healthy, highly motivated and well equipped personnel (Makarov, 2000; Kharchenko, 2011, 2012). Russian scientists have used lately the term "departure of conscience" when they analyse the causes of aviation events conditioned by the insufficient development of the appropriate cultural values in a person that makes decisions (Leychenko, 2006).

One of the possible approaches to the solution of these problems is formalization and mathematical presentation of the STS operators' activities on the base of the systemic analysis. Taking into account in the act of DM by the H-O within sociotechnical systems, besides the separate (Shmelova, 2011; Kharchenko, 2011):

1. **Professional Features** $\overline{F}_{\exp}$**:** Theoretical training (knowledge), thinking (skills), heuristic thinking (skills of the second level), practical qualification (abilities), acquired H-O during professional activity.
2. **Non-Professional Features** $\overline{F}_{np}$**:**

$\overline{F}_{ip}$ – is set of H-O *individual-psychological* features (IOF): perception, thinking, experience, imagination, memory, nature, intention, health;

$\overline{F}_{pf}$ – is set of H-O *psycho-physiological* features (PPF): features of the temperament, nervous system, emotional type, gain coefficient; and

$\overline{F}_{sp}$ – is set of H-O *socio-psychological* features (SPF): concentration on the problem, attention, commitment, economical interests, social priorities, moral, enables to predict the H-O's actions on the basis of modelling the foresighting "large-scale" outcomes of individual actions (Kharchenko, 2011; Shmelova, 2011).

The analysis of social-physiological features conducted by the authors allowed to make a conclusion that the activities of operator are influenced by the own image, the image of corporation as well as by interests of a family. At the same time respondents – operator pay special attention to interests of their families, their own economic status and professional promotion. The analysis of priorities determined the following model of advantages: social and economic priorities of a person, political views and legal norms of a person, spiritual and cultural orientations of a person as to the influence of individual-psychological factors the most significant of them are health and experience. In the conditions when a situation develops in the direction to a catastrophic occurrence, then the temperament and ability to receive information become more significant factors (Shmelova, 2011; Kharchenko, 2011). The research into the influence of individual-psychological and social-psychological features on the professional activity of human-operator of STS allowed obtaining data on such structural components of a personality of a specialist, such as motives or behaviour, values and priorities, hierarchy and development of these dynamic categories on the stages of DM by H-O.

Heuristic 3: The authors suggest to put the well-known fast reflexion theory of V.O. Lefevre (fast reflexion Lefervre, 2001) into the basis of modelling the behaviour of H-O in the process of his/her DM as a part of STS. Leferve's reflexion is the reflexion of consciousness and shows itself in the intuitive thinking. As Lefevr2 (2011) states, human's mentality has the so-called "information processor", which has the function of automative generation of images simultaneously with its subjective world. The operation of this processor creates the specific spectre of human reactions, which appear very quickly (almost immediately) and cannot be controlled consciously. Lefevr introduces the notion of automative (intuitive) choice, which happens in unexpected situations (conditions of uncertainty and risk), when the influence of external (social) environment can be the controlling power in the moment of making a choice.

Such behaviour received the name of *reflexive models of bipolar choice*, when in the moment of choice, the H-O can be in one of the alternatives:

positive pole A

negative pole B.

In order to take into account the complex of the factors that influences H-O of the STS and based on the heuristics of socio-technical concept of human-operators DMM, algorithms and techniques, the *general model* of H-O's decision-making process (DMP) has been worked out by authors (Figure 1) (Shmelova & Sikirda, 2001).

FUNDAMENTAL DECISION-MAKING MODELS IN THE CONTEXT OF SOCIO-TECHNICAL CONCEPT

Presently the most widely-spread scientific approach is the one that marks out three main DMM: Rational Decision-Making Model, Bounded Rationality Model and Intuitive Model (Allard, 2003). In this connection authors propos the following interpretation of the classical DMM in the context Socio-Technical Concept.

Rational Decision-Making Models

It is well-known, that *Rational decision-making model* defines the solution as a result of an ordered process of thinking and based on the notion of discretionarity. Discretionary called (Kahneman, 1994) such person's actions, which are based on deliberation, which are based on gathered information about current state of the

Figure 1. The general model of H-O's decision-making process

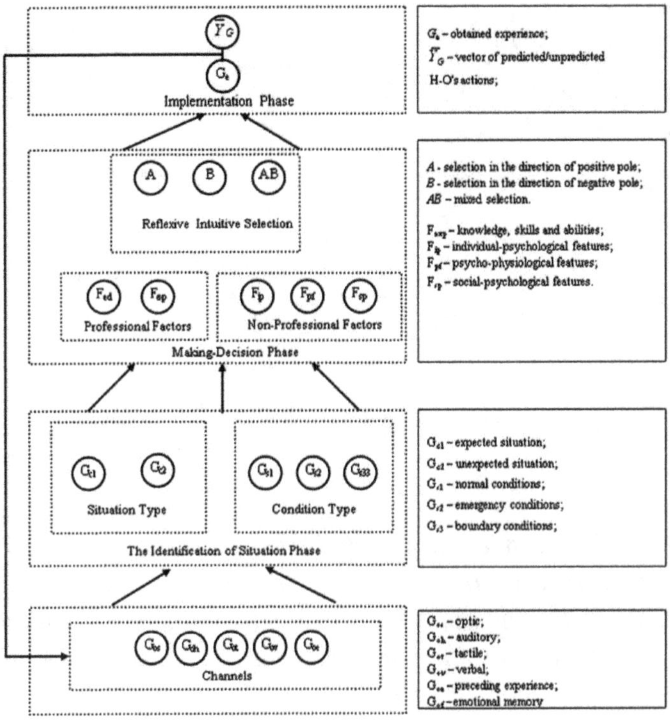

environment and comparing the available alternatives to them explicitly. In this case the selection's criteria of the preferred embodiment usually set in advance, and the solution (the chosen alternative) is the most competitive (optimal) for specific circumstances. We also need to consider the fact that in the conditions of STS the special role is given to the sources and means of operator's perception of information; the amount, ways and forms of flow of this information rapidly grow and get more complicated.

In this connection, according to the rational model for H-O as a socio element, under the systematic *DMP* (Simon, 2009; Rizun, 2013, 2014; Rizun & Taranenko, 2014a) in *socio-technical concept* could be interpreted using 5 significant steps (author' edition see on Figure 2).

Bounded Rationality Decision-Making Models

The *bounded rationality DMM* was proposed by Herbert Simon (Simon, 2009). Instead discretionary it presupposed using the *heuristics,* which could be described as follows:

Figure 2. DMP of rational model in the context of socio-technical concept

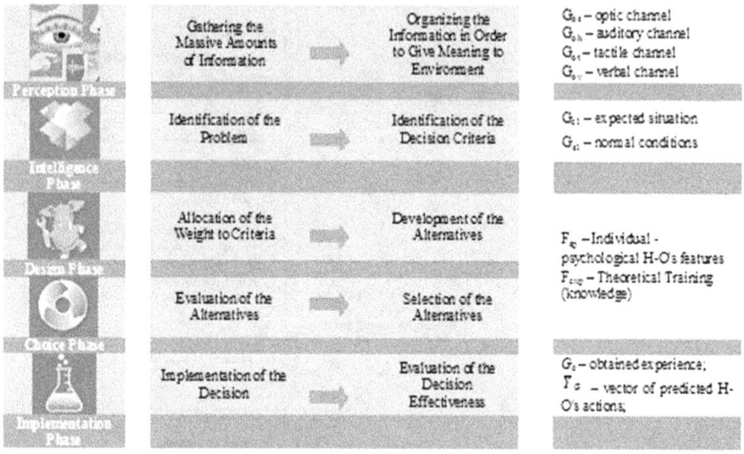

a set of prescriptions connected with desired (forbidden) patterns of perception, behaviour, that emerged as a result of personal professional experience and practical problems solving skills, and which have the similar initial conditions and prerequisites, and which are used in situations, when the rational DMM of cannot be used as a "soft" option of rationality in a situation of individual choice.

Thus, bounded rationality is an information behavioral prerequisite for modeling for *individual* solutions. Proponents of bounded rationality (Luck, 2009) very often correlate it with "adaptive toolbox" – as the tried and tested strategy set which people use to solve the problems and to meet the demands of environments. These strategies are commonly named heuristics – simple making-decision patterns which are used as simplification rules for stopping the options search.

In contrast to complete rationality in DM, bounded rationality model for H-O as an element of the in *socio-technical system* implies the following steps (Simon, 2009; Rizun, 2013, 2014; Rizun, 2013; Rizun & Taranenko, 2014a), which in author's version are presented in the Figure 3.

Intuitive Decision-Making Models

Intuition represents: "a quick apprehension of a decision situation based on past experiences and the reinforcement associated with these experiences, which is devoid of conscious thought" (Lunenburg, 2010; Luck, 2009); "a cognitive conclusion based on a decision maker's previous experiences and emotional inputs" (Pratt, 2007);

Figure 3. DMP of bounded rationality model in the context of socio-technical concept

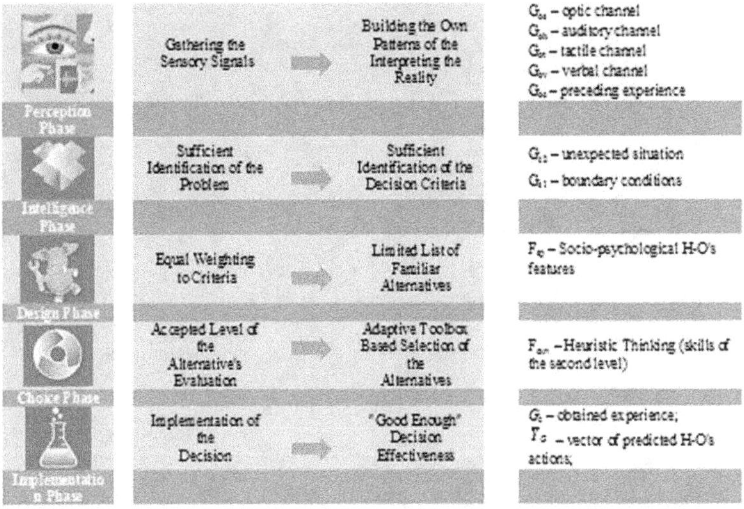

"a tacit form of knowledge that orients decision making in a promising direction" (Policastro, 1999).

Intuitive DMM is based on the usage of "internal intuition", common sense, associative and logical thinking, professional and emotional experience, impressions. The DM in the equally (or differently) as influenced by the knowledge, experience, motivational structure, purpose, life principles and values of the decision, impressions, assumptions, as well as limited timeframe decision. But we have to take into account the following statement: we need to "recognize that our initial, automatic response may be incorrect" (Hastie & Dawes, 2001). That's why it is important that we make decisions using *informed* intuition. Author's version of informed intuitive model steps for H-O as an element of the in *socio-technical* is presented in the Figure 4 (Rizun, 2013, 2014; Rizun & Taranenko, 2014a).

IDENTIFICATION OF FUNDAMENTAL DECISION-MAKING MODELS OF THE SOCIO ELEMENT OF THE MODERN CYBER-SPACE

Identification of the Rational Decision-Making Model

Identification of fundamental intellectual processes, which lie on the basis of DM behaviour of the socio element, can be based on the *analogies* with the *devices* (elements), provides technical support for these solutions. This approach allows

Figure 4. DMP of informed intuitive model in the context of socio-technical concept

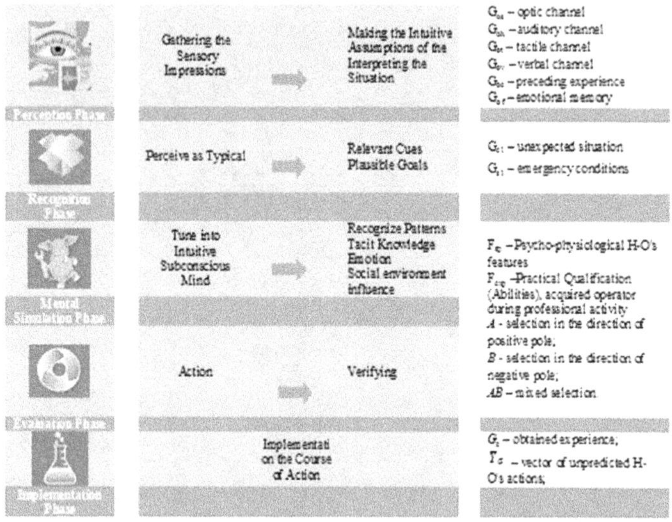

performing a "technical" classification and formalization of the specificity of the H-O behaviour based on his/her existence with modern cyber-space.

In real conditions the H-O takes part in controlling STS and relies on the meters' readings, trying to keep them within the limit of allowed maximum value of the controlled magnitude, not allowing the excess of this level. Depending on the psycho-physiological features of a human control can be provided either in a rather quiet regime, which has a slight deviation from the allowed one, or with a significant amplitude of deviations.

On the other hand, the automatic control theory known, that the PID-controller is a link in the contour of the control with feedback loop, used for maintaining the setpoint values of the measured parameter Rizun, 2013; Rizun & Taranenko, 2014a, 2014 b). Mathematical form of the PID algorithm is:

$$u(t) = K_p \cdot \varepsilon + \frac{1}{T_I} \cdot \int_o^t \varepsilon dt + T_d \frac{d\varepsilon}{dt} \tag{1}$$

The authors suggest the concept of interpretation of H-O making-decision processes

as a process of correspondence (maintenance) to the adjusted (required) set point level of the set problem solving (optimal, sufficient solution) (Rizun, 2013).

In this view the author was proposed the hypothesis that:

the basic processes of the Rational DM could be adequately simulated and identified by the transient processes of the proportional-integral-derivative (PID) controller.

The flowchart STS with the H-O, identified in the system on the basis of analogy with the classical PID-controller, can have the following structure (Figure 5).

In the flowchart: $r(t)$ – the problem, given at the input STS; $u(t)$ – solution of the problem, given at the input, made by the operator (PID-controller); $d(t)$ – technical errors in the process of the decision transfer (external disturbance); $y(t)$ – result of the influence of the decision, made by operator, on the system (measured output of the controlled object); $f(t)$ – errors of measurements of the results of influence (measurement noise); P – controlled object; $\varepsilon(t)$ – signal about the correspondence of the set problem $r(t)$ to the actual results of the decision made by the operator $y(t)$:

$$\varepsilon(t) = r(t) - y(t) \quad (2)$$

On the first step the following interpretation of the terms of the mathematical expression of the PID-controller as of the analogy of the STS H-O was developed:

- T_I – *integral gain*: the measure of the level of the cumulative set of the professional: *Theoretical Training (Knowledge)* and non-professional *individual-psychological features*, which allows to solve the current problem with the required quantitatively evaluated professional level of problem domain *awareness* level.
- T_d – *derivative gain*: the measure of the level of the cumulative set professional: *Practical Qualification (Abilities)* and non-professional *psycho-physiological features*, which allows to solve the current problem with the required quantitatively evaluated professional level of *pace* of making logical operations.

Figure 5. Example of the flowchart STS with H-O, identified as a classical PID-controller

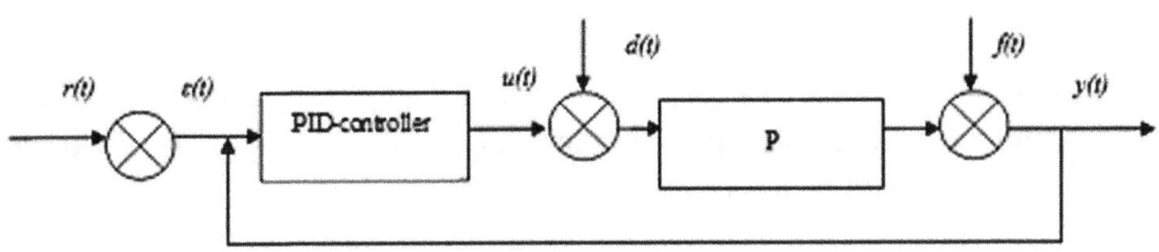

K_p – *proportional gain*: the measure of the level of the cumulative set professional: *Thinking (Skills)* and non-professional *socio-psychological features*, which allows to solve the current problem with the required qualitative evaluated professional level of purposefulness, persistence and commitment.

$\varepsilon(t)$ – error (mismatch): the measure of difference between the current variant of problem solution by human operators and optimal (sufficient, acceptable) solution (desired set point).

According to the classical theory of *STS* of control for the interpretation of stages of the DMP by the H-O it is suggested to use the following technological analogies of transitive characteristics of PID-controller (Figure 6) (Rizun, 2014).

T_r – perception time (Perception Phase); t_{max} – time for identification of the problem and development of the decision criteria and alternatives (Intelligence and Design Phases); μ_{max} – characteristics of the volume of the initially developed decision criteria and alternatives; T_p – time of extrapolated decision search (Choice Phase) with the set error ε_s (allowed deviation from the optimal decision); T_{cl} – period of damping oscillations, which is a process of the made decision realization (Implementation Phase).

Thus, the authors propose to *accept the hypothesis:* the basic processes of the Rational DMM could be adequately simulated and identified by the transient processes of the PID-controller.

With the objective of a deeper study of the specificity of behavior of H-O of the *STS* in the process of DM it is suggested to consider the structural scheme of DMM, which is based on the following rules (Figure 7):

- The *X-axis* is taken as the scale of measuring the degree of decision rationality, and
- The *Y-axis* is a scale of measuring the degree of problem's uncertainty.

Then we can suggest that each of the DMM of the H-O of the *socio-technical* systems, besides from their classic interpretation (main diagonal of the matrix), can have specific variants (strategies of model's application).

For example, the models of the strategy of Rational Decision-Making can be used in the conditions of full uncertainty as well as:

- In the conditions of weak *uncertainty* (in case when there are no strict limits of time and the deep analysis of effects of the made decisions in any of the possible risk situations is required); and

Figure 6. Example of the transitive process of DM for the problem $r_i(t)$ by the H-O, identified as a classic PID-controller

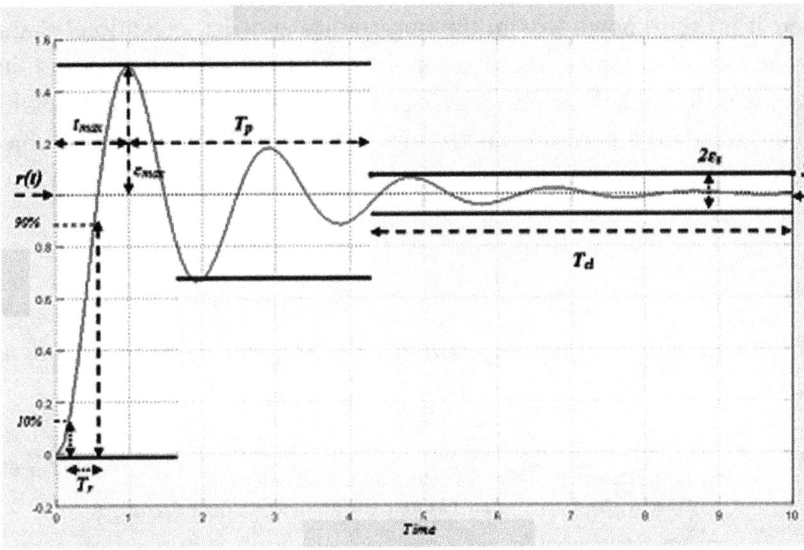

Figure 7. Structural scheme of decision-making models

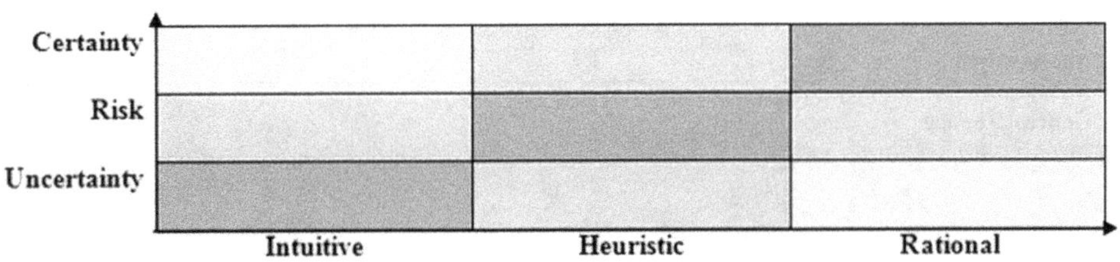

- In conditions of *risk*, when making a quick decision with completely predictable effects can allow to win some time for choosing the nontrivial solution.

Then, depending of the type of the problem, as well as on personal characteristics of the H-O it is suggested to use the following classification of strategies of applying the Rational Decision-Making Model, built on the analogy with PID-controller (PI, PD-controller) (Table 1) (Rizun, 2013; Rizun & Taranenko, 2014a, 2014b).

Identification of the Bounded Rationality and Intuitive Decision-Making Models

However, it is well-known that in the uncertainty and risk conditions characterized by the lack of complete information, inaccuracy of goals setting and limited time resources, to replace the *Rational* logical thinking, which is more suitable for performing standard algorithms and routine DMP, comes the need for subjective creative approach to solving problems based on personal experience and "gut feeling" (Table 1).

Table 1. The Taxonomy of types of strategies of rational decision-making model application

Human-Operator's Features	**Professional: Thinking (Skills)** **Non-Professional: Individual-psychological**	**Professional: Theoretical Training (Knowledge)** **Non-Professional: Socio-psychological**	**Professional: Practical Qualification (Abilities)** **Non-Professional: Psycho-physiological**
Analogy with Automatic Control Theory		PI- controller 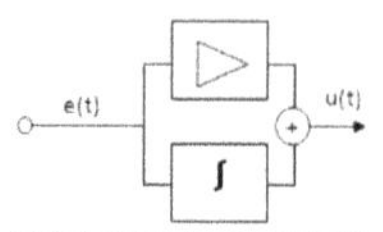	PD- controller 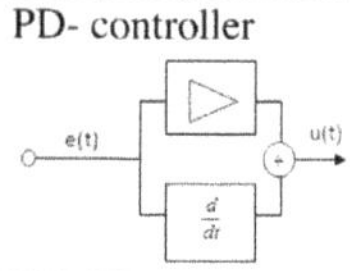
	PID- controller 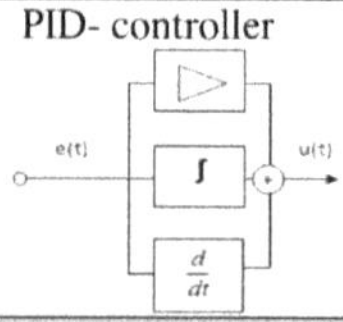		
Decision-Making Conditions	Certainty	Risk	Uncertainty
Strategy	Strategy of maximum utility (MU)	Strategy of maximum benefit strategy (MB)	Strategy of minimum cost (MC)
Style	Optimal Rational – decision-making style (DMS), which allows to make decisions that will satisfy the set conditions (optimal ratio of expenses and quality)	Qualities Rational DMS, which allows to make decision with the maximal quality of all the possible alternatives' work-out	Rapid Rational DMS, which allows to make decisions with the maximal speed
Problem's type	Logically determined problems of various degree of complexity and urgency	Strategic problems that require long deep theoretic analysis and justification	Simple, well-mastered, but requiring high preciseness of conduction, routine problems

That is why, continuing the analogy between the DMM of the H-O of the socio-technical systems and the analogies with the devices (elements) provides technical support for H-O solutions (Automatic Control Theory), the authors were proposed the following:

On the one hand, form the Automatic Control Theory (ACT) it is well-known that the signal passing through the nonlinear elements (NE), "enriched" by additional harmonic components defined by the parameters of this elements. On the other hand, a person, using for making-decision in unprogrammed, precarious and often stressful situation, their own experience and intuition (subjective vision, feeling of the current and future situation) alters input information, "enriched" it's by new subjectively formed components (decision) on the basis of their own knowledge, professional experience (skills) of the decisions adoption and implementation and the current state of psycho-physical activity.

Then, the phenomenon of *Bounded Rationality and Intuitive DMM* could be conceptualized by the author as:

the set of DM techniques in order to transform the input with missing, untrustworthy, inconsistent, "noisy" or not capable of interpretation cyber- information of the given particular decision problem into the right course of action, "burdened by individual heritage" or "enriched" in the form of own propensity, hobby, tendency, expectation, axioms and judgment, presumption or bias and their justification. (Rizun, 2013)

In this regard, the authors nominated the following hypothesis:

The basic processes of the Bounded Rationality and Intuitive DMM can be adequately identified by the specificity of the nonlinear elements. (Rizun, 2013)

Bounded Rationality Decision-Making Models

"Arbitrary"

The NE "two-position relay" for STS could be interpreted by authors as follows:

1. Degree level of the risk exception in the output signal is determined by value of the parameter |b|.
2. The decision, have made using this style, is invariant with respect to the different circumstances of time and place.
3. NE "two-position relay" is proposed to define by term "*Arbitrary*", thus emphasizing the style's features, which are the result of *automatism* of cumulative set of the *skills* and *individual-psychological features*, and which determine

Figure 8. Two-position relay

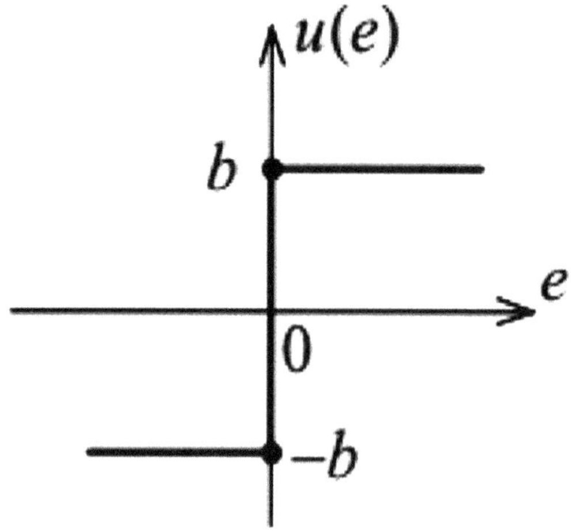

the reproducing the one unique familiar solution in the situation, which was identified as a typical (quantitatively evaluated as parameter |b|).

4. Applicable for solving the problems:
    a. On the one hand, cannot be solved using a logical algorithm; and
    b. On the other hand – do not require creative and flexible approach with taking into account the peculiarities of the problem.

"Principles"

The nonlinear elements "saturation" for STS could be interpreted by authors as follows:

Figure 9. Saturation

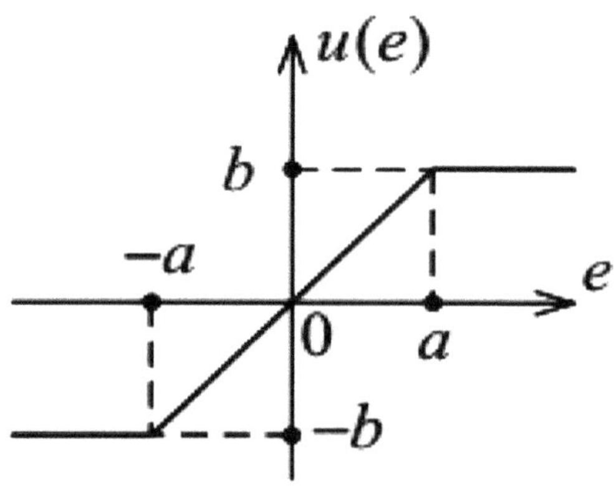

1. On the phases and levels of the DM, which were identified by H-O as non-compliance with the rational ($-a < x(t) < a$), the "*heuristics*" are used as a "framework" for limiting the depth of analysis of the problem.
2. As identification of the fact of the reducing the DM risk, this style provides the justified use of Rational DMM.
3. Degree level of the risk exception in the output signal is determined by value of the parameter |b|, the borders of possibility of using a rational DMM – by value of the parameter $|a|$.
4. NE "*saturation*" is proposed to define by term "*Principles*", thus emphasizing the style's features of cumulative set of the *knowledge* and *socio-psychological features* 1), which are the result of presence the individual's deeply science-based theories and principles (heuristics) of DM.
5. Applicable for solving the problems:
   a. On the one hand, must be solved in the conditions of incomplete information, semi-structural problem statement and fuzzy goal; and
   b. On the other hand – require purposefulness, persistence and commitment with taking into account the peculiarities of the problem.

"Anchoring"

The nonlinear elements "backlash" for STS could be interpreted by authors as follows:

1. Degree of depth analysis of the problem in the output signal is determined by the parameter $|a|$.
2. If it is necessary (and possible) individual can to provide the maximum compliance with rational DMM.

Figure 10. Backlash

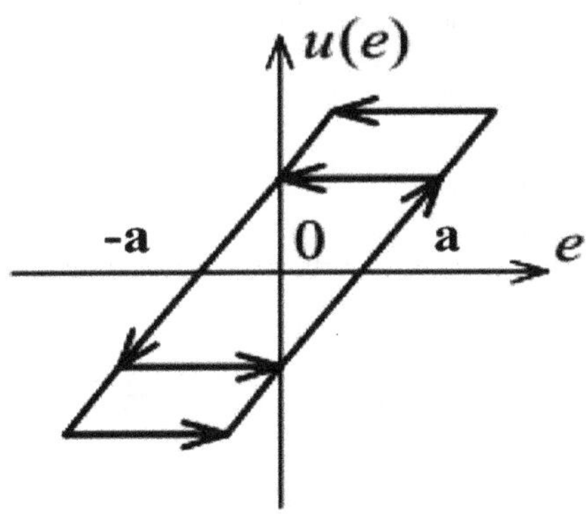

3. Nonlinear element *"backlash"* is proposed to define by term *"Anchoring"*, thus emphasizing the style's features of cumulative set of the *abilities, habits* and *psycho-physiological features* as a professionally reinforced preferences (justified bias), which allow to make the single correct decision by the force of will in the critical situation (in the conditions of increased risk / weak uncertainty).
4. Applicable for solving the problems:
    a. On the one hand, required the ability to identify the complex of characteristics of the original problem – the degree problem uncertainty/risk and the level of the time constraints;
    b. On the other hand – predetermines the need for flexible use of proven presumptions of DM.

Thus, depending on the type of the problem, as well as on personal characteristics of the H-O) it is suggested to use the following classification of strategies of applying the Bounded Rationality DMM, built on analogies with the NE (Table. 2) (Rizun, 2013, 2014; Rizun & Taranenko, 2014a, 2014b).

Table 2. The taxonomy of types of strategies of bounded rationality decision-making model application

Human-Operator's Features	**Professional: Theoretical Training (knowledge)**	**Professional: Heuristic Thinking (skills of the second level)**	**Professional: Practical Qualification (Abilities)**
	Non-Professional: Individual-psychological	**Non-Professional: Socio-psychological**	**Non-Professional: Psycho-physiological**
Decision-Making Conditions	Certainty	Risk	Uncertainty
Strategy	Principles (PR)	Arbitrary (AR)	Anchoring (AN)
Style	Bounded Rational DMS	Automatic DMS	Experimentally Bounded Rational DMS
Problem's Type	Problems, which require combination of theoretical analysis and justification	Problems, which do not require creative and flexible approach	Problems, which require combination of limiting depth analysis of the problem and early making desicion
NE	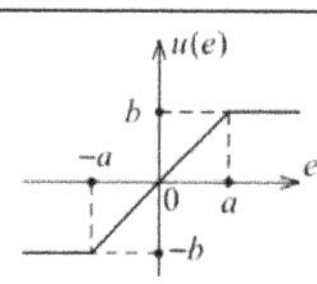	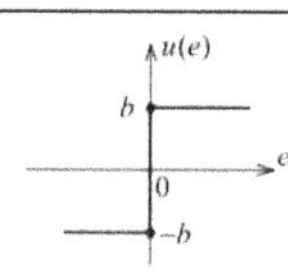	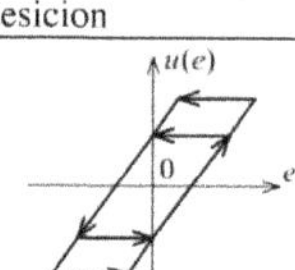

Intuitive Decision-Making Models

"Common Senses"

The NE "deadband" for STS could be interpreted by authors as follows:

1. Latitude of the intuitive perception of the situation by "big picture" in the output signal is determined by parameters of the nonlinear element – by so-called deadband $|a|$.
2. NE "*deadband*" is proposed to define by term "*Common senses*", thus emphasizing the presence of own axioms and judgments as a result of cumulative set of the *knowledge* and *socio-psychological features*, in the implementation of which observed the phenomenon of "enrichment" of the input signal by accumulated in the course of professional activities axioms and judgments, allows to think in terms of "big picture" and to perform the modeling of possible outcomes, using the strategy of mental simulations.
3. Increasing the size of the dead zone leads to a substantial simplification of the decision process (the minimum number of available operator heuristic).
4. The pursuit of maximum satisfaction of rational decision-making model (with a certain combination of the basic parameters of the problem and the type of operator) does not provide an optimal solution, but "enough" solution can be obtained intuitively internally limited time.
5. Applicable for solving the problems:
    a. On the one hand, under uncertainty and risk, as well as a limited time, and based on common sense, recognize the key aspects that need analysis, intuitively perceiving the other as a "big picture"; and
    b. On the other hand – retains the elements of discretion (rationality).

Figure 11. Deadband

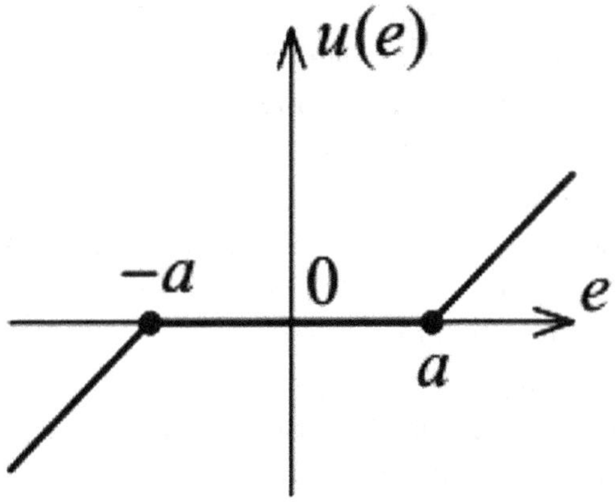

Figure 12. Deadband with saturation

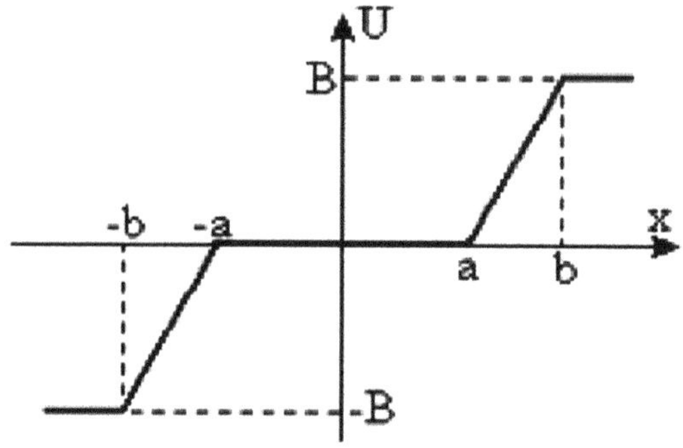

"Empirical"

The NE "deadband with saturation" for STS could be interpreted by authors as follows:

1. On the phases and levels, in which were identified the high uncertainty of problem is realized the "*Arbitrary*" strategy by adopting a unique solution.
2. Moving to the "*Common senses*" strategy by using based on own empirical experience axioms and judgments, which make the possibility on stages, characterized by an average degree of uncertainty, to evaluate the situation, using the theory of "big picture";
3. Latitude of the intuitive perception of the situation by "big picture" in the output signal $Y(t)$ is determined by parameters of the NE – the so-called deadband $|a|$, the degree of development of "behavioral bias", which eliminate the rationality fact – by parameter $|b|$.
4. NE "*deadband with saturation*" is proposed to define by term "*Empirical*", thus emphasizing the presence of individual strategies in the bases of the *Own Experience* and of cumulative set of the *skills* and *socio-psychological features*, in the implementation of which observed the phenomenon of "enrichment" of the input signal by using the "author" behaviour strategies, which have the flexible response to the specifics of the situation as a whole and its individual stages of the decomposition analysis. These "author" behaviour strategies are the result of the presence of great deal of relevant experience – complex of the tacit knowledge, emotional memory and "behavioural prejudice".
5. DMP retains the elements of the simplified discretionary (rationality), acceptable in the decision within the limited time.
6. Applicable for solving:
    a. On the one hand, complex nonstructural problems characterized by dynamic conditions and ambiguous goals; and

b. On the other hand – problems, having an average degree of uncertainty and time constraints.

"Preference"

The NE "three-position relay without deadband" for STS could be interpreted by authors as follows:

1. Latitude of the intuitive perception of the situation by "too big picture" in the output signal *Y(t)* is determined by parameters of the NE – the so-called deadband $|a|$, the degree of development of "preferences", which eliminate the rationality fact – by parameter $|b|$.
2. NE "*three-position relay without deadband*" is proposed to define by term "*Preference*", thus emphasizing the presence of adaptive intuitive automatism as a result of cumulative set of the *habits* and *psycho-physiological features*, characterized by the possibility of thinking by "too big picture" and phenomenon of "enrichment" of the input signal by adaptive application their own propensity, hobby, tendency, expectation.
3. Applicable for solving the problems:
   a. On the one hand, in situations characterized by increased uncertainty and tight time limit; and
   b. On the other hand – absent the elements of the discretionary (rationality).

Figure 13. Three-position relay without deadband

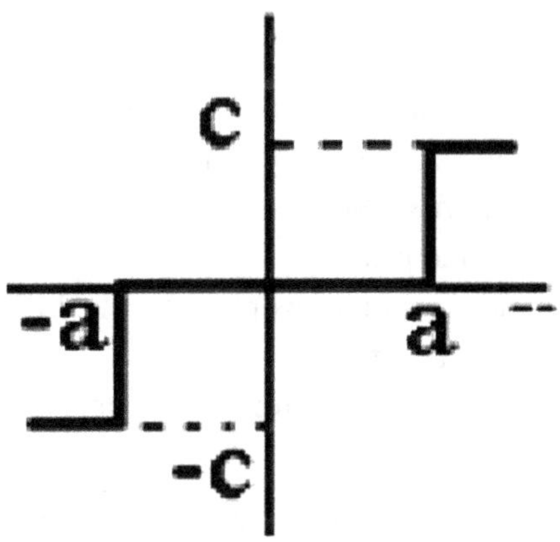

Figure 14. Two-position relay with deadband

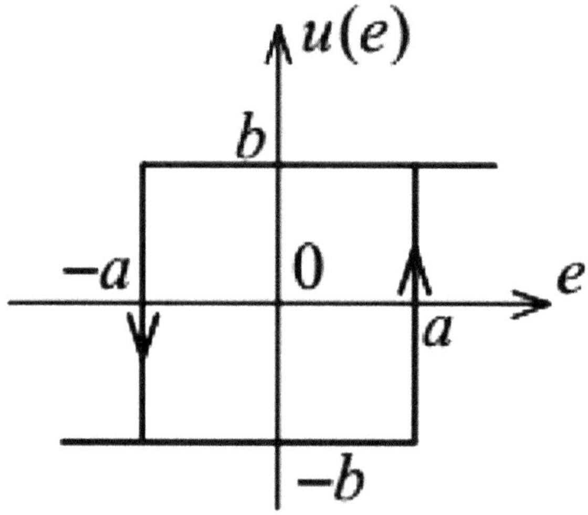

"Trial and Error"

The NE "*two-position relay with deadband*" for STS could be interpreted by authors as follows:

1. Latitude of the intuitive perception of the situation by "big picture" in the output signal *Y(t)* is determined by parameters of the nonlinear element – the so-called deadband $|a|$, the degree of development of " exhaustive trial ", which eliminate the rationality fact – by parameter $|b|$.
2. NE "*two-position relay with deadband*" is proposed to define by term "*Trial and Error*", thus emphasizing the presence of style's features, based on the cumulative set of the *abilities* and *psycho-physiological features* and first of all on the "Gut Feeling" and results of emotional memory, and which realizes by two solutions – Failure or Success.
3. Applicable for solving the problems:
    a. On the one hand, in situations characterized by extreme conditions of complete uncertainty; and
    b. On the other hand – need for immediate decision-making.

Thus, depending on the type of the problem, as well as on personal characteristics of the H-O it is suggested to use the following classification of strategies of applying Intuitive Rationality DMM, built on analogies with the NE (Table 3).

Table 3. The taxonomy of types of strategies of intuitive decision-making model application

Human-Operator's Features	**Professional: Theoretical Training (knowledge)**	**Professional: Thinking (skills)**	**Professional: Heuristic Thinking (skills of the second level)**	**Professional: Practical Qualification (Abilities)**
	Non-Professional: Individual-psychological	**Non-Professional: Socio-psychological**	**Non-Professional: Psycho-physiological**	
Decision-Making Conditions	**Certainty**	**Risk**	**Uncertainty**	
Strategy	Common senses (CS)	Empirical (EM)	Preference (PR)	Trial and Error (TE)
Style	Own Judgment DMS	Own Experience DMS	Semi-Automatic DMS	"Gut Feeling" DMS
Problem's Type	Problems, characterized by an average degree of uncertainty and time constraints	Problems, characterized by dynamic conditions and ambiguous goals, as well as having an average degree of uncertainty and time constraints	Problems, characterized by increased uncertainty and tight time limit	Problems characterized by extreme conditions of complete uncertainty and the need for immediate DM
NE	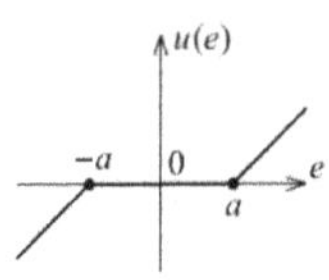	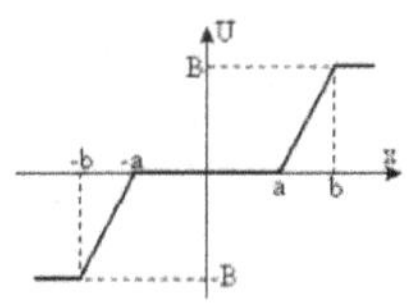	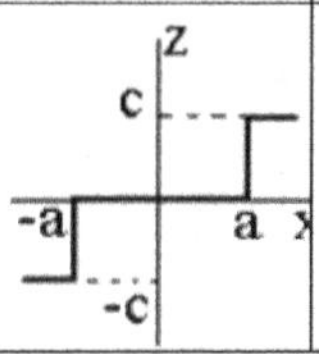	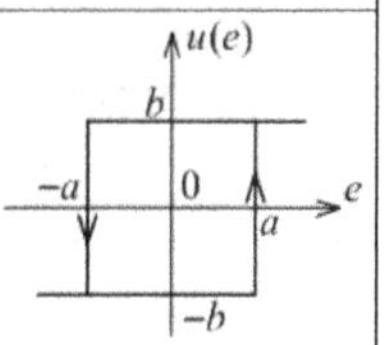

Thus, the authors propose to accept the following hypothesis: *the basic processes of the Bounded Rationality and Intuitive Decision-Making Models can be adequately identified by the specificity of the nonlinear elements.*

The taxonomy of the specific indicators of *activities*, depending on the DMM (Beach, 1978; Christopher, 2009) and the main conditions of STS for H-O as a socio element is formulated and presented by authors in Table 4 (Rizun, 2013; Rizun & Taranenko, 2014a, 2014b)).

MODELS OF THE COLLECTIVE DECISION-MAKING IN THE CONTEXT OF SOCIO-TECHNICAL AND SOCIAL NETWORK ANALYSIS CONCEPTS

In connection with the fact that H-O is considered as an element of socio-technical system, his professional activity is quite often connected with a collective decision-making.

Table 4. Taxonomy of the indicators of human-operator activities depending on DMM

Indicators	The Conditions of Operator's Activity		
	Normal Conditions	Boundary Conditions	Emergency Conditions
		"Free" Conditions	
	Decision-Making Models		
	Rational	Bounded Rationality	Intuitive
Capacities			
Completeness of information	Complete	Incomplete	Arguable
Professional futures	Thinking	Heuristic Thinking	Heuristic Thinking, Hobbies and Abilities
Non-Professional futures	Individual-psychological	Socio-psychological	Psycho-physiological
Perception channels	Sensory Information	Sensory Signals	Sensory Impressions
Structuring of the problem	Structured	Semi-Structural	Nonstructural
Speed	Deliberate	Rapid	Instant
Goal	Clear	Fuzzy	Competing
Indeterminacy	Certainty	Risk	Uncertainty
Techniques			
Problem Processing	Problem Solving	Problem Resolution	Problem Finding
Accuracy	High	Enough	Low
Tools	Value	Heuristic	Judgment
Options Difference	Weighted	Acceptable	Indistinguishable
Situational Awareness	Details	Patterns, Clusters	Big Picture
Accountability	Statistics, Analytical Tools	Pre-Formulated Rules	Perception, Mental Simulations, Feelings
Experience	Not necessarily	Embedded in Direct Habit	Embedded in Direct Experience
Emotional Level	Emotion Free	Legitimate Role for Emotions	Emotionally Associative
Effort	Requires Cognitive Effort	Requires Developed a Range of Successful Adaptive Strategies	Appears Effortless
Choice			
Metrics	Numerative	Good enough	Failure or Success
Time Horizon	Present	The Near Future	Future
Level of Detail	Detailed	Grouped	Aggregate
Decisions Formulation	Prescriptive	Suggested	Descriptive
Nature	Objective	Subjective	"Gut Feeling" Based
Conscious Level	Conscious	Traditional	Unconscious
Creativity	Creativity Free	Creativity Free	Creative
Decisions	Optimal	Satisfied	More Relevant

Collective decision-making involves H-O generating problems, providing potential solutions, voting for solutions and for trusting to each other, and aggregating individual votes and ultimately deriving a final collective decision. As collective DM usually requires more time than the individual one, such decisions are usually made in conditions of uncertainty or risk, but with the influencing of the *reflexive models of bipolar choice.*

Thus collective *DMP* in *socio-technical concept* could be interpreted using following 5 significant steps (author' edition see on Figure 15).

The classic formulation of the problem of mathematical foundations of Collective Multi-Criteria DMP can be represented as follows:

1. **Trust-Based Phase:** Suppose that in the process of collective choice of experts involved N (H-O). In that case, if the H-O h_i wants to express their opinion about the relative trustworthiness of H-O h_j, it can be represented by the conditional probability of the *trust-phase*:

Figure 15. Collective DMP in the context of Socio-Technical Concept

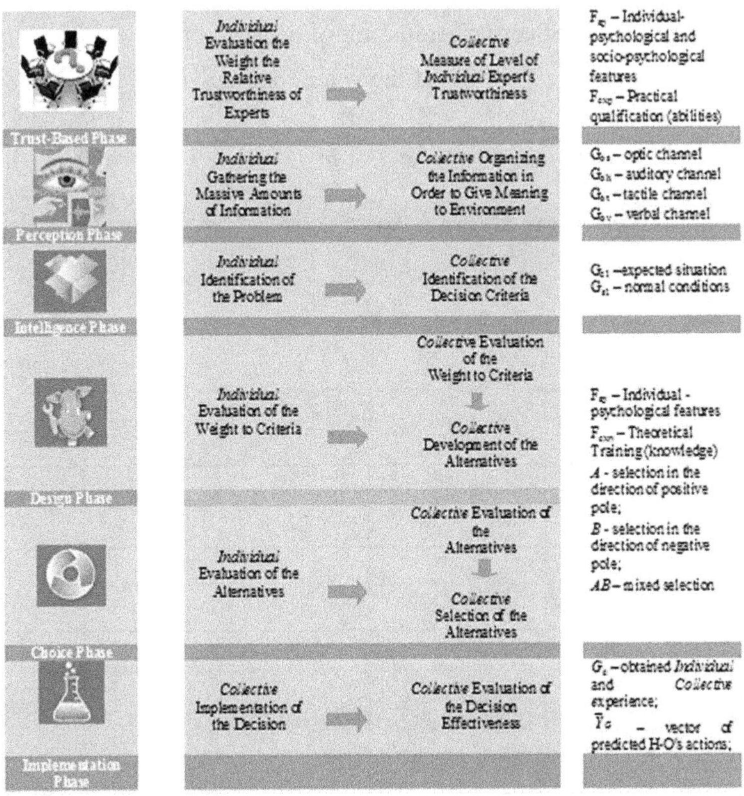

$$w_{h_i,h_j}^{trust} = P(h_j \text{ is good in domain } d_l | \; h_i \text{ is the expert in domain } d_l) \quad (3)$$

This relationship states that h_i trusts, according to some probability, that h_j will make a good decision given h_i's previous understanding of domain d_i by h_j

The first phase resulted is a *Trust-Based* matrix of the form shown in Table 5.

Thus, the Collective Measure of Level of Individual Expert's Trustworthiness $U_{h_1}^{trust}$ for H-O h_j may be defined as:

$$U_{h_1}^{trust} = \sum_{j=1}^{N} w_{h_i,h_j}^{trust} \quad (4)$$

2. **Intelligence and Design Phases:** Developed a set of *M* criteria $Q = \left\{q_r\right\}_{r=1}^{m}$ and *K* alternatives of solving the problem $A = \left\{a_t\right\}_{t=1}^{K}$. Each Alternative is represented by a set of objective valuation (price, quantity, quality, time, etc.) characteristics $O = \left\{o_{tr}\right\}_{t=1 r-1}^{K \; m}$ for each criterion (Table 6).

For the purpose of collective evaluation of each criterion weight q_r each *H-O* h_j puts his/her estimator $w_{h_i,q_j(p)}^{votedForCr}$ of the *votedForCl-phase*:

$$w_{h_i,q_j(p)}^{votedForCr} = P(q_j \text{ is good criteria for the problem } p | \; h_i \text{ is aware of } p) \quad (5)$$

Table 5.

Trust	$H - O_1$	$H - O_2$	...	$H - O_n$
$H - O_1$	–	w_{h_1,h_2}^{trust}	...	w_{h_1,h_n}^{trust}
$H - O_2$	w_{h_2,h_1}^{trust}	–	...	y_{2n}
...	...	...	...	...
$H - O_n$	w_{h_n,h_1}^{trust}	w_{h_n,h_2}^{trust}	...	–

Table 6.

Alternatives	q_1	q_2	...	q_m
a_1	o_{11}	o_{12}	...	o_{13}
a_2	o_{21}	o_{22}	...	o_{23}
...	...	...	...	...
a_n	o_{n1}	o_{n2}	...	o_{nm}

The second phase resulted is a *criteria's vote-based* matrix of the form shown in Table 7.

The above vote $w_{h_i,q_j(p)}^{votedForCr}$ state that *H-O* h_i believes that, according to some probability, criteria q_j is a good for evaluation the alternatives of the problem *p*.

Thus, Collective Measure of Criteria Weight $U_{q_j}^{votedForCr}$ can be determined by the multiplicative compression:

$$U_{q_j}^{votedForCr} = \sum_{i}^{N} U_{h_i}^{trust} w_{h_i,q_j(p)}^{votedForCr} \quad (6)$$

Table 7.

Criteria	q_1	q_2	...	q_m
h_1	$w_{h_1,q_1(p)}^{votedForCr}$	$w_{h_1,q_2(p)}^{votedForCr}$	...	$w_{h_1,q_m(p)}^{vvotedForCr}$
h_2	$w_{h_2,q_1(p)}^{votedForCr}$	$w_{h_2,q_2(p)}^{votedForCr}$	...	$w_{h_2,q_m(p)}^{votedForCr}$
...	...	...	...	...
h_n	$w_{h_n,q_1(p)}^{votedForCr}$	$w_{h_n,q_2(p)}^{votedForCr}$	...	$w_{h_n,q_m(p)}^{votedForCr}$

3. **Choice Phase:** For the purpose of collective evaluation of each alternative a_t each H-O h_j his/her estimator $w^{votedForAl}_{h_i,a_j(p)}$ of the *votedForAl-phase:*

$$w^{votedForAl}_{h_i,a_j(p)} = P(a_j \text{ is good alternative for the problem } p | h_i \text{ is aware of } p) \quad (7)$$

The third phase resulted is *alternative-based* matrix of the form shown in Table 8.

The above vote $w^{votedForAl}_{h_i,a_j(p)}$ state that *H-O* h_i believes that, according to some probability, a_j is a good solution for problem *p*.

Thus, Collective Measure of Alternative Weight $U^{voted}_{q_j}$ can be determined by the multiplicative compression:

$$U^{voted}_{a_t} = \sum_{j=1}^{m} U^{votedForCr}_{q_j} O_{tj} \cdot \sum_{i}^{n} w^{votedForAl}_{h_i,a_t(p)} \quad (8)$$

Thus, on the collective *DM* on each particular object in each phase influenced by the collective *DM* result by the previous phases. In other words, the measure of "*importance*" (weight) of the current estimated object (operator criterion, alternative) is determined by the influence of the measures "importance" (weight) of objects related to the early stages. This recursive relationship can be represented as a graph.

In this regard, *to estimate collective DM in socio-technical concept the instruments of the Social Network Analysis (SNA) are suggested.*

The Analysis of Networks and networked systems, however, has a long history of graph theory in discrete mathematics (Ahuja, 1993; Bollobas, 1998; West, 1996).

Table 8.

Alternative	a_1	a_2	…	a_k
h_1	$w^{votedForAl}_{h_1,a_1(p)}$	$w^{votedForAl}_{h_1,a_2(p)}$	…	$w^{votedForAl}_{h_1,a_m(p)}$
h_2	$w^{votedForAl}_{h_2,a_1(p)}$	$w^{votedForAl}_{h_2,a_2(p)}$	…	$w^{votedForAl}_{h_2,a_m(p)}$
…	…	…	…	…
h_n	$w^{votedForAl}_{h_n,a_1(p)}$	$w^{votedForAl}_{h_n,a_2(p)}$	…	$w^{votedForAl}_{h_n,a_m(p)}$

From the late 1990s onwards, research on Social Networks has branched onto a number of fields, and has been generally carried out under the umbrella term of complex networks, a new emerging area in which networks are studied in several domains, using data from a wide variety of sources. The classes of networks studied include computer, biological, financial, medical, physical, and transportation networks, among many others.

In a graph, we call a unit – whether an human, company – a *vertex* or node. A tie between two nodes indicates the presence of the relationship connecting them. Absence of a tie indicates absence of a relationship. A tie with a direction is called an arc, and a tie without direction is called an *edge*. One could also note the value or volume of flow as the weight of a tie and thus obtain a network that would then be a *weighted digraph.*

A primary use of graph theory in SNA is to identify "important" actors. The concepts of centrality and prestige seek to quantify graph-theoretic notions about an individual actor's prominence within a network by summarizing structural relations among the *g* nodes. Group-level indexes of centralization and prestige assess the dispersion or inequality among all actors' prominences. The three most widely used centrality measures are *degree, closeness*, and *betweenness* (Freeman 1979, 2000, 2004).

The main goal of research using SNA is to understand the general properties of the data, represented by networks, often by analyzing large datasets collected with the aid of technology. The data is often abstracted at the level at which the networks are treated as large graphs, often with little or no concern on whether the nodes (*actors*) represent human, companies, or other entities. Such an abstraction is possible because in many ways the problems addressed in complex network research are similar across different domains. Relevant problems include understanding of the structure of the networks (i.e., by identifying underlying properties of the relationship and edge structures), the evolution of such structures (i.e., how the networks change over time), and how information propagates within the networks.

One of the most important indicators of evaluation nodes Social Network, reflecting the degree of importance of the effect of a recursive one object to the related objects is PageRank indicator. *PageRank* indicator was proposed in (Brin & Page, 1998; Page, Brin, Motwani, & Winograd., 1999) as a ranking mechanism for the Google search engine that assigns to each page a global importance score based on link analysis. The basic idea is that if page *u* has a link to page *v*, then the author of *u* is implicitly saying that page *v* is somehow important to *u*. Thus, page *u* is conferring some amount of importance onto *v*, and this amount is determined by the importance of *u* itself and the number of outlinks in *u* over which it is divided.

In our case, we can transform the classical definition of the *PageRank* as the following:

PageRank is a global importance score of the multiphase collectively evaluated objects (expert, criteria, alternative), based on the number of significant relationships with the another objects from the previous (f-1) phases of evaluation and the importance (weight) of the objects containing those relationships $W = \left\{ w_j^{f-phase} \right\}$, *where* $w_j^{f-phase}$ *– weights of this objects on the f-phase.*

According to this random walk model, the importance *P(i)* of a *objects i* could be determined as:

$$P^{(j)}(i) = \sum_{q \to p} \frac{P^{(j-1)}(i)}{d(i)} \quad (9)$$

where $d(i) = \sum_{j \in \Pi(i)} w_j^{f-phase}$ is the *weighted* degree of object *i* (*weighted*-degree of a node is the sum of the weights of all links attached to node i);

The authors suggested that the instruments of *collective multi-criteria DM* and *PageRank* of the graph node have similar mathematical and methodological bases (the both lies in the definition of "importance" of nodes, which depends on the "importance" of their neighbors). To confirm that the authors suggest the examples of Collective Selection of the Alternatives using the two methodologies, describe above (Table 9a and Table 9b).

Table 10 contains results of *collective multi-criteria DM* and *PageRank*. For comparison of these results the column of distribution of places of received Rank indices is added.

The application of software for statistical analysis of data in accordance with the SNA theory allows having a visual reflection of differences between PageRank of Alternative, presented by *trust-based*, *criteria's vote-based*, *alternative-based* and *vote-based decision* social sub-networks (Figures 16, 17).

Thus, the comparison of results of calculations of multiplicative Collective Measure of Alternative and *PageRank* confirms the significance of the *suggestion* about existence of the analogy between mathematical bases of the *collective multi-criteria DM* and the of the SNA theory.

Table 9a. The example of the results of the collective multi-criteria DMP

Trust	h_1	h_2	h_3	h_4	Criteria	k1	k2	k3	k4
h_1	-	0,8		0,2	h_1	0,25	0,5	0,75	1
h_2		-	0,4	0,3	h_2	0,25	0,75	0,5	1
h_3		0,5	-	0,4	h_3	0,5	1	0,25	0,75
h_4	0,3		0,8	-	h_4	1	0,5	0,75	0,25
$U_{h_1}^{trust}$	**0,3**	**1,3**	**1,2**	**0,9**	$U_{q_j}^{votedForCr}$	**1,9**	**2,775**	**1,85**	**2,725**

Table 9b.

Alternative	a1	a2	a3	a4	Collective Measure of Alternative	a1	a2	a3	a4
h_1	1	4	3	2	k1	3	2	2	4
h_2	1	2	3	4	k2	4	3	3	4
h_3	2	3	1	4	k3	5	3	2	3
h_4	3	4	2	3	k4	6	1	4	5
$U_{q_j}^{votedForAl}$	**0,16667**	**0,30954**	**0,21426**	**0,30954**	$U_{q_j}^{voted}$	**7,0666**	**6,3142**	**5,726**	**11,723**

Table 10. Results of comparison of collective multi-criteria DM and pagerank

Alternatives	$U_{q_j}^{voted}$	Rank	PageRank	Rank
a1	**7,0666**	**2**	**0,111342**	2
a2	**6,3142**	**3**	**0,100529**	**3**
a3	**5,7260**	**4**	**0,092004**	4
a4	**11,723**	**1**	**0,127944**	**1**

Figure 16. Example of the social sub-networks for collective multi-criteria decision-making

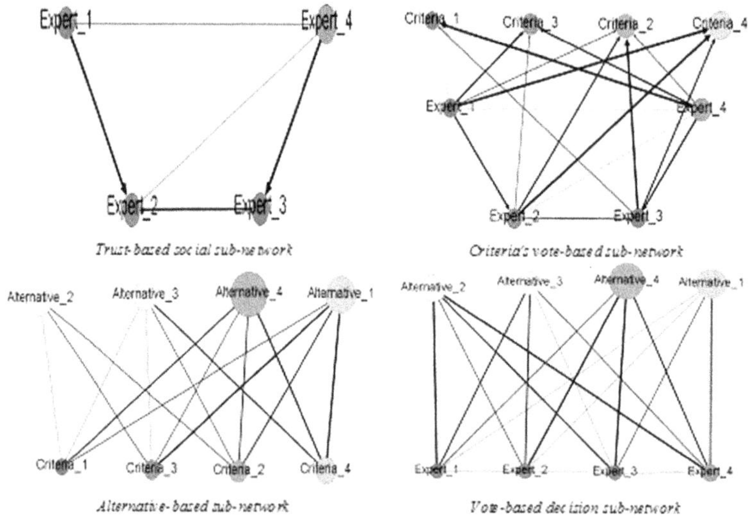

Figure 17. The example of the graphic interpretation of alternative significance on the basis of pagerank values

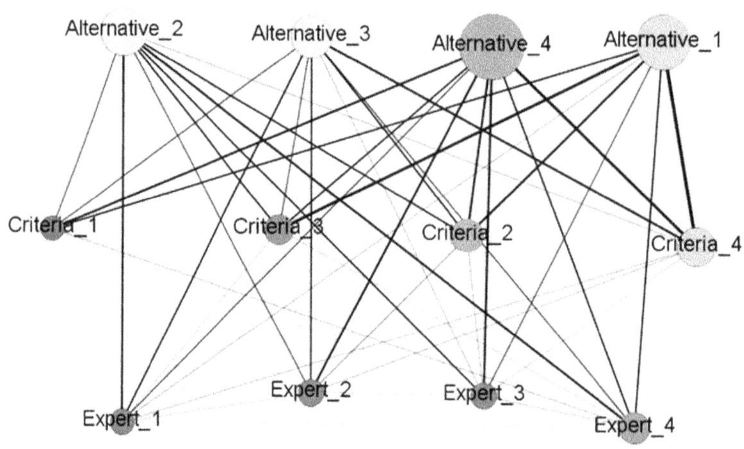

CONCLUSION

The conclusions which were received as a result of the experiments allow developing and improving the theory of DM in terms of the study and classification of specificity of the H-O's DMP in the context of the *socio-technical concept*:

1. The general model of H-O Decision-Making process on the concepts of socio-technical concept based by authors on the following heuristics:
    a. Primary modes of operation by the STS's human operators are the following: *normal*; *emergency*; *boundary*; *"free"* conditions.
    b. *Features* which affect the DM by a H-O are divided into *professional* features: level of knowledge, skills, abilities, preceding experience as well as the features of *non-professional* nature: psycho-physiological, individual-psychological, social-psychological.
    c. Decision-making of the H-O in unexpected situations (conditions of uncertainty and risk) influence of external (social) environment which causes the behavior received the name of *reflexive models of bipolar choice.*
2. The fundamental intellectual processes, which lie on the basis of DM behavior of the H-O and based on the analogies with the devices (elements), provides technical support for these solutions, are identified. This approach, proposed by authors, allows performing a "technical" classification and formalization of the specificity of the H-O behavior based on his/her existence with modern cyber-space:
    a. The basic intellectual processes of the *Rational DMM* can be adequately identified by the transient processes of the *PID-controller*.
    b. The basic intellectual processes of the *Bounded Rationality* and *Intuitive DMM* can be identified by the transient processes of the *NE*.
3. The instruments of *collective multi-criteria DM* and *SNA theory* have similar mathematical and methodological bases.

The taxonomy of the most typical automatic control theory elements and their compliance with certain decision-making models with a point of view of decision-making processes specificity and on human-operator behavior in the context of the socio-technical concept was obtained

Thus, proposed by authors of this chapter *concept of the behavioral models system* is a modern and scientifically sound methodological basis for the formation of employees the entirely new DM skills and competences in the conditions of occurrence of contemporary cyber-organizations – a socio-technical systems.

REFERENCES

Ahuja, R. K., Magnanti, T. L., & Orlin, J. B. (1993). Network Flows: Theory, Algorithms, and Applications. Prentice Hall.

Allard, C. R. (2003). *Antecedents of Effective Decision Making: A Cognitive Approach. Research Memoranda 045.* Maastricht: METEOR, Maastricht Research School of Economics of Technology and Organization.

Bai, J. (2007). Intuitive Decision Theory Analysis and the Evaluation Model. Management Science and Engineering, 1(2).

Beach, L. R. (1978). A Contingency Model for the Selection of Decision Strategies. *Academy of Management Review*, (3): 439–449.

Bertsch, V., Treitz, M., Geldermann, J., & Rentz, O. (2007). Sensitivity analyses in multi-attribute decision support for off-site nuclear emergency and recovery management. *International Journal of Energy Sector Management*, *1*(4), 342–365. doi:10.1108/17506220710836075

Brin, S., & Page, L. (1998). The Anatomy of a Large-scale Hypertextual Web Search Engine. In *Proc. of the Seventh World Wide Web Conference*. doi:10.1016/S0169-7552(98)00110-X

Christopher, L. (2009). *Intuitive and rational approaches to decision making in education.* (Doctoral Dissertation). University of Hull.

Elizabeth, A. S. (2001). The Role of Tacit and Explicit Knowledge in the Workplace. *Journal of Knowledge Management*, *5*(4).

Flueler, T. (2006). *Decision Making for Complex Socio-Technical Systems: Robustness from Lessons Learned in Long-Term Radioactive Waste Governance (Environment & Policy)*. Springer. doi:10.1007/1-4020-3529-2

Hammond, K. R. (1996). *Human Judgment and Social Policy: Irreducible Uncertainty, Inevitable Error, Unavoidable Injustice*. New York: Oxford University Press.

Hastie, R., & Dawes, R. M. (2001). *Rational Choice in an Uncertain World: The psychology of judgment and decision making*. Thousand Oaks, CA: Sage Publications.

Human Factors Guidelines for Safety Audits Manual . (2002). First Edition. Doc. 9806-AN/763. Canada, Montreal: International Civil Aviation Organization.

Kahneman, D. (1994). New Challenges to the Rationality Assumption. *Journal of Institutional and Theoretical Economics*, (150), 18–36.

Keating, C. B., Fernandez, A. A., Jacobs, D. A., & Kauffmann, P. (2001). A methodology for analysis of complex sociotechnical processes. *Business Process Management Journal*, *7*(1), 33–50. doi:10.1108/14637150110383926

Kharchenko, V. P. (2011). Methodology for Analysis of Decision Making in Air Navigation System. *Proceedings of the National Aviation University*. doi:10.18372/2306-1472.48.51

Kharchenko, V. P. (2012) A methodology for analysis of flight situation development using GERT's and Markov's networks. *World Congress "Aviation in the XXIst century. Safety in Aviation And Space Technologies.*

Kharchenko, V. P. (2012). Modeling of behavioral activity of air navigation system's human-operator in flight emergencies. *Proceedings of the National Aviation University*. doi:10.18372/2306-1472.51.2150

Kharchenko, V. P., Shmelova, Y. V., & Sikirda, N. N. (2011). The Human Operator's Decision-Making graphic analytical models. *Proceedings of the National Aviation University*.

Kormyshev, V.M. (2011). *Definition of Competence in Information Systems*. Academic Press.

Kovalev, Yu. N. (2012). Modeling of personality and activities of air navigation system operator.*World Congress "Aviation in the XXIst century. Safety in Aviation and Space Technologies.*

Lamb, C. J., & Lachow, I. (2006). *Reforming Pentagon Strategic Decision-making*. Institute for National Strategic Studies.

Lefebvre, B. A. (2001). *Functions of Fast Reflexion in Bipolar Choice. In Reflexing processes and operations* (Vol. 1, pp. 34–46).

Leychenko, S.D., Malishevskiy, A.V., & Mikhalic N.F. (2006). *Human Factors in Aviation*. Kirovograd: YMEKS.

Lunenburg, F. C. (2010). The Decision Making Process. *National Forum of Educational a MD Administration and Supervision Journal*, *27*(4).

Makarov, R. N., Nidziy, N. A., & Shishkin, G. K. (2000). *Psychological Foundations of Didactics in Flight*. MAPCHAK.

March, J. G. (1991). Exploration and Exploitation in Organization Learning. *Organization Science*, *2*.

Nielsen, H. (2011). Bounded Rationality in Decision Making. Dobbs Ferry, NY: Manchester University Press.

Page, L., Brin, S., Motwani, R., & Winograd, T. (1999). *The PageRank Citation Ranking: Bringing order to the web. Technical report, Computer Science Department*. Stanford University.

Pavlenko, M. A. (2007). The Method of Analysis of the Automated Control Systems of Air Traffic Operator. *Information Processing Systems, No.*, *1*(59), 78–81.

Pavlovskaia, O.O. (2005). Methods for Determining the Parameters of a Linear Mathematical Model of the Operator. *Computer Technology, Management, Electronics, 5*(7), 41-44.

Policastro, E. (1999). Intuition. In M. A. Runco & S. R. Pritzker (Eds.), *Encyclopedia of creativity* (Vol. 2, pp. 89–93). San Diego, CA: Academic Press.

Pratt, M. G., et al. (2007). Exploring Intuition and its Role in Managerial Decision Making. Academy of Management Review, 32(1).

Pupkoff, K. (1974). The Optimization of Connection Between Human and Techniques in Man-Machine Systems. *Preprints of JFAC-JFORS Symposium*, 419-426.

Reason, J. (1990). *Human Error*. Ashgate.

Rizun, N. (2012). The Questions of the Testee Identification as PID-Controller in the Automated System of Professional Certification. *Vimiryuvalna obchislyuvalna tehnika tehnologichnih, 3*(41), 133-141.

Rizun, N. (2013). Computer Modelling of Cognitive Processes.*Proc. of the Federated Conference on Computer Science and Information Systems* (FedCSIS).

Rizun, N. (2013). Development of Methods and Models of Inaccuracy Minimization of the Machine-to-Human Interaction in Automated Systems of Professional Readiness Level Diagnostics. *Nauchniy Visnyk NMU, 2*.

Rizun, N. (2014). Methodology of Intellectual Express-Evaluation of the Decision-Making Effectiveness. Economics & Business Communication Challenges. *International Week. Research Papers*, 209-229.

Rizun, N., & Taranenko, Y. (2014a). Simulation Models of Human Decision-Making Processes. *Management Dynamics in the Knowledge Economy*, *2*(5), 241-264.

Rizun, N., & Taranenko, Y. (2014b). Simulation Modelling of the Decision-Making Activity of the Automated Control Systems Operators Using Nonlinear Elements. *Ambient Technology & Creativity Support Systems*, *188*, 157-178.

Saroj. (2009). *Decision Making: Meaning and Definition.* Retrieved from http://www.excellentguru.com/index.php

Shmelova, T.F. (2012). Analysis of Human-Operator's Decision-Making in Air Navigation System. *Radio Electronic and Computer Systems*, *9*(52), 287-292.

Simon, H. A. (1997). *Models of bounded rationality: Empirically Grounded Economic Reason.* Cambridge, MA: MIT Press.

Simon, H. A. (2009). *Economics, Bounded Rationality, and the Cognitive Revolution.* Northampton, MA: Edward Elgar Publishing.

Stanovich, K. E., & West, R. F. (1999). Discrepancies Between Normative and Descriptive Models of Decision Making and the Understanding/Acceptance Principle. *Cognitive Psychology*, *38*(3), 349–385. doi:10.1006/cogp.1998.0700 PMID:10328857

Strelkov, Y. K. (2001). *Engineering and Professional Psychology.* Academic Press.

Taranenko, Y. (2008). Formation of the Optimal Characteristics for Oscillating Systems vibro-frequently Data Densitometers. *Math. Electronics Universities*, *7*(51), 47–58.

Timochko, A.I. (2007). Modeling of the Decision Maker Activities in the Management Systems Networks. *Radiotekhnika*, *151*, 85-91.

Towler, M. (2010). *Rational Decision Making: An introduction.* New York, NY: Wiley.

Tversky, A., & Kahneman, D. (1974). Judgment under Uncertainty: Heuristics and Biases. *Science*, *185*(4157), 1124–1131. doi:10.1126/science.185.4157.1124 PMID:17835457

Wagner, R. K. (1986). *Tacit Knowledge and Intelligence in the Everyday World in Practical Intelligence: Nature and Origins of Competence in the Everyday World.* New York: Cambridge University Press.

Weick, K. E. (2001). *Making Sense of the Organization.* Oxford, UK: Blackwell Ltd.

Young, J.F. (1979). *Robotics.* Mechanical Engineering.

Zhabreev, V.S. (2007). *Models and Evaluation of Human Factors in the Large Systems: a Tutorial.* Academic Press.

KEY TERMS AND DEFINITIONS

Bounded Rationality and Intuitive Decision-Making Model: The set of human-operator's thinking process techniques in order to transform the input with missing, untrustworthy, inconsistent, "noisy" or not capable of interpretation cyber-information of the given particular decision problem into the right course of action, "burdened by individual heritage" or "enriched" in the form of own propensity, hobby, tendency, expectation, axioms and judgment, presumption or bias and their justification.

Collective Decision-Making: The thinking techniques, which presupposed involving the human-operators of the socio-technical system in the processes of generating problems, providing potential solutions, voting for solutions and for trusting to each other, and aggregating individual votes and ultimately deriving a final collective decision.

Decision-Making Model: Set of the human-operator's thinking process techniques in order to transform the input of the given particular decision task into the right course of action.

Decision-Making Process: The process of choice the course of action with influence of the professional as well as individual-psychological, psycho-physiological and socio-psychological features of the human-operator, which will contribute optimal / adequate / acceptable/ justifiable results in optimal / acceptable / limited time to full-time / part complete / incomplete / inconsistent information.

Human-Operator: An element of a complex socio-technical system, which performs the main system's parameter's control and monitoring functions with the help of the machines tools and technology, own experience and skills as well as the communication with the other human, and is able to make decisions under certainty, uncertainty and risk.

Rational Decision-Making Model: The human-operator's logical linear multi-step thinking process techniques for stable permanent maintaining the maximal compliance between the achieved as a result of the decision and required level of the problem solution.

Socio-Technical Systems: The modern automated human-machine control systems, which characterized, on the one hand, by the extensive usage of computer technology, and on the other hand, high dependence on the human operators behaviour – decision-making models, algorithms and techniques.

Chapter 10
Aggregators News Sources of Main Portals in Poland

Jan Kreft
Jagiellonian University, Poland

Mariana Petrova
Jagiellonian University, Poland

ABSTRACT

Aggregation of media content comes as a popular form of media market activity which is supposed to facilitate access to information in the conditions of its excessive amount. It can be defined as selection and hierarchization of information and determination of its value. Aggregation can be performed with the direct participation of editorial teams, or it can be performed automatically, with the use of algorithms. The research conducted in April 2014 – April 2015 was to indicate the sources of information published by the news portals dominating on the Polish market, namely: Onet.pl and WP. pl. The research referred to the main section Wiadomości (News). During the research 1829 sources of information were identified in Onet.pl and 2272 sources of information in WP.pl. Considering Onet.pl, there were 549 pieces of the portal news and 1239 pieces of news obtained from different sources. WP.pl presented 582 pieces of the portal news and 1690 pieces of news received from other sources. Thus, Onet.pl presented its original content more frequently. The scale of using external sources allows us to state that both portals are, first of all, the aggregators of medial contents, and they use external information sources, however the scale of the practice is differentiated. Generally, it is possible to state that only every third message placed in the most important section of both portals was created in their editorial offices.

DOI: 10.4018/978-1-5225-1656-9.ch010

INTRODUCTION

In new media space, users are confronted with the excessive contents, a phenomenon which is metaphorically referred to as information or content overload. Its manifestations are commonly known. Content overload refers to all the platforms, let us just mention social media – for example, popular tweets, no more than 1/3 of which is worth reading; at least it is possible to come to such a conclusion, based on the survey ordered by Harvard Business Review (Hudson, 2012).

The social consequences of information overload also come as the subject of the research. The problem may be defined as one of the most important issues of contemporary media, first of all considering problems with short-term memory, already recognised in neurological research and challenges that must be faced by users when its capacity reaches the limits, and the Anti Information Deficiency syndrome appears along with the information overload syndrome (Ericson, 2013), lack of concentration, increased perceptible stress, weakened intellectual efficiency and the fear of missing out.

Information overload is related to the paradox defined by Barry Schwarz, and it has been confirmed by the research carried out in 2008 in Northwestern University's Media Management Center. It consists in the passiveness resulting from helplessness in the view of information overload. Hence, we are allowed to conclude that more choice does not contribute to more satisfaction (Vahlberg, et.al, 2008). As Fred Hirsch (2005) concludes, it comes as an element of "the tyranny of small decisions." Renata Salecl (2011) comes to a similar conclusion, writing about the perception of life as a composition constructed from decisions and choices. In media environment information is frequently multiplied.

In his "Data Smog" David Shenk (1997) discusses the problem of impoverishment through excessive information. He also discusses the development of new information technologies which allow us to access media and to confront a complicated process of acquiring valuable information and its interpretation. He writes about media environment where the ability to avoid useless information becomes a virtue, and meta-information, that is namely: knowledge how to match various kinds of information in the most efficient way, is the most precious kind of information itself.

In such conditions, the ability to manage information is of crucial significance. Information overload requires solutions which can be interpreted as optimal for individual needs. Aggregation of media content comes as such a solution, at least at a declarative level.

Aggregation of media content comes as a combination of conveying texts, photographs, videos or audio records in one form, in accordance with one basic rule: collection and systematization of media content, its hierarchization, the use of internal links, packaging, recording, rewriting and illustrating content in the Internet. An aggregator of media content, usually news, is a coordinator in the environment of media producers, which combines the contents with the simultaneous determination of their hierarchy, according to quickly changing criteria of importance, popularity and value of the information.

Evaluation of information has crucial significance for aggregation. Viewed from such a perspective, aggregation is divided into aggregation performed by editors (editors select the content) which is based on the activity of network users (e.g. Digg and Reddit) and automatic aggregation which uses algorithms (e.g. Google News). In the latter case, the automats use the context analysis, and they group information covering similar issues. In other cases, editors may, for example, complete automatically segregated information and add headings or their own editorial articles.

Facing excessive amounts of media content created by users and by media organisations, aggregation is considered as an unavoidable consequence which follows the development of media technologies, as well as a privilege for its beneficiaries, namely: the access to media content which is selected to suit their requirements.

There have been a lot of activities initiated in the Internet, aiming at providing help for users to facilitate and catalogue information accessible in the net. In 1993 bookmarks appeared in browsers. At first they served as catalogues for the links collected by particular users (Humphries, 2000). Then the most popular social networking services started their advance. Bookmarks placed in those sites became the foundation for the concept of Web 2.0 and UGC (User Generated Content). Let us have an example of the Digg company. Established in 2004, the company developed a recommendation system for the news exchange and ranking. Reddit appeared in 2005 and Newsvine in 2006. The most modern portal, however, is Pinterest.com which initially introduced itself as the image curation service. The project appeared on the international arena in 2012, with a voluntary registration and without any invitation system. In terms of the number of site views Pinterest.com has taken the third position, after Facebook and Twitter (Palis, 2012). In accordance with the survey provided by Experian Hitwise, an analytical company, in 2012 Pinterest.com had more users than LinkedIn, Tumblr and Google+ (Kennaly, 2012). The fact comes as the best proof for the demand for the photo-content filtration in the net.

Another portal which uses other methods to retain the required content is Clipboard. It allows users to save whole pages or their parts, such as texts, photographs or website elements. All the saved data are in the user's profile and each saved fragment is provided with relevant links and tags which users may ascribe manually themselves. Moreover, it contains well-recognised social elements: users may share their saved content, make friends and look for new acquaintances.

Among other similar projects we can list Likehack.com – the service of automatic search and retaining content based on the activity of users in social networks. Recently the company has significantly invested into the project which is now functioning in its alfa version, where over 500 thousand users have their accounts (Rusbase, 2012).

Diversification of aggregation practice allows us to formulate a classification of news aggregators, divided into four categories:

- Feed aggregators (bars),
- Specialty aggregators,
- User-curated aggregators, and
- Blog aggregators (Isbel, 2010).

The first category includes such web portals as Yahoo!, News, My Yahoo, Google News. They display headlines of the news, sometimes with the first lines of the contents and the links to the whole articles from the original websites. Specialty aggregators collect information on the required subjects from various sources at one place (e.g. Everyblock, Outside.In or aggregators focused on particular subjects as Techmeme and Taegan Goddard's Political Wire). Similarly, to feed aggregators, specialty aggregators usually display the headlines of the news with several first sentences and links to the source website and its name. Such aggregators comprise several subjects and sources. User-curated aggregators come as sites which are provided with links sent by users and fragments of texts taken from various websites. Frequently, such links come from a wider range of sources than links in news aggregators. The links refer to various posts entered into blogs and multimedia content, such as videos in YouTube and links to some traditional sources of information. The fourth type of aggregators reminds a traditional news aggregator, however, the websites use contents provided by the third parties in order to create a blog on particular subjects (e.g. Gawker Media, The Huffington Post). Bloggers provide a synthesis of information from various sources, describing one story and providing the links to the original content

Considering the point of view represented by the users, the main advantages of aggregators are:

- Saving time.
- **Saving Money:** A lot of mass media try to limit the access to their content, using paid subscription. However, it involves a number of negative effects, therefore it is easier to become an aggregator which has already settled all the issues with the owners of the copyrights.
- Increased accuracy of the search results.
- Filing and categorisation of the content.
- One entry point.

- **Limited Number of Advertisements:** In exchange we can obtain mere, clean text and the required information.
- **Comfort:** There is no need to browse through huge amounts of text; it is enough to obtain the essence of the whole material and decide which things deserve our attention.

Considering the point of view represented by organisations which create media content, an important question is the influence of social network sites on readership of news content. According to the survey provided by ComScore, 47.8% (that is 181.5 million) of Europeans visit news sites (an increase by 9% in comparison to the data from last year); in the case of 10% of such visits, viewing news is preceded by a visit on Facebook. The research also indicates that this percentage is gradually growing, and that fact proves the growing role of social networking sites in the process of searching and using the required information (Mohamud, 2012).

It is worth noticing that media content is not the only element which undergoes the process of aggregation. The audience which plays the role of a direct purchaser of media content (thanks to the subscription and purchase of goods) and a receiver of advertisements and other promotion messages, undergo that process too (Kreft, 2015). Such aggregation of users fosters aggregation of revenues. Through some sources which are considered as "trusted" by users, they undergo aggregation by advertisers – hence, users will have to receive advertisements on the subjects which could possibly turn out to be interesting for them.

Generally, the basic operational principle of the largest aggregators consists in the collection and systematization of news; therefore, users may look for the required news in the informational space which has been created with the use of key-words. Non-news sources which can also contain the same key-words may be eliminated from searching at once.

All aggregators offer an opportunity to follow the latest news to their receivers; the main websites usually display currently updated news of the day. It should be emphasized that in Yahoo! News the most important news for the home page are selected by the editorial team, whereas in Google News and Yandex.Novosti the formatting of the website is automatic. All the news which come to the aggregator are divided into themes (e.g. politics, society, sports, etc.) and then into particular articles. Such a process is implemented with the use of special algorithms that cluster news streams and text analysis (it is required to format the headlines). As a result, each subject is surrounded by separate chapters-articles. The most important news is selected from all the information and then added to the relevant (Shafer, 2001). The principles of the ranking may be different but the "weight" of the article depends on its "size" that is on the amount of news in the particular chapter, the number of references made by other mass media, on the extent to which the news provided by other media are different for other texts and on the time of news publishing. In that way the preferences of particular employees do not affect the ranking of particular texts as there is no space for edition itself.

The algorithm of Google News is also called the StoryRank (analogically to PageRank), and it was developed in 2001 by Krishna Baharat, the Head of the R&D Centre of Google in Bangalor (Glaser 2010). Google News had been operating in the mode of Beta testing since April 2002 until April 2006, and then the company officially announced the launch of that service. During the official launch Krishna Baharat explained the mission of the corporation as the use of the computer to make regulations of news from the whole world possible in the real time, to see in the bird's eye view what is being announced on almost every single issue.

The Internet offers several dozen aggregators in their regional versions, which follow the local news and can make rankings of publications on the events from various countries in over twenty languages. Aggregation constantly changes its forms. For example, Google tends to create news service available for the whole world. In its page, Google News displays not more than 200 characters of the original text. An announcement is provided with a link to a full version of the text available on the site which has been referred to. The viewed site may provide an access to the full text of the publication, or it can offer paid subscription to users, which may also allow them to read other texts. The aggregator of Google News offers its users several ways of site personalisation. It is possible to subscribe for a newsletter and obtain notifications on some interesting subjects via e-mail. Additionally, there is a possibility to choose the view of the Google News site, to personalize it, to add or remove some theme modules or to arrange their sequence. Users who have no time to set their preferences are offered a service of automatic recommendation for theme modules. Mobile phone owners can also access the mobile version of Google News.

Selection of media content provided by aggregators is usually performed with the consideration of its numerous sources. Let us just mention Yahoo! News which selects information from approximately 7 000 sources, among others: from Reuters, Associated Press, AFP, USA Today, The Chicago Tribune, Los Angeles Times, Financial Times and BusinessWeek Online. An aggregator collects the content in one, easily accessible location in the on-line space (Madsen, 2011). The content is usually categorised by topics and provided with proper links to the source information and, more and more frequently, it depends on the preferences declared by users.

On one hand, the advantages of aggregation for users are emphasized, such as: saving time and money (a possibility of avoiding payments for the content protected with subscription fees), high accuracy of searching, a possibility to receive the latest news, comfort, fast indexation of information, access to the content provided in an open format. On the other hand, however, aggregation is interpreted as an important manifestation of deterioration in media quality, their unification and appropriation of value by new-media organisations which "freeload" on the producers of media content; it may be also considered as the worsening of journalism standards and domination of financial aims over any other targets of media organisation.

Aggregators also affect the way in which people function in news environment. The habit of reading only one newspaper or one Internet website has been fading away. The ratio of loyalty towards one particular publication has been decreasing. Advanced Internet users have been for a long time considering that visiting several websites a day is natural, as well as selecting a source for each theme, which seems to be the most adequate at a given moment. In that way, the most important organisations of mass media lose their revenues because the outflow of their receivers results in the decrease of income from newspaper sales. Such a situation, in turn, results in the outflow of receivers and advertisers (Bakker, 2011).

Considering horizontal news portals, aggregation of media content is performed on a large scale. These are the conclusions presented by the PEW (Special Report) which has analysed 152 aggregators and producers of original media content.

LITERATURE REVIEW

Considering the multi-dimensional character of aggregation, it is analysed from different points of view. The research subject covers, first of all, the relations between aggregators and press publishers and other traditional media (Chiou & Tucker, 2010). The scope of using media contents available in the Internet is also analysed (Kaplan, 2010), as well as the consequences of providing hyperlinks, for example, in the context of the demand for providing so-called spreadable media (Jenkins et.al., 2013), and in the context of the concept concerning link economy (Dellarocas et.al., 2013). The legal consequences of using media contents created by other entities by aggregators within the process of "mastering of content" also undergo the analysis (Picard, 2014).

Ignoring undisputable advantages of receivers and small publishing companies, numerous editorial teams have expressed their dissatisfaction from the fact that there are various aggregators. Their negative attitude particularly results from the fact that the Internet traffic provided by news aggregators is too low; therefore, the price obtained from the use of the news products they provide is unsatisfactory. The publishing companies indicate that the revenue obtained for providing readers with particular articles (which they have noticed thanks to the aggregators) does not cover the lost profits, which are taken by website owners because of the decrease in the number of direct views at their own home pages. The decrease in the popularity of websites is affected not only by the increasing competition among publishers but also by the outflow of receivers who select other sources of information. In accordance to the survey provided by Outsell, almost half of Internet users read only the headlines of the articles provided by aggregators. Among the readers of Google News 44% of them do not click on the links redirecting to the original content, and they do not go to the websites of publishing companies, stating that reading headlines comes as a sufficient source of information. In that way media do not generate any income (Doctor, 2009).

The influence of media content aggregation on individual receivers' behaviour is also analysed (Dellarocas et.al., 2015), including changes of market behaviour, such as for example the influence of using one information channel on the loyalty to a particular publisher (Bakker, 2011), allocation of users and advertisers' attention to the most popular sites by aggregation (Hong, 2012), and the influence of novelty and popularity of media content on users' attention. There are also some models provided which present the use of the hyperlinks which consider opinions concerning the attractiveness of the content (Roos, et.al., 2013).

The scientific achievement referring to the influence of aggregation on journalism is also significant (Westley, 2011). Such influence is usually interpreted as a threat to traditionally understood journalism, and its intensity is difficult to be identified. The starting point for these considerations can be a statement provided by Shayne Bowman and Chris Willis who claim that the highly respected profession of a journalist has found itself in an unusual moment in history, when its hegemony as a guard of news has been threatened not only by technologies and competition but also, possibly, by its receivers to whom it serves (Bowman, Willis, 2003). The possibility of aggregation has forced the managers of media organisations to face a dilemma: to concentrate on better journalism or to focus on better business (Quinn, Quinn-Allan, 2005). Robert Picard (2009) provides a justification for the decreasing role of journalists in such a context, presenting it in his article Why journalists deserve low pay. There also some analyses carried out on the decreasing role of journalists in obtaining and selecting news and their significance in contextualization of information (Anderson, 2013).

Considering the aspect of media organisation management, aggregation of media content is analysed as a form of rationalisation of the costs referring to information acquiring, a possibility of replacing the content which is expensive in production with some cheaper content. The consequences of treating aggregators as the first and the most important place to obtain information from are also analysed, as well as the mutual relations between aggregators and press publishers who pursue the increase in traffic on the Internet sites of their newspapers and magazines by using aggregators (by providing relevant links). The substitutional and complementary role of aggregation comes as an element of strategic challenges connected with differentiation of media contents and the tendency to increase the quality of the provided content (Jeon & Esfahani, 2012). Aggregation is also analysed in the context of intermediation and disintermediation (Chircu & Kauffman, 1999).

Considering operation of horizontal news portals, the scale of the mutual dependence of aggregators and press publishers who pursue the increase of traffic on the Internet sites of their newspapers and magazines, with the use of aggregators (Yin, 1994).

The decision made by the Associated Press to break up the cooperation with the Google News, and the pressure exerted by European media content creators to restrain the free use of their content by content aggregators (especially the Google News) came as an important impulse for the research on the relations between content aggregators and news agencies.

If information is treated, first of all, as a product sold on the media market, in accordance with its unique, the use of the offer presented by news agencies becomes an investment which should be profitable.

From the point of view presented by the most important users of the offer presented by news agencies, the basic factor referring to their selection is the relation of the quality of the provided service to its costs, in comparison to the alternative costs referring to the acquisition of similar news products. Thus, the costs related to the acquiring of verifiable, reliable current news (so-called hard news) come as an important factor which affects strategies applied by agencies looking for new sources of income, and organisations using their services and cutting their own costs related to the acquiring of high quality content.

METHODOLOGY AND HYPOTHESES

The conducted research is based on an assumption that recognising information sources in basic news sections of the Internet portals, it is possible to define common features of the analysed media organisations.

Presented in the article, the basic research problem resulting from such an assumption is to provide an answer to the following research question: what information sources are used by the news portals dominating in Poland?

In accordance with the methodology suggested by R.K. Yin (1994), the article focuses on the crucial variables, omitting other data and elements. The analysis of the collected research material follows the general analytical concept, and its explanation depends on the following research hypotheses:

H.1: News portals dominating in Poland are aggregators of media content.
H.2: News portals dominating in Poland are aggregators which use external information sources.

H.3: As aggregators of media content, news portals dominating in Poland use external information sources more frequently than their own, original ones.

To verify the hypotheses, the research was carried out (which started on 1st April 2014 and ended on 1st April 2015). The research consisted in the indication of the media content sources used in the main news sections of the following news portals: Onet.pl and WP.pl in their section "News." The observation was performed in weekly cycles – the news was analysed on every third day during the week and then a one-week break was taken. It means that the information on the news sources was collected during 123 days. The observation was performed once a day. The analysed "News" sections were updated during those days, thus both portals published more news in the analysed sections.

Based on the analysed data, the following set was collected. It presents the frequency of the use of the particular information sources by Polish news portals. The proportions of the use of the original media content and external materials obtained from other information centres can be also observed.

Considering Onet.pl., the materials of the Autorzy Onet project were considered as the original content of the portal. The project has been intended to increase the number of readers (Autorzy, 2015). Any user of Onet.pl can join the project and the best materials are published on the homepage of the portal twice a day. Their authors can obtain their remuneration – in cash, or they can choose promotion. During the time of the research Onet.pl published the list of 111 names of the authors who were ready to create content for Onet.pl for free.

For the requirements of the research, the journalistic genres present in the "News" section were unified. In some rare cases it referred to elaborated information, announcements, articles, comments, short captions under photographs, hypertexts or blog content.

Findings

Onet.pl and WP.pl are the leaders among other news portals in Poland. Approximately 15.2 – 15.6 million users visited their websites during the research period, and the range of the portals reached the level of 73% (Megapanel).

Onet.pl which was initially named OptimusNet (founded in 1996) originally came as a catalogue of websites (Historia). In 1998 the first content service in HTLM appeared, namely: Poradnik Podatkowy (Tax Advisor), and then the Republika WWW platform which enabled its users to open e-mail accounts on OnetPoczta and their own websites. In 1999 the subsequent news portals appeared: Pogoda (Weather), Wiadomości (News) and Rozrywka (Entertainment). In 2002 another service was started: Domeny (Domains) where users were able to purchase domains such as .net, .com, .org, .info, .biz.

In 2001 Onet.pl was incorporated into a media group, namely: International Trading and Investment Holdings S.A. which also owned the TVN station (History). In 2005 a new version of the browser OnetSzukaj was provided. It enabled quick filtration of the portal resources. In the same year Onet started to publish press articles (Der Spiegel, La Vanguardia, The Guardian, New York Times, The Independent, Le Monde, L'Express, Le Figaro), and in 2007 it entered agreements with the CNN television station, Fashion TV and Financial Times, a daily newspaper. In the same year the IVONA, a new speech synthesizer, was implemented into the news services. It enabled users to listen to the news.

During that time Onet entered into approximately 200 agreements concerning content supply, including an agreement with the PAP. Most agreements did not refer to any payment from Onet – instead the portal offered its contractors the access to over 9 million users (Fura, 2010).

During the research Onet.pl offered a number of various sections Wiadomości (News), Sport, Biznes, Moto, Kobieta (Woman), Facet (Fellow), Dziecko (Child), Gry (Games), Technowinki (Technical News) and e-mail services, virtual disc services, VoD (a website offering videos, films, series and cartoons), Zumi (a localizer), Sympatia (a dating website) Biznes.pl (a website offering business tools), Plejada (a show-business website), Very Unique Magazine (an Internet magazine), Zapytaj, Polish Digital Community (a website offering IT news and information), Digart.pl (a website for digital art creators), Blog.pl (a location for bloggers), Medonet (a medical website), Bistro24 (ordering food by the Internet) and RodziceRadzą (a social network for parents).

During the research period, Onet.pl usually placed 13 basic pieces of news, which were updated with various frequency. Since October 2014 the number of news pieces changed, and it reached the level of 21. During 123 days when the observations were carried out, 1829 information sources were analysed.

Based on the collected data, it is possible to state that Onet.pl obtained information from 177 sources: press agencies, TV stations, newspapers, etc. In order to read the selected news, a user must click on it, as it is at WP.pl. Onet.pl does not provide any links to the original sources, however, contrary to WP.pl., at the end of each news piece it provides the links which redirect users to the homepages of media organisations which first published the particular piece of news.

During 123 days of the portal observation, 1829 information sources were analysed at Onet.pl, including 549 original news pieces and 1239 news pieces obtained from external sources.

Table 1. The number of the observation days and analysed pieces of news in a year

News Presented in the Main News Section of Onet.pl	Days with the Number of Presented News Pieces	The Number of News Pieces
13	71	923
16	3	48
17	42	714
20	3	60
21	4	84
Total:	123	1829

Source: the author's own study

Table 2. Sources of media content in "News" section at Onet.pl

All news	1829
The number of the sources	1788
Onet.pl original news	549
Other information sources stated	1239

Source: the author's own study

Wirtualna Polska (wp.pl) appeared in 1995 as an aggregator of websites, similarly to Onel.pl. It evolved into a horizontal portal which combines the functions of commercial platform for advertisement and communication. WP.TV. is available for cellular phones and other mobile appliances. The owner of the portal is Grupa Wirtualna Polska which belongs to O2 partnership controlled by the European Media Holding.

Grupa Wirtualna Polska offers, among others, its homepage (WP.pl, o2.pl) and the news section, including WP Wiadomości (News), WP Pogoda (Weather), WP Program TV (TV Guide) and Wawalove.

During the research period, the "News" section offered by WP.pl usually included 13 basic news pieces which were updated with various frequency. Since September 2014 the function of town selection was introduced to the "News" section, along with the selection of opinions. It enabled users to access additional information from the selected towns or to read opinions presented by various publicists (Opinions: Kraków, Trójmiasto, Wrocław, Katowice, Poznań, Warszawa). After one of the cities was selected, the number of news pieces presented on the homepage of WP.pl could rise up to 26.

During the research period 2272 news pieces from the "News" section of WP.pl were analysed; the total number of the indicated information sources was 3044 (in some cases more than one source were indicated).

The research indicates that the total number of information sources in the "News" section reached the level of 2272. After the analysis of the collected data, it is possible to state that there were 582 original news pieces of WP.pl (25.6% of the total number of the analysed news), and 1690 news pieces obtained from external sources.

During the research time period, WP.pl used 182 information sources.

Table 3. The number of the observation days and analysed pieces of news in a year

The Number of News Pieces Presented Daily in the "News" Section of Wp.pl	The Number of Days with the Number of Presented News Pieces	The Total Number of News Pieces
12	23	276
13	38	494
16	1	16
17	2	34
18	1	18
19	10	190
23	1	23
25	1	25
26	46	1196
Total:	123	2272

Source: the author's own study

Table 4. Information sources at WP.pl (the "News" section)

All news	2272
The number of the indicated sources	3044
WP.pl original news	582
Other information sources stated	2462

Source: authors

CONCLUSION

Polish market is dominated by two news portals, namely: Onet.pl and WP.pl. In their most important "News" sections, the portals present the latest news which are updated with various frequency.

The research carried out between April 2014 and April 2015 aimed at the indication of the information sources. In order to achieve the aim, three research hypotheses were formulated. The first hypothesis referred to the fact that the news portals dominating in Poland were content aggregators; the second one assumed that the aggregators used external sources of information; the third hypothesis referred to the fact that the portals used external sources of information more frequently than their own sources.

During the research there were 1829 sources of information identified at Onet.pl and 2272 sources of information at WP.pl.

The research lasted for one year, and it allowed the author to verify the assumed hypotheses. During the research period both portals used various news sources. Onet.pl published the news pieces obtained from 177 sources and WP.pl – from 182 sources. Considering differentiation of the information sources (given in absolute values), both portals do not differ significantly from each other.

Considering Onet.pl, there were 549 original news pieces of the portal published and 1239 news pieces obtained from external sources. WP.pl presented 582 original news pieces and 2462 news pieces obtained from external sources. Such a comparison indicates that Onet.pl used its own information sources proportionally more often. The scale of using external sources allows the author to state that both portals are first of all media content aggregators, and they use external information sources most frequently, although the scale is differentiated.

The obtained results lead us to the following conclusions: each time users refer to both portals as to the information sources, they should be inclined to think about the original primary sources of information. Both news portals dominating on the Polish market still remain media content aggregators in their main "News" sections. The information is obtained from various sources, and it is selected in the editorial offices of both portals. Generally, only 1/3 of the published news has been created in the editorial offices of both aggregators.

REFERENCES

Anderson, C. W. (2013). Towards a sociology of computational and algorithmic journalism. *New Media & Society*, *15*(7), 1005–1021.

Bakker, P. (2011). *New journalism 3.0-aggregation, content farms, and Huffinization: The rise of low-pay and no-pay journalism.* In The Future of Journalism Conference 2011.

Bardoel, J., & Deuze, M. (2001). "Network journalism": Converging competencies of old and new media professionals. *Australian Journalism Review*, *23*(2), 91.

Bowman, S., & Willis, C. (2003). *We media. How audiences are shaping the future of news and information.* Academic Press.

Chiou, L., & Tucker, C. E. (2010). *News, Copyright and Online Aggregators.* Discussion paper, NET Institute Working Paper No. 11-18.

Chircu, A. M., & Kauffman, R. J. (1999). Strategies for Internet middlemen in the intermediation/disintermediation/reintermediation cycle. *Electronic Markets*, *9*(1-2), 109–117. doi:10.1080/101967899359337

De Bauw, H., & Vandenweghe, V. (2011). Brussels Court of Appeal Upholds Judgment Against Google News and Google Cache. *Eubelius*. Retrieved January 5, 2016 from http://www.eubelius.com/en/spotlight/brussels-court-appeal-upholds-judgment-against-google-news-and-google-cache

Dellarocas, C., Sutanto, J., Calin, M., & Palme, E. (2015). Attention allocation in information-rich environments: The case of news aggregators. *Management Science*, mnsc.2015.2237. doi:10.1287/mnsc.2015.2237

Doctor, K. (2009). News Users 2009. Outsell. Retrieved from http://www.outsellinc.com/store/products/886?refid=pr886

Ericson, J. (2013). *Information overload: how the Internet inhibits short-term memory*. Retrieved 21 January, 2016 from http://www.medicaldaily.com/information-overload-how-internet-inhibits-short-term-memory-257580

Fura, M. (2010). *Chcemy rozmawiać z wydawcami, jak rozliczać się za treści w sieci.* Gazeta Prawna 2010. Retrieved January 3, 2016 from http://www.gazetaprawna.pl/blogi/michal_fura/archiwa/P3.html

Glaser. (2010). Google News to Publishers: Let's Make Love Not War. *PBS*. Retrieved 2015 November, 2015 from http://www.pbs.org/mediashift/2010/02/google-news-to-publishers-lets-make-love-not-war035 Historia Onet.pl

Hirsch, F. (2005). *Social limits to growth*. Routledge.

History. (n.d.). *ITI holdings*. Retrieved December 27, 2016 from http://www.itiholdings.com/index.php/History/History/

Hong, S. (2012). Online news on Twitter: Newspapers' social media adoption and their online readership. *Information Economics and Policy*, *24*(1), 69–74. doi:10.1016/j.infoecopol.2012.01.004

Hudson, A. (2012). *The age of information overload*. Retrieved 28 December, 2015 from http://news.bbc.co.uk/2/hi/programmes/click_online/9742180.stm

Humphries, L. (2000). *Extras-itlist and other bookmark managers*. Retrieved from http://www.llrx.com/extras/itlist.htm

Isbell, K. (2010). *The rise of the news aggregator: Legal implications and best practices*. Berkman Center Research Publication.

Jenkins, H., Ford, S., & Green, J. (2013). *Spreadable media: Creating value and meaning in a networked culture*. NYU Press.

Jeon, D. S., & Esfahani, N. N. (2012). *News Aggregators and Competition among Newspapers in the Internet*. Academic Press.

Kaplan, D. (2010). *Mark Cuban: Google, Content Aggregators Are Vampires; Newspapers Are Zombies*. PaidContent.org. Retrieved 12 December, 2015, from https://gigaom.com/2010/02/02/419-onmedia-mark-cuban-google-content-aggregators-are-vampires-newspapers/

Kennaly, M. (2012). Pinterest now the THIRD most popular social media network beating LinkedIn and Tumblr. *Daily Mail*. Retrieved 23 December, 2015 from http://www.dailymail.co.uk/news/article-2127078/Pinterest-popular-social-media-network-beating-LinkedIn-Tumblr.html?ito=feeds-newsxml

Kreft, J. (2016). From Readers to Users and Creators. Migration of Printed Media Readers to the Internet: Research Results of 2007-2015. *Social Sciences*, *90*(4), 32–42. doi:10.5755/j01.ss.90.4.14250

Madsen, P. (2011). *Aggregating agendas: Online news aggregators as agenda setters*. Paper presented at 36th Annual Conference of the Midwest Association for Public Opinion Research, Chicago, IL.

Megapanel PBI/Gemius, IV.2014 – IV.2015.

Mohamud, A. (2012). Nearly 50 Percent of Internet Users in Europe Visit Newspaper Sites. *comScore*. Retrieved 25 December, 2015 from http://www.comscore.com/por/Insights/Press-Releases/2012/1/Nearly-50-Percent-of-Internet-Users-in-Europe-Visit-Newspaper-Sites

Autorzy Onet, czyli publikuj i zarabiaj na swoich treściach. (n.d.). Retreived December 12, 2015, from antyweb.pl/autorzy-onet-czyli-publikuj-i-zarabiaj-na-swoich-tresciach/

Palis, C. (2012). *Pinterest traffic growth soars to new heights: Experian report.* The Huffington Post.

PEW. (n.d.). *Special Reports – Economic Attitudes*. Retrieved January 10, 2016, from http://www.stateofthemedia.org/2010/special-reports-economic-attitudes/nielsen-analysis

Picard, R. G. (2009). Why journalists deserve low pay. *The Christian Science Monitor*, 19.

Picard, R. G. (Ed.). (2014). *Media product portfolios: Issues in management of multiple products and services*. Routledge.

Quinn, S., & Quinn-Allan, D. (2005, January). The World-wide spread of journalism convergence. In *Proceedings of the 2005 Journalism Education Association Conference*. Griffith University, School of Arts.

Roos, J. M., Mela, C. F., & Shachar, R. (2013). *Hyper-media search and consumption*. Academic Press.

Rusbase. (2012). *Управляющий партнер Altair Capital Management инвестировал в Likehack 4,5 млн руб.* Retrieved 23 December, 2015 from http://rusbase.vc/news/upravlyayushij-partner-altair-capital-management-investiroval-v-likehack-45-mln-rub/

Salecl, R. (2011). *The tyranny of choice*. Profile Books.

Shafer. (2002). *Automated News. The new Google News site, news untouched by human hands*. Retrieved 17 December, 2015 from http://www.slate.com/articles/news_and_politics/press_box/2002/09/automated_news.html

Shenk, D. (1998). *Data smog: Surviving the information glut*. Harper San Francisco.

Vahlberg, V., Shearer, E., Peer, L., Williams, V., & Figueroa, B. (2008). From "Too Much" to "Just Right". Engaging Millenials in Election News on the Web. Media Management Center, Northwestern University. *Tulostettu*, *5*, 2009.

Westley, B. (2011). How a Narrow Application of 'Hot News' Missapropriation Can Help Save Journalism. *The American University Law Review*, *60*(3), 4.

Yin, R. K. (1994). *Case study research. Design and methods*. Thousand Oaks, CA: Sage.

Chapter 11
Competences as a Core Factor Impacting Market Research Usage in Poland:
Luxury of Thriving Business or Necessity of Struggling One – State of Market Research Industry in Poland

Zofia Bednarowska
Jagiellonian University, Poland

Michał Andrzej Chrzanowski
Jagiellonian University, Poland

ABSTRACT

Almost two-thirds of Polish companies declare that the most important source of information are their own observations and personal experience (Kaper & Kuziak, 2006, p. 6). Only 32% of companies use benchmarking as an external source of information, while over two thirds of the companies do not measure their effectiveness at all (Bednarowska, 2015). They do not mention market research, which is an essential element in the management and development of enterprises. Not only they

DOI: 10.4018/978-1-5225-1656-9.ch011

provide knowledge about the environment, but also support knowledge management in the enterprise. In order to proceed with acquiring knowledge, it is necessary to overcome a group of internal and external determinants that prevent many Polish companies from proceeding with market research. However, one of the crucial one are competences of decision makers and power sponsors of market research projects.

1. INTRODUCTION

1.1. Knowledge Management

In a knowledge-based economy companies have to adapt to the rapidly changing reality of the market, which is driven by innovation. As Peter Drucker predicted, knowledge replaced equipment, capital, materials and labour to become the most important element in production and, as a consequence, competitive advantage is determined by knowledge resources (Wang & Wang, 2012: 8899, Drucker, 1993). It has been noticed that *the foundation of organizational competitiveness in the contemporary economy has shifted from physical and tangible resources to knowledge* (Wong & Aspinwall, 2005: 64).

In this context it is crucial to understand how organizations identify, create, share and accumulate knowledge. This is studied by researchers from a relatively new discipline – knowledge management (Darroch, 2005: 101). It is actually both a field of science and a very practical imperative of mature organizations. Private companies will be examined here as an example of organizations.

According to research carried out by Jenny Darroch (2005), an organization capable in knowledge acquisition, knowledge dissemination and responsiveness to knowledge is more innovative and performs better (2005: 112). According to the study, the most influential factor of company innovation and performance is the company's capability in terms of the following:

- Organization values employee attitudes and opinions.
- Organization has a well-developed financial reporting system.
- Organization is sensitive to information about changes in the marketplace.
- Organization works in partnerships with international customers.
- Organization gets information from market surveys (2005: 112).

This chapter will focus on one of those capabilities – a company's ability to acquire knowledge about its marketplace and clients through market research information. How does this information translate into knowledge and help the company develop?

It is worth specifying that knowledge and information are two different concepts. Knowledge is defined as an asset or resource, unlike information or data, is not easily understood, classified, shared and measured. It is invisible, intangible and difficult to imitate. Expanding the knowledge base within an organization is not the same as expanding its information base (Terra & Angeloni, 2003: 2). In this paper, knowledge is understood as a familiarity or understanding of facts by an employee, gained by receiving and processing information. This approach emphasizes the role of the employee's competences to process information – necessary in the creation and accumulation of knowledge in the company.

1.2. Market Research as a Contribution to Knowledge Management

Innovative companies need sources of information in order to access the latest trends. The foremost source is up to date market information. One of the key data sources in market information systems (MkIS) are market research studies and ready to buy market data (e.g. online and offline systems which enable users to manage data themselves, such as Tableau). Their use can reduce uncertainty in decision making, as well as help to build market knowledge and market orientation (Mazurek-Łopacińska & Sobocińska, 2010: 11). Market recognition, research and development in general constitute adaptive functions of companies - the last stage of the development of the social structure in the organization, according to the model of open systems (Katz & Kahn, 1966). Therefore, they may contribute to the development of the organization.

Furthermore, market data aims to provide answers about the company itself, in order to learn and understand it more thoroughly. It is common knowledge that reliable sources of information should be valued in any decision making process. Market research includes not only the study of the demand for certain products (goods or services), but also the analysis of the product and package, of customers (end users), of prices and other elements of market impact and the competitive situation. The results are used in strategic management and operational planning, and especially in sales and impact on the market (Lichtarski, 2005: 265).

Market research is not the only source of information in knowledge management. There are several internal and external sources of information, including own knowledge and the experience of an organization's employees, trade contacts and offers, fast developing business intelligence based on information technologies, official statistics, competitors' materials (catalogues, websites) and professional journals (Bartosik-Purgat & Schroeder, 2014: 105). Here, emphasis will be placed on market research as an information source. It requires the most effort and extensive resources, but proves to be a more reliable source of information than manager intuition in reducing the risks resulting from a turbulent competitive market (Moszkowicz & Bembenek, 2013: 59). Paradoxically, it is the latter that Polish companies trust more often (Bartosik-Purgat & Schroeder, 2014: 108).

1.3. Case Study of Poland

This paper will investigate the situation of Polish companies and their adoption of knowledge management strategies, namely – usage of market research information.

Poland underwent a massive economic and political transformation in 1989, transitioning from a centrally planned communist economy dominated by nationalized companies to a free market. As a result, Polish companies exist in the free market no longer than 27 years, and the experience of Polish managers is even shorter. Thanks to EU integration Poland joined efforts with other countries in developing a European knowledge-based economy (OECD, 1996). This makes the Polish market an interesting case of in terms of a massive adoption of new managerial strategies, including knowledge management. In this paper, two sides of the demand-supply chain are analyzed – market research agencies (supply) and Polish companies (demand). Polish market research agencies (called this way in order to differentiate them from companies that are their clients in a B2B relationship) were established shortly after the transformation at the beginning of the 1990s, and their development is closely connected to the maturity of Polish companies understood as the extent of knowledge acquisition and management.

Almost two-thirds of Polish companies declare that the most important sources of information are their own observations and personal experience (Kapera & Kuziak, 2007: 6). The budgets of many Polish companies do not cover the regular purchase of data about consumers and the market, or, perhaps more surprisingly, even the purchase of data in situations crucial for the companies, such as entering new markets and introducing new products, where business action would benefit from the support of business information. Global companies, such as Procter & Gamble or Heineken USA, are much more aware of the significance of market data and they invest large sums of money to reduce uncertainty (Aviv 2001). Oftentimes, the marketing budget needs to cover other important cost positions, such as advertising (Pun & Heese, 2015). This makes it even more difficult to include market research in the decision making process of the companies. During the Great Recession of 2007 one of the first cost reductions in Polish companies was cutting market research expenses (PTBRiO, 2013: 39). This reflects the view that market research is considered more of a luxury in a profitable company than a necessity.

Market research can be one of the most complicated ways of obtaining data. Yet, if the research is carried out correctly, it may serve as a self-assessment of an organization's market position, and a comparison to the competitors that can help estimate company value. Polish private companies, when asked about their source of information for measuring their performance, show little activity in using it in everyday practice. 32% of companies surveyed use benchmarking as an external source of information, while over two thirds of Polish companies do not measure their effectiveness at all (Bednarowska, 2015). Other research shows that only 8,9% of companies applied market research in the last three years, when asked about marketing activities (Johann 2013). How is it possible to evaluate a company's performance and work on improvement without that kind of critical knowledge?

The main hypothesis of the research and analysis postulates that poor usage of market research is determined by a range of internal and external factors.

Our strategy of the literature search and analysis was based on the assumption that the reason behind the research problem is manifold, including array of the internal and external factors. Various aspects of organizations have been analyzed, in order to identify those that may be critical in the low usage of market research.

In the case of internal factors, the research was based on the theoretical framework proposed in the Wong & Aspinwall study (Wong & Aspinwall, 2005). In the study several factors of knowledge management adoption were identified: management leadership and support, culture, information technology, strategy and purpose, measurement processes, organizational infrastructure, process and activities, motivational aids, resources, training and education, human resources (2005: 66). Factors proposed by Wong & Aspinwall constituted a starting point for a review of existing research studies on Polish management and market research usage, in order to identify factors contributing to the current situation. Additionally, those factors were compared with reports prepared by the Polish Society of Market and Opinion Research (PTBRiO) covering the state of the marketing research industry in Poland. As a result, several key factors were identified as crucial for understanding the mechanisms of market research usage among Polish companies.

Research regarding external factors was conducted in a similar manner, focusing on the development of the Polish market research industry.

2. INTERNAL FACTORS

The first group of factors refers to the most important areas of the organization, highlighted in organizational theory and knowledge management theory. The factors were as follows: organization structure, decision-making process, organization management, and organizational culture, which all constitute elements of strategic management. If the management (including management leadership) is poor, then there is a higher risk of low usage of market research.

2.1. Organization Structure

It is organization structure that may impact the decision of using market research and afterwards – the choice of market research company. Organizational maturity, size of the organization, technology applied in a company, social system of the organization all shape organization structure (Lichtarski, 2005: 301) and determine what decisions are taken there. Unsurprisingly, one of the biggest determinants of the use of market research may be the size of the company. As data shows, bigger companies tend to have bigger research budgets (Churchill, 2002: 32-33). A more open attitude to market research is also driven by the level of centralization in the company, as well as by the position of the marketing department within company structure (Churchill, 2002: 31). In industry literature, there are discussions which organizations are more flexible and whether marketing department structure or level of their centralization may impact work efficiency (Grover & Vriens, 2006: 19). A significant factor may be the fragmentation of marketing departments, as well as their presence in various departments of the company, as it hinders internal communication within the company. If they are placed in a non-strategic, less decisive department, then it is less likely to enforce the decisions within the whole company – such as buying market research studies.

2.2. Decision Making Process

Companies which are buying market research studies are consumers in the B2B relation. Their tendency to buy market research is influenced by their model of decision making processes, as well as the extent of their rationality (Hatch, 2002: 270-278). It is widely recognized that B2C consumers do not always follow a rational decision-making process (Solomon et al., 2007: 259). And similar situations may occur in the case of B2B customers, even if it is a company decision. Since buying market research studies oftentimes takes place when the company is in a state of uncertainty and ambiguity, this has an even stronger impact the outcome of the decision.

The level of complexity of the decision making process within the organization determines a policy of research supplier choice. According to research by the Polish Society of Market and Opinion Research (PTBRiO), 40% of private companies in Poland outsource studies to one research agency, or choose on the basis of a tender (PTBRiO, 2009: 8). The decision to make use of the research is also influenced by a list of corporate priorities – one of the biggest obstacles is cost, which may prevent the company from deciding to go ahead with a market research study (PTBRiO, 2013: 39).

2.3. Organization Management and Organizational Culture

Management in Polish companies may determine the decisiveness of companies in the field of marketing research. The Polish style seems to be often based on questionable information (Kapera & Kuziak, 2006), without reference to full market information, competitive environment and prompts for development.

The low level of usage of market research in everyday management is strongly linked to introducing knowledge management, which is still scarcely applied in Polish companies, as its results are only visible in the long-term (Lichtarski, 2005: 381). Knowledge management includes learning about the company itself in order for it to grow. It is mainly organizational research that can support the management with insights and evidence in order to introduce necessary changes, improve processes and ways of decision making within the company (Wheeler 2009).

Both the reason and the result of low usage of market research may be identified by the low level of innovation of Polish companies. This condition has been proven - Poland is among the least innovative countries in the EU (Balcerowicz & Wziątek, 2009: 9). This may also have a negative impact. In the European Commission report "Research and Innovation performance in the EU" (2014) Poland had been characterized as "a low knowledge capacity system with a specialisation in low knowledge-intensive sectors" (2014: 48). These sectors have a low tendency to circulate knowledge within their companies, and an even lower tendency to acquire knowledge. According to the report authors Polish industry sectors composition around low research intensive sectors is connected with country's weaknesses of R&I performance (2014: 225). Poland has a low intensity of business researchers, which reflects the minor role the business sector plays in the national R&I system (2014: 219). Only 3.4% of small and medium-sized companies are reported to introduce marketing or organizational innovations (2014: 219).

Organizational culture is important in building a strategy (Aniszewska, 2007: 13). Since market research is a cognitive part of the culture of the organization (Oczachowski, 2011: 23-24), companies with their own market research department may conduct market research studies more often. Even the process of building a strong marketing team (or market research team) is reflected in final purchase decisions.

Organizational culture differs by industry, and this is also apparent in the application of market research. A PTBRiO study shows different degrees of usage of market research within different industries - companies from the FMCG sector perform research the most frequently, and at the same time trust in the power of their information (PTBRiO, 2012: 18). Companies from the telecommunications industry are also eager to use external data, while construction companies do it less frequently (PTBRiO, 2013: 57). The explanation of this phenomenon may be a relationship between the importance of organizational culture in the company and the industry. The most well developed organizational culture is visible in the service industry, which is built on personal relationships with the client (Aniszewska, 2007: 43). Companies from this industry are among the most frequent clients of market research agencies.

Even if some companies are convinced to conduct market research, they do not do it on a regular basis. Market monitoring is particularly important in dynamically changing industries. The latest data or long-term forecasting may significantly strengthen the competitiveness of the company, yet 54% of companies use only ad hoc research studies (PTBRiO, 2013: 20). It also shows, indirectly, the reluctance to use complex, multi-stage (also more expensive) research studies, but on the other hand - their greater information value.

The classic Geert Hofstede theory (Hofstede, 2000) lists four dimensions of organizational culture. Polish culture is characterized by a high rate of masculinity, as well as a high rate of power distance and strong uncertainty avoidance, and also a relatively low level of individualism. All in all, it leads to "spontaneous" management and a poor ability to delegate tasks (Żakowska, 2007: 16). Strong uncertainty avoidance is not conducive to ordering market research, especially when performing it for the first time. The low level of individualism and a high rate of power distance can mean braking innovative proposals in the field of research and its utilization becomes difficult.

2.4. Decision-Makers and Market Researchers Competences

Competences of the managers and various stakeholders impact the use and reception of market research data in the strategic management of the company. The decision to order market research and use its results depends on company structure and involves managers from different levels of the company ladder – e.g. market research specialists, directors of marketing, the board of directors, or CEO. Many of those groups involved have a conflict of interest, especially regarding the distribution of the company budget.

Oftentimes, the use of market research data relies on the ability of market research companies and marketing specialists to convince members of the boards that it will contribute to the company's profit. The authors suggest that the usage of more complex data sources like market research studies requires specific competences of both sides – namely strategic thinking and the use of information about market trends. According to the Deloitte study on Polish decision-makers, there is a significant gap between the level of their competences and the level of competences required for companies to thrive and compete (Deloitte, 2015: 10). A survey study on employers (meaning middle level managers) and employees (meaning the general Polish population of adults) analyses several key competences. Strategic thinking and the use of information about market trends are the most important qualities in decision-makers (Deloitte, 2015: 7). At the same time, 70% of employees negatively assessed the competences of employers, and 30% of managers hold the same opinion about board members (e.g. 30% of employers attributed to board members average, low or extremely low competence of strategic thinking).

The ability to process complex information included in market research reports is another important feature of the decision-maker in the context of market research usage. There are no research studies conducted specifically about market research reports, therefore general data gathered in the PIAAC project will be reviewed. PIAAC stands for Programme for the International Assessment of Adult Competencies and is a worldwide study by the Organisation for Economic Co-operation and Development (OECD). The main aim of the program is to measure skills of literacy, numeracy and problem solving skills in technology-rich environments.

The following analysis was conducted on the Polish dataset from 2015 on respondents working in the private sector and supervising more than five people (OECD dataset 2015). The main limitation of that research is the inability to distinguish decision-makers in the context of market research usage. Nevertheless, it provides some substantial information about the competences of managers in Polish companies.

The PIAAC study measures several skills used by respondents during their work. The authors identified skills similar to the ones that are needed to process market research reports:

- Reading diagrams, maps or schematics.
- Reading professional journals and publications.
- Usage of simple algebra or formulas.
- Usage of advanced math or statistics.

Figure 1. Usage of skills among managers
Source: own analysis of PIAAC data (OECD dataset 2015)

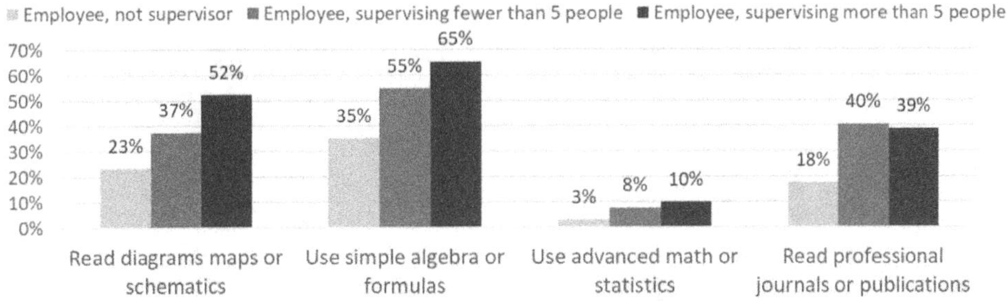

It is clear that the level of proficiency in these skills can help to differentiate managers of the higher level from managers of the lower level. The bigger number of employees supervised by a respondent, the higher likelihood that they read diagrams, maps or schematics (52% of supervisors of more than five people), use simple algebra or formulas (65%), use advanced math or statistics (10%) and read professional journals or publications (39%). Only 10% of managers possess the skill to use advanced math or statistics, which can be an obstacle in processing both basic descriptive statistics, but also more complex statistical models used in market research.

The main value of the PIAAC data lies in the measurement of literacy and numeracy skills. The table in the Appendix presents chosen levels of proficiency in those skills (OECD, 2013: 64, 76).

In the context of processing market research reports, at least a third level of proficiency is required both in literacy and numeracy. It includes the abilities to process multiple-page reports, construing meaning across larger chunks of text, interpreting and basic analysis of data and statistics in texts, tables and graphs. More complex presentation of the results (e.g. statistical models) would require a 4th level of proficiency. At that level a person is able to conduct tasks that require analysis and more complex reasoning about quantities, data, statistics, chance, spatial relationships, proportions and formulas. Tasks at this level may also require understanding arguments or communicating well-reasoned explanations for answers or choices (OECD, 2013: 64).

There are 64% managers who possess sufficient literacy proficiency to process basic market research reports (62% in the case of numeracy proficiency). Only 14% of managers have a skill set needed for processing more advanced market research results (18% in the case of numeracy proficiency). What this data shows is that a

Figure 2. Literacy and numeracy proficiency of Polish managers
Source: own analysis of PIAAC data (OECD dataset 2015)

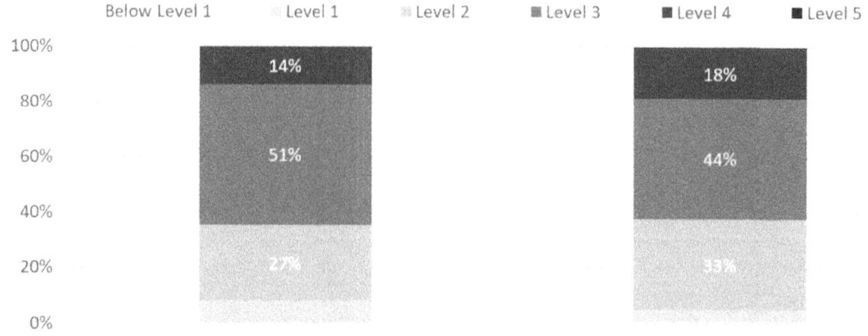

staggering 86% of Polish managers do not have the skills to take full advantage of complex sets of data, such as comprehensive market research reports. This leads the authors to take the grim view that had most managers received a complex report, including insights and recommendations, they would not be able to single-handedly process all of the information and implement the proposed changes in their companies.

Results are similar in a group of respondents who declare that they have to read diagrams or schematics at least once a week and those who have to read reports, articles, magazines and journals with the same frequency.

Those results suggest that around 45% of managers do not have skills required for processing market research reports. This greatly impacts the potential of market research industry development in Poland.

3. EXTERNAL FACTORS

3.1. Research Market Industry Maturity

Companies are working in a competitive environment, with multiple sectors: social, cultural, legal, economic and technological (Hatch, 2002: 79-81). In terms of the economic dimension, one of the most impactful factors is research market industry maturity. The general global market research is evaluated as stable, and market growth as quite modest, according to the European Society for Opinion and Marketing Research (ESOMAR, 2015). The forecasts for the global market are optimistic - the Research Industry Index (RII) suggests growth (AMA, 2015). Poland is in the top 10 fastest growing markets in Europe, yet it is the new EU member countries that

Figure 3. Literacy and numeracy proficiency of respondents who frequently use diagrams, maps, schematics and reports
Source: own analysis of PIAAC data (OECD dataset 2015)

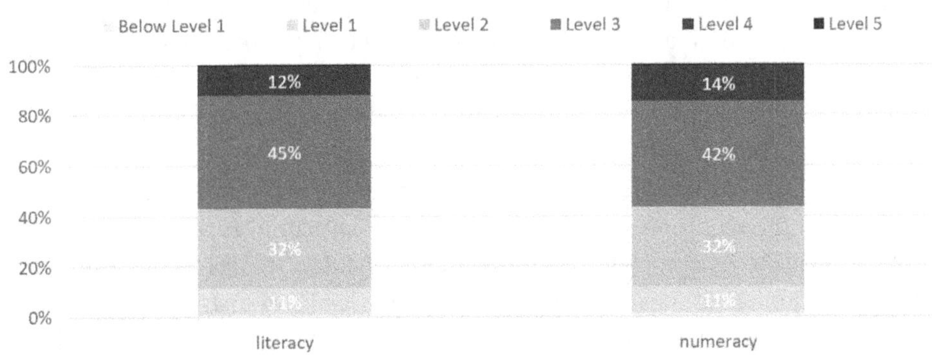

are now developing faster. Poland is the biggest market among new EU member states, making it the eighth biggest market in Europe, based on turnover, and 20th in the world (ESOMAR, 2015). Yet, the Polish market research industry is said to be immature, in comparison to Western Europe and North America (ESOMAR, 2012).

Market research is a complex service, as it consists of several stages mostly not transparent or immediately comprehensible to a client. Hence, the trust placed in a market research company's work is based mainly on the quality of reports which are the final products of the service. Establishing confidence in market researchers is crucial, since it translates into better decisions, better use of research and mutual satisfaction with cooperation (Grover & Vriens, 2006: 3-5).

However, market research buyers criticize the Polish market research industry for low quality of their products, which undermines the general trust in market research value (PTBRiO, 2012). The buyers indicate several aspects of the researchers' work that need to be improved: the understanding of research goals (79%), usability of the reports (85%), ability to summarize findings (84%), structure of the reports (76%), misspellings (75%), language of the reports (68%), and graphic layout (52%) (PTBRiO, 2012). In conclusion, they refer to the most significant features of the reports. On the other hand, the existing highly aware clients expect more and more from the results, not only reports but full development scenarios for companies to make the right strategic decisions (Mazurek-Łopacińska & Sobocińska, 2010: 11). There is a substantial area for improvement here. Better usability and clarity of market research reports would help managers with lower levels of skills to process information included in them, and translate it into knowledge which is actually usable for the company. Those shortcomings of market research reports influence the trust in presented results and lower the need for market research services.

The market's immaturity is also caused by low awareness of the benefits stemming from using market research – this applies particularly to smaller companies. There is also relatively low awareness of the possibilities of using ready data, that is information which is already available, e.g. transaction statistics generated by website systems and CRM analysis (Kaniewska-Sęba et. al., 2006: 60-61). This leads to a low level of integration of different information sources and marketing advice. And this is also an area where Polish market agencies are becoming more active (PTBRiO, 2013: 39). However, a key challenge is "creative intelligence, necessary to synthesize these vast data flows into information, knowledge, insight, foresight, and strategy" (Moran, 2013).

A major role in the educational process of building market research awareness is played by the requirements of European Union structural funding. Each of the projects needs to be completed with research evaluation studies for the use of the European Union (European Commission, 2006). Thanks to that, many companies have conducted market research studies, sometimes for the first time. This leads to an increase of the importance and rank of social and market research in Poland (PTBRiO, 2013: 47).

3.2. B2B and B2P Service in Poland

There is a gap between the interests and standards of market research agencies and their customers' needs (Davis et. al., 2012). There is a conflict of interest in the relationship because research agencies usually aim at maximizing the price, while companies are interested in the most innovative information, at the lowest possible cost (Kaniewska-Sęba et. al., 2006: 297).

The very process of acquiring the B2B service is lengthy and very demanding. The value of transactions in the B2B market is much greater than in the B2C market, but there are less transactions completed, and winning a new client involves a lot of work and considerable financial outlay (Kaniewska-Sęba et. al., 2006: 24). Providing services for the B2B market is challenging. The maintenance of the B2B relationships is one of the most desirable values and is considered the most important source of development of sustainable competitive advantages for companies (Luo & Kumar, 2013: 141). Studies have shown that the most important part of the relationship are interpersonal skills and the willingness to reduce uncertainty regarding the research study (Moorman et al., 1993).

A lack of clear targets of the ordered marketing research studies, or the lack of vision of the final outcome may easily result in the study's failure. Usually, the failure is attributed to the market researchers, rather than the clients. Researchers tend to embrace research problems in the wrong way and clients are not able to shape their business problem into a research problem (Mazurek-Łopacińska & Sobocińska, 2002: 35).

This problem occurs when the client company has not implemented marketing-oriented strategies. A marketing-oriented company should be a company which uses interactive marketing, resulting in customer satisfaction and in maintaining a strong position in a competitive market (Nowotorska-Romaniak, 2014: 158). It requires the company to specify clear long-term goals, to define a target segment of customers, and strategic planning of actions leading to achieving goals with efficiency. Orientation on how to best meet the needs of buyers requires the company to conduct regular market research exploring those needs (2014: 159).

If a company is not marketing-oriented and the information about customers is not shaping company strategy, then there is no real need for market research. Market research products should be able to solve clients' problems by means of information about their customers or other market stakeholders. Designing such market research products requires the translation of the marketing problem into a research problem (PTBRiO, 2016: 5). A precise, well-constructed marketing goal is easier to translate into a research problem than a vague one. Moreover, the process of designing market research involves a few important decisions to be made, concerning survey methods, participants recruitment criteria, the scope of research tools and methods to be used. These decisions should be based on the client's strategic plan and require that the client possesses at least a minimal level of pre-existing knowledge about the customers and the market. The final requirement for that process is that the client intends on implementing the market researcher recommendations.

In this view, the use of market research is an indicator of marketing-oriented companies. Several studies have shown that only a small portion of companies use market research in Poland (e.g. Nowotorska-Romaniak, 2014: 161). At the same time, the majority Polish companies declare that they have enough information about the market and their customers (Rawski, 2014: 230). The study shows that 83% of Polish companies declared that the market information resources they have are sufficient. The following graph (Graph 4) presents the percentage of Polish companies which declare that they have sufficient information about the market.

According to Rawski's study (2014: 231) more than 80% of Polish companies declare that they have sufficient market information about: needs and preferences of buyers (95%), market segmentation (88%), market share (84%), size of the market (83%), environment analysis (82%), brand awareness (79%), competitors (78%) and evaluation of marketing (78%). However, this does not mean that they are using information resources they own. Therefore, the authors conclude that one of the reasons behind the low usage of market research is simply that companies do not feel they would benefit from market research.

Figure 4. Assessment of the level of sufficiency of the market information resources by Polish companies (% of them declaring sufficient information)
Data source: Rawski (2014: 231)

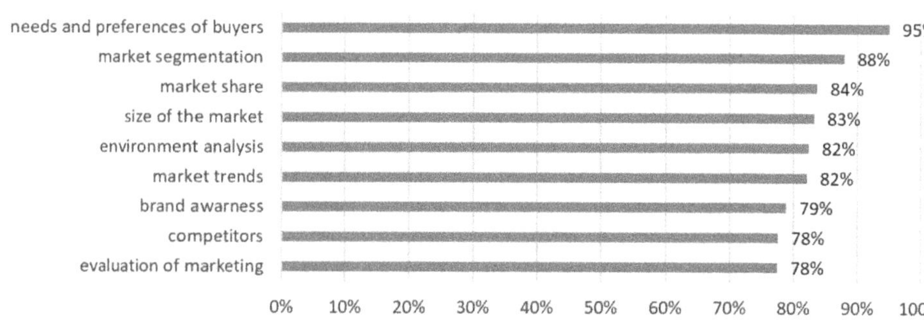

3.3. Research-Based Policy Culture

Generally, a supportive economic environment is the key to the development of any industry, in particular the newly established private market research industry. The use of market research is closely related to the development of the Polish free market after communism collapsed in 1989. First market research agencies were established at the beginning of the nineties, whereas first worldwide market research was already being conducted at the end of the 19th century (Olejniczuk-Merta, 2010:22). Since the early nineties, we can observe a steady growth in the market research expenses of Polish companies (PTBRiO, 2011-2016).

Introduction of foreign capital enforced more intensified use of market research in several industries in Poland. A good example is the Polish banking industry, where several mergers and acquisitions by foreign banks occurred after Poland's access to the European Union in 2004. It was followed by a significant increase in the value of financial market research expenses (Grzegorczyk, 2012: 187). Adaptation of the Western-European style of management culture resulted in an increase in the use of market research.

Figure 5. Market research industry growth in comparison to GDP per capita in Poland

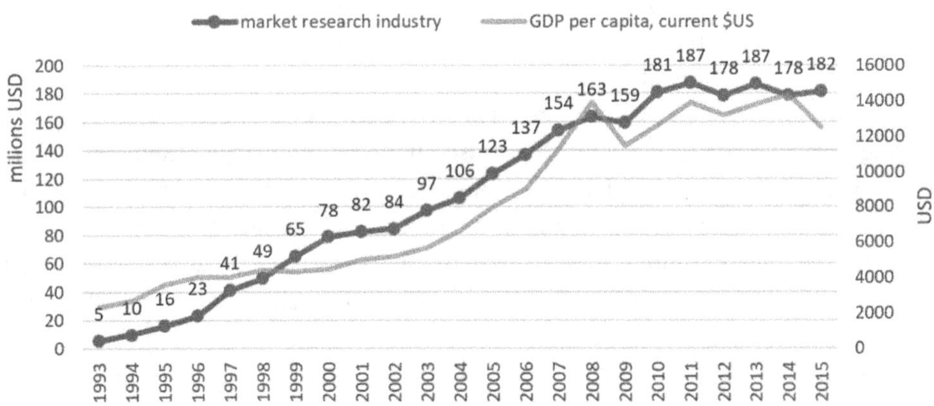

4. CONCLUSION

The study focused on the private companies' capabilities to acquire and process market information as part of their knowledge management. According to cited papers, those capabilities are crucial for innovative company performance in a knowledge-based economy. Poland was chosen as a case study as an example of a developing country with a short history of free market, where management strategies are being adopted from Western countries. Among knowledge sources, marketing research was selected as one of the most important sources of information about the market environment and, at the same time, the most demanding in terms of resources and managerial competence. Wong & Aspinwall`s study on knowledge-management adoption (Wong & Aspinwall, 2005) was used as a starting point for identifying factors that affect low market research usage among Polish companies.

The aforementioned study identified the key factors that influence knowledge management adoption in Western European countries. The authors then applied that same approach to Polish companies. Several aspects of Polish companies and Polish market research industry were reviewed for the purpose of identifying possible factors that interfere with the progress of knowledge management adoption and market research usage.

Market research is widely recognized as an essential element in the management and development of companies. This chapter identifies a battery of concepts proving that the use of market research supports the decision-making process within companies, and, as a result – company development as a whole. Not only does it provide knowledge about the competitive environment and reduces uncertainty during critical business decisions, but also supports knowledge management and the growth of the company. However, the level of usage of market research in the Polish market is fairly low, as is the awareness of its benefits. The key hypothesis was verified, and a group of internal and external determinants was identified that prevent many Polish companies from proceeding with market research and acquiring vital knowledge.

The internal determinants (ones which originate in the companies) concern the organization structure of the company, the type of decision making process, organization management and organizational culture style.

The organizational structure of private companies can take many forms, a fact that sometimes makes the companies difficult to compare. A major factor in the decision to conduct a market study is the type of company (service or production), and the particular sector in which it operates. Companies from the service sector are the ones most often willing to carry out studies (PTBRiO, 2013). The location of the marketing or research department (or other departments responsible for decisions regarding research) is crucial in this context – it influences the speed and efficiency of decision making (the number of persons who need to be consulted about the decision, and who need to be consulted about the scope of the research). Therefore, a complicated and multi-layered structure can delay decisions. Fragmented marketing departments can be the cause of similar scenarios (if they are located in several different departments). Consequently, the decisions require coordinating the work of many people. The organizational structure and decision making process also affect the way research is carried out in a company.

Another significant issue to mention is the importance of an open attitude of power sponsors to market research and market data in general. It stems from knowledge of the study's topic and experience with purchasing research reports or data.

Organizational management and organizational culture are further factors influencing the decision on whether or not to use research and data. The lack of a stable, well-established organizational culture can result in the lack of a need to enhance knowledge about the company by both the employees and employers, and therefore a low level of concern about acquiring this kind of knowledge from outside sources. These factors differ in intensity depending on the type of industry sector and company size.

Unfortunately, ad hoc research still tends to dominate, even in the case of Polish companies which have good internal structure, an efficient decision making process and a willingness to enhance their knowledge about their consumers, the competitive environment and the company itself. Panel studies and omnibus surveys constitute a small share of the market, compared to other countries (ESOMAR, 2015).

However, as the authors' analysis shows, the most prominent factor for poor market research usage are the competences of decision-makers and market researchers. Based on an international study on respondents working in the private sector and supervising more than five people (OECD dataset 2015), a Polish dataset from 2015 has been analysed. This analysis identifies skills needed to process market research reports and to verify their proficiency among Polish managers, who are the potential recipients of market research products. The analysis showed that managers at higher levels of the company structure tend to have higher levels of those skills. The analysis was narrowed down to those declaring they have to use the skills at work at least once a week, such as reading diagrams, reports, articles, magazines and journals.

Other skills. crucial to acquire knowledge from market research, are literacy and numeracy. The results in both research studies show that merely 64% of managers possess sufficient literacy proficiency to process basic market research reports, and 62% have the required skills in terms of numeracy. What is more, in the case of more advanced research results, only 14% of managers have the necessary levels of literacy, and 18% of numeracy proficiency.

A list of external factors that hold companies back from conducting market research or using data includes market research agencies' experience, as well as a generally poor standard of B2B and B2P services, together with a slowly developing research-based policy culture in Poland.

The Polish market research industry is still relatively young and inexperienced, which affects the perception of market research, and therefore the frequency of using market research products. The global market research industry is considered to be stable, and the Polish economy is one of the largest in Europe (ESOMAR, 2015) – all this bodes well for the Polish market research industry. However, in order for more entities to appreciate the value of research and market data, it is crucial that both sides – market research companies and their clients – work together on building mutual trust. Research by PTBRiO (2013) indicates that research products are often criticized by clients for their poor quality and low relevance to their business problems. Researchers on the other hand point to poor knowledge of research on the part of their clients, which hinders cooperation, especially in the case of smaller companies – a great deal of them do now know what should be researched or how to use the findings.

To some extent, this situation is caused by the perceived low level of service quality in both the B2B and B2P markets, and not just in the research industry. The European Union's requirements are certainly a positive factor in the growth of research, as they impose an obligation to evaluate a company's performance – and often also to conduct market research – before a company can receive financing, or acquire company information before key decisions are taken. As shown in the analysis above, for many companies it is frequently the first time they come in contact with market research.

The main hypothesis of the chapter was confirmed. The low level of the usage of market research as a source of information and building knowledge in the company is caused by a range of internal and external factors. The factors were listed due to the secondary data analysis completed and state of art. As the analysis shows, the core factor impacting market research usage are competences of their recipients.

The study had a preliminary character, it is possible to draw some conclusions and make recommendations for further analysis. Future research recommendation is to conduct research dedicated to Polish companies that would cover all mentioned factors and bring conclusive results regarding the significance of each of them to the explanation of the problem.

REFERENCES

AMA. (2015). *Research Industry Index*. Retrieved from http://www.marketingresearch.org/stay-informed/research-industry-index

Aniszewska, G. (2007). *Kultura organizacyjna w zarządzaniu strategicznym. w: Kultura organizacyjna w zarządzaniu. Aniszewska, Grażyna. (red.)*. Warszawa: Polskie Wydawnictwo Ekonomiczne.

Aviv, Y. (2001). The effect of collaborative forecasting on supply chain performance. *Management Science, 47*(10), 1326–1343. doi:10.1287/mnsc.47.10.1326.10260

Balcerowicz, E., & Wziątek, A. (2009). *Determinanty rozwoju innowacyjności firmy. Ekspertyza przygotowana dla Polskiej Agencja Rozwoju Przedsiębiorczości*. CASE - Centrum Analiz Społeczno-Ekonomicznych. Retrieved from http://www.parp.gov.pl/files/74/75/76/487/493/7614.pdf

Bednarowska, Z. (2015). *An investment, not an expense. Why marketing research and market analysis are essential for a company's success*. Retrieved from www.research-pmr.com/a44/an-investment-not-an-expense-why-marketing-research-and-market-analysis-are-essential-for-a-companys-success

Churchill, G. A. (2002). *Badania marketingowe: podstawy metodologiczne*. Warszawa: Wydaw. Naukowe PWN.

Davis, D. F., Golicic, S. L., Boerstler, C. N., Choi, S., & Oh, H. (2013). Does marketing research suffer from methods myopia? *Journal of Business Research, 66*(9), 1245–1250. doi:10.1016/j.jbusres.2012.02.020

Deloitte. (2015). *Liderzy na dziś – Liderzy na jutro. Jakie kompetencje przywódcze mają członkowie zarządów i rad nadzorczych?* Retrieved from http://www2.deloitte.com/content/dam/Deloitte/pl/Documents/Reports/pl_Rady-Nadzorcze-Liderzy-na-dzis.pdf

ESOMAR. (2012). *Global Market Research 2012. An ESOMAR Industry Report in cooperation with KPMG Advisory*. Amsterdam: ESOMAR.

ESOMAR. (2015). *Global Market Research 2015. An ESOMAR Industry Report in cooperation with BDO Accountants & Advisors*. Amsterdam: ESOMAR.

European Communities. (2006). Evaluation Methods for the European Union's External Assistance. Guidelines for Project and Programme Evaluation. *Reproduction (Cambridge, England), 3*.

Grover, R., & Vriens, M. (2006). The Handbook of Marketing Research: Uses, Misuses, and Future Advances. Thousand Oaks, CA: Sage Publications, Inc.

Grzegorczyk, W. (2012). *Marketing Activities of Polish Banks*. Advanced Research in Scientific Areas.

Hatch, M. J. (2002). *Teoria organizacji, przekł. Łuków, P.* Warszawa: Wydawnictwo Naukowe PWN.

Hofstede, G. (2000). *Kultury i organizacje: zaprogramowanie umysłu. Tłum. Durska, Małgorzata*. Warszawa: Polskie Wydawnictwo Ekonomiczne.

IMF. (2016). *Too Slow for Too Long Cataloging-in-Publication Data*. Retrieved from http://www.imf.org/external/pubs/ft/weo/2016/01/pdf/text.pdf

Johann, M. (2013). *Strategie marketingowe stosowane w praktyce polskich przedsiębiorstw*. Katedra Zarządzania Strategicznego Kolegium Nauk o Przedsiębiorstwie. Retrieved from http://www.sgh.waw.pl/katedry/kzs/Konferencja/10.%20Maria%20Johann.pdf

Kaniewska-Sęba, A., Leszczyński, G., & Pilarczyk, B. (2006). Badania marketingowe na rynku business-to-business. Kraków: Oficyna Ekonomiczna. Oddział Polskich Wydawnictw Profesjonalnych.

Kapera, K., & Kuziak, M. (2007). *Pozyskiwanie i wykorzystywanie informacji rynkowych. Raport z badania ankietowego przedsiębiorstw.* Katedra Marketingu UEK Kraków. Retrieved from http://ankieta.uek.krakow.pl/report/Raport_Badanie_AE_Krakow_2006.pdf

Katz, D., & Kahn, R. L. (1966). *The social psychology of organizations*. New York: John Wiley & Sons.

Kirsch, I., & Thorn, W. (2013). *Technical Report of the Survey of Adult Skills*. PIAAC.

Luo, A., & Kumar, V. (2013). Recovering Hidden Buyer-Seller Relationship States to Measure the Return on Marketing Investment in Business-to-Business Markets. *Journal of Marketing Research, 50*, 143-160.

Mazurek-Łopacińska, K., & Sobocińska, M. (2002). Badania marketingowe. Podstawowe metody i obszary zastosowań. Wrocław: Wydawnictwo Akademii Ekonomicznej.

Mazurek-Łopacińska, K., & Sobocińska, M. (2010). Kierunki rozwoju badań marketingowych w dobie społeczeństwa i gospodarki opartych na wiedzy. In *Badania marketingowe: nowe wyzwania*. Wrocław: Wydawnictwo Uniwersytetu Ekonomicznego. Prace Naukowe Uniwersytetu Ekonomicznego we Wrocławiu.

Moorman, C., Deshpandé, R., & Zaltman, G. (1993). Factors Affecting Trust in Market Research Relationships. *Journal of Marketing*, *57*(1), 81-101.

Moran, R. (2013). *The Futures of Marketing Research*. Retrieved from http://www.sagepub.com/upm-data/43895_Epilogue.pdf

Oczachowski, D. (2011). Kultura kognitywna organizacji a prowadzenie badań marketingowych. In *Badania marketingowe: metody, nowe podejścia i konteksty badawcze*. Wrocław: Wydawnictwo Uniwersytetu Ekonomicznego. Prace Naukowe Uniwersytetu Ekonomicznego we Wrocławiu.

OECD. (2013). *OECD Skills Outlook 2013 - First results from the survey of adult skills*. http://doi.org/10.1787/9789264204256-en

Olejniczuk-Merta, A. (2010). Trzy generacje badań marketingowych. In *Badania marketingowe: nowe wyzwania*. Wrocław: Wydawnictwo Uniwersytetu Ekonomicznego. Prace Naukowe Uniwersytetu Ekonomicznego we Wrocławiu.

PTBRiO. (2009). *Wizerunek branży badawczej 2009*. Prezentacja wyników badania IX-X 2009. PTBRiO.

PTBRiO. (2011). *Rocznik PTBRiO 2010/2011*. Retrieved from http://ptbrio.pl/index.php/publikacje/rocznik-ptbrio

PTBRiO. (2012a). *Wizerunek branży badawczej 2011*. PTBRiO. Retrieved from www.ptbrio.pl/images/stories/publikacje/wizerunek_branzy_badawczej_raport_wersja_pelna.pdf

PTBRiO. (2012b). *Rocznik PTBRiO 2011/2012*. Retrieved from http://ptbrio.pl/index.php/publikacje/rocznik-ptbrio

PTBRiO. (2012c). *Raport idealny*. Retrieved from http://www.ptbrio.pl/index.php/publikacje/raporty

PTBRiO. (2013a). *Rocznik PTBRiO 2012/2013*. Retrieved from http://ptbrio.pl/index.php/publikacje/rocznik-ptbrio

PTBRiO. (2013b). *Badania marketingowe*. Rocznik Polskiego Towarzystwa Badaczy Rynku i Opinii. Edycja XVII, 2012/13.

PTBRiO. (2014). *Rocznik PTBRiO 2013/2014*. Retrieved from http://ptbrio.pl/index.php/publikacje/rocznik-ptbrio

PTBRiO. (2015). *Rocznik PTBRiO 2014/2015*. Retrieved from http://ptbrio.pl/index.php/publikacje/rocznik-ptbrio

PTBRiO. (2016). *Rocznik PTBRiO 2015/2016*. Retrieved from http://ptbrio.pl/index.php/publikacje/rocznik-ptbrio

Pun, H., & Heese, H. S. (2015, August). A note on budget allocation for market research and advertising. *International Journal of Production Economics*, *166*, 85–89. doi:10.1016/j.ijpe.2015.04.013

Żakowska, M. (2007). Kultura organizacyjna polskiego biznesu według kryteriów zaprogramowania kulturowego. In *Kultura i gospodarka: ku antropologii życia gospodarczego we współczesnej Polsce*. Tychy: Śląskie Wydawnictwo Naukowe.

KEY TERMS AND DEFINITIONS

Artificial Intelligence: Artificial Intelligence refers to computational understanding of intelligent behavior which works as a system that perceives its environment and takes actions that maximize its chances of success.

Belief: The assumption that is held to be true, by an individual or a group, regarding concepts, events, people, and things.

Bounded Rationality and Intuitive Decision-Making Model: The set of human-operator's thinking process techniques in order to transform the input with missing, untrustworthy, inconsistent, "noisy" or not capable of interpretation cyber-information of the given particular decision problem into the right course of action, "burdened by individual heritage" or "enriched" in the form of own propensity, hobby, tendency, expectation, axioms and judgment, presumption or bias and their justification.

Cognition: The mental processes involved in judging, knowing, learning, perceiving, recognizing, remembering, thinking, and understanding that lead to the awareness of the world.

Collective Decision-Making: The thinking techniques, which presupposed involving the human-operators of the socio-technical system in the processes of generating problems, providing potential solutions, voting for solutions and for trusting to each other, and aggregating individual votes and ultimately deriving a final collective decision.

Collective Intelligence: Shared or group intelligence that emerges from the collaboration, collective efforts, and competition of many individuals and appears in consensus decision making.

Competency: The important skill that is needed to do a task.

Critical Thinking: The unbiased examination of assumptions underlying current beliefs to assess their correctness and legitimacy, and thus to validate or invalidate the beliefs.

Crowdsourcing: Process of obtaining needed services, ideas, or content by soliciting contributions from a large group of people, especially an online community.

Cyborganization: Cyborganization refers to a hybrid of a human and computer shaped form of organization that shaped by technology-mediated coordination, social computing and organizational algorithms.

Decision-Making Model: Set of the human-operator's thinking process techniques in order to transform the input of the given particular decision task into the right course of action.

Decision-Making Process: The process of choice the course of action with influence of the professional as well as individual-psychological, psycho-physiological and socio-psychological features of the human-operator, which will contribute optimal / adequate / acceptable/ justifiable results in optimal / acceptable / limited time to full-time / part complete / incomplete / inconsistent information.

Holacracy: Holacracy refers to an organizational form which grounds on a peer-to-peer workplace that of flexible structure, self-organized holons, unique decision making process, more autonomy-based etc.

Human Computation: A computer science technique in which a machine performs its function by outsourcing certain steps to humans, usually as microwork. This approach uses differences in abilities and alternative costs between humans and computer agents to achieve symbiotic human-computer interaction.

Human-Operator: An element of a complex socio-technical system, which performs the main system's parameter's control and monitoring functions with the help of the machines tools and technology, own experience and skills as well as the communication with the other human, and is able to make decisions under certainty, uncertainty and risk.

Innovation: The application of better solutions that meet new requirements, unarticulated needs, or existing market needs. This is accomplished through more-effective products, processes, services, technologies, or business models that are readily available to markets, governments and society.

Learning: The measurable change in behavior through experience, instruction, and study.

Morphing: Morphing means continuous ongoing transformation of an organization that of structural change, strategical change etc.

Process: The sequence of linked procedures which consume one or more resources to convert inputs into outputs.

Rational Decision-Making Model: The human-operator's logical linear multi-step thinking process techniques for stable permanent maintaining the maximal compliance between the achieved as a result of the decision and required level of the problem solution.

Second Machine Age: The Second Machine Age refers to the new age that grounds on software-driven technological developments and human-machine collaboration.

Skill: The ability and capacity acquired through the deliberate, systematic, and sustained effort to execute the complex activities.

Socio-Technical Systems: The modern automated human-machine control systems, which characterized, on the one hand, by the extensive usage of computer technology, and on the other hand, high dependence on the human operators behaviour – decision-making models, algorithms and techniques.

Thinking: The activity of utilizing the individual's mind to consider something.

Third Industrial Age: The Third Industrial Age refers to "the industrialization of information age" that grounds on digitalization and renewable, distributable and smart energy for every continent of the world.

Transient Competitive Advantage: Transient Competitive Advantage is a strategy that grounds on short-lived opportunities and focuses to gain short-term strategic advantage of competition among rivals.

APPENDIX

Table 1. The level of proficiency in literacy and numeracy skills

Level	Literacy	Numeracy
3	Texts at this level are often dense or lengthy, and include continuous, non-continuous, mixed, or multiple pages of text. Understanding text and rhetorical structures become more central to successfully completing tasks, especially navigating complex digital texts. Tasks require the respondent to identify, interpret, or evaluate one or more pieces of information, and often require varying levels of inference. Many tasks require the respondent to construct meaning across larger chunks of text or perform multi-step operations in order to identify and formulate responses. Often tasks also demand that the respondent disregard irrelevant or inappropriate content to answer accurately. Competing information is often present, but it is not more prominent than the correct information.	Tasks at this level require the respondent to understand mathematical information that may be less explicit, embedded in contexts that are not always familiar and represented in more complex ways. Tasks require several steps and may involve the choice of problem-solving strategies and relevant processes. Tasks tend to require the application of number sense and spatial sense; recognising and working with mathematical relationships, patterns, and proportions expressed in verbal or numerical form; and interpretation and basic analysis of data and statistics in texts, tables and graphs.
4	Tasks at this level often require respondents to perform multiple-step operations to integrate, interpret, or synthesise information from complex or lengthy continuous, non-continuous, mixed, or multiple type texts. Complex inferences and application of background knowledge may be needed to perform the task successfully. Many tasks require identifying and understanding one or more specific, non-central idea(s) in the text in order to interpret or evaluate subtle evidence-claim or persuasive discourse relationships. Conditional information is frequently present in tasks at this level and must be taken into consideration by the respondent. Competing information is present and sometimes seemingly as prominent as correct information.	Tasks at this level require the respondent to understand a broad range of mathematical information that may be complex, abstract or embedded in unfamiliar contexts. These tasks involve undertaking multiple steps and choosing relevant problem solving strategies and processes. Tasks tend to require analysis and more complex reasoning about quantities and data; statistics and chance; spatial relationships; and change, proportions and formulas. Tasks at this level may also require understanding arguments or communicating well-reasoned explanations for answers or choices.
5	At this level, tasks may require the respondent to search for and integrate information across multiple, dense texts; construct syntheses of similar and contrasting ideas or points of view; or evaluate evidence based arguments. Application and evaluation of logical and conceptual models of ideas may be required to accomplish tasks. Evaluating reliability of evidentiary sources and selecting key information is frequently a requirement. Tasks often require respondents to be aware of subtle, rhetorical cues and to make high-level inferences or use specialised background knowledge.	Tasks at this level require the respondent to understand complex representations and abstract and formal mathematical and statistical ideas, possibly embedded in complex texts. Respondents may have to integrate multiple types of mathematical information where considerable translation or interpretation is required; draw inferences; develop or work with mathematical arguments or models; and justify, evaluate and critically reflect upon solutions or choices.

(PIAAC 2013: 64, 76). OECD (2013). OECD Skills Outlook 2013: First Results from the Survey of Adult Skills

Compilation of References

Achrol, R., & Kotler, P. (1999). Marketing in the Network Economy. *Journal of Marketing*, *63*, 146–163.

Agamben, G. (1998). *Homo Sacer: Sovereign Power and Bare Life* (1st ed.). Stanford, CA: Stanford University Press.

Ahuja, R. K., Magnanti, T. L., & Orlin, J. B. (1993). Network Flows: Theory, Algorithms, and Applications. Prentice Hall.

Akyuz, H. I., & Samsa, S. (2009). The effects of blended learning environment on the critical thinking skills of students. *Procedia: Social and Behavioral Sciences*, *1*(1), 1744–1748. doi:10.1016/j.sbspro.2009.01.308

Allard, C. R. (2003). *Antecedents of Effective Decision Making: A Cognitive Approach. Research Memoranda 045*. Maastricht: METEOR, Maastricht Research School of Economics of Technology and Organization.

Almirall, E., Lee, M., & Wareham, J. (2012). Mapping Living Labs in the Landscape of Innovation Methodologies. *Technology Innovation Management Review*, *2*(9).

Alston, K. (2001). Re/thinking critical thinking: The seductions of everyday life. *Studies in Philosophy and Education*, *20*(1), 27–40. doi:10.1023/A:1005247128053

AMA. (2015). *Research Industry Index*. Retrieved from http://www.marketingresearch.org/stay-informed/research-industry-index

Amazon Mechanical Turk. (2016). Available at https://www.mturk.com/mturk

Anderson, C. (2012). *Makers: the new industrial revolution*. New York: Crown Business.

Anderson, C. W. (2013). Towards a sociology of computational and algorithmic journalism. *New Media & Society*, *15*(7), 1005–1021.

Anderson, J., & van Weert, T. (Eds.). (2002). *ICT in Education*. UNESCO, Division of Higher Education.

Andreou, C., Papastavrou, E., & Merkouris, A. (2014). Learning styles and critical thinking relationship in baccalaureate nursing education: A systematic review. *Nurse Education Today*, *34*(3), 362–371. doi:10.1016/j.nedt.2013.06.004 PMID:23830067

Angeli, C., & Valanides, C. (2009). Instructional effects on critical thinking: Performance on ill-defined issues. *Learning and Instruction*, *19*(4), 322–334. doi:10.1016/j.learninstruc.2008.06.010

Aniszewska, G. (2007). *Kultura organizacyjna w zarządzaniu strategicznym. w: Kultura organizacyjna w zarządzaniu. Aniszewska, Grażyna. (red.)*. Warszawa: Polskie Wydawnictwo Ekonomiczne.

Antoncic, B., & Zorn, O. (2004). The mediating role of corporate entrepreneurship in the organizational support performance relationship: An empirical examination. *Managing Global Transitions*, *2*(1), 5–14.

Arend, B. (2009). Encouraging critical thinking in online threaded discussions. *The Journal of Educators Online*, *6*(1), 1–23.

Armstrong, K. (2004). *A short history of myth* (Vol. 1). Canongate Books.

Arnkil, R., Järvensivu, A., Koski, P., & Piirainen, T. (2010). *Exploring Quadruple Helix Outlining User-Oriented Innovation Models*. Työraportteja 85/2010 Working Papers. University of Tampere.

Arum, R., & Roksa, J. (2011). *Academically adrift. Limited learning on college campuses*. Chicago, IL: The University of Chicago Press.

Atay, S., & Karabacak, Ü. (2012). Care plans using concept maps and their effects on the critical thinking dispositions of nursing students. *International Journal of Nursing Practice*, *18*(3), 233–239. doi:10.1111/j.1440-172X.2012.02034.x PMID:22621292

Atwal, G., & Williams, A. (2009). Luxury Brand Marketing – The Experience is Everything! *Journal of Brand Management*, *16*(5), 338–346.

Autorzy Onet, czyli publikuj i zarabiaj na swoich treściach. (n.d.). Retreived December 12, 2015, from antyweb.pl/autorzy-onet-czyli-publikuj-i-zarabiaj-na-swoich-tresciach/

Avaaz. (2015). *Avaaz*. Retrieved November 24, 2015, from http://www.avaaz.org/en/index.php

Aviv, Y. (2001). The effect of collaborative forecasting on supply chain performance. *Management Science*, *47*(10), 1326–1343. doi:10.1287/mnsc.47.10.1326.10260

Bai, J. (2007). Intuitive Decision Theory Analysis and the Evaluation Model. Management Science and Engineering, 1(2).

Baid, H., & Lambert, N. (2010). Enjoyable learning: The role of humour, games, and fun activities in nursing and midwifery education. *Nurse Education Today*, *30*(6), 548–552. doi:10.1016/j.nedt.2009.11.007 PMID:20044181

Bailin, S. (2002). Critical thinking and science education. *Science & Education*, *11*(4), 361–375. doi:10.1023/A:1016042608621

Bajgier, S. M., Maragah, H. D., Saccucci, M. S., Verzilli, A., & Prybutok, V. R. (1991). Introducing Students to Community Operations Research by Using a City Neighborhood as a Living Laboratory. *Operations Research*, *39*(5), 701–709. doi:10.1287/opre.39.5.701

Baker, S., & Holt, S. (2004). Making Marketers Accountable: A Failure of Marketing Education? *Marketing Intelligence & Planning*, *22*(5), 557–567.

Bakker, P. (2011). *New journalism 3.0-aggregation, content farms, and Huffinization: The rise of low-pay and no-pay journalism*. In The Future of Journalism Conference 2011.

Balcerowicz, E., & Wziątek, A. (2009). *Determinanty rozwoju innowacyjności firmy. Ekspertyza przygotowana dla Polskiej Agencja Rozwoju Przedsiębiorczości*. CASE - Centrum Analiz Społeczno-Ekonomicznych. Retrieved from http://www.parp.gov.pl/files/74/75/76/487/493/7614.pdf

Ballantyne, D., & Varey, R. J. (2006). Creating Value-in-Use through Marketing Interaction: The Exchange Logic of Relating, Communicating and Knowing. *Marketing Theory*, *6*(3), 335–348.

Ballon, P., Glidden, J., Kranas, P., Menychtas, A., Ruston, S., & Van der Graaf, S. (2011). Is there a Need for a Cloud Platform for European Smart Cities? In P. Cunningham & M. Cunningham (Eds.), *Proceedings of the eChallenges e-2011 Conference*. Dublin: International Information Management Corporation (IIMC).

Ballon, P., Pierson, J., & Delaere, S. (2005). *Test and experimentation platforms for broadbandinnovation: Examining European practice*. The 16th European Regional Conference by the International Telecommunications Society, Porto, Portugal.

Bampo, M., Ewing, M., Mather, D., Stewart, D., & Wallace, M. (2008). The Effects of the Social Structure of Digital Networks of Digital Networks on Viral Marketing Performance. *Information Systems Research*, *19*(3), 273–290.

Banich, M. T. (2009). Executive function: The search for an integrated account. *Current Directions in Psychological Science*, *18*(2), 89–94. doi:10.1111/j.1467-8721.2009.01615.x

Barak, M., & Dori, Y. J. (2009). Enhancing higher order thinking skills among inservice science teachers via embedded assessment. *Journal of Science Teacher Education*, *20*(5), 459–474. doi:10.1007/s10972-009-9141-z

Bardoel, J., & Deuze, M. (2001). "Network journalism": Converging competencies of old and new media professionals. *Australian Journalism Review*, *23*(2), 91.

Barton, K., & McCully, A. (2007). Teaching controversial issues where controversial issues really matter. *Teaching History*, *127*, 13–19.

Basheer, A.M. & Ibrahim, A. (2010). Mobile Marketing: Examining the Impact of Trust, Privacy Concern and Consumers' Attitudes on Intention to Purchase. *International Journal of Business and Management*, *5*(3), 28–41.

Bauman, Z. (2000). *Liquid Modernity*. Cambridge, UK: Polity Press.

Bauman, Z., & Lyon, D. (2013). *Liquid Surveillance: A Conversation*. Cambridge, UK: Polity.

Baum, L. M., & Newbill, P. L. (2010). Instructional design as critical and creative thinking: A journey through a Jamestown-era Native American village. *TechTrends*, *54*(5), 27–37. doi:10.1007/s11528-010-0434-z

Beach, L. R. (1978). A Contingency Model for the Selection of Decision Strategies. *Academy of Management Review*, (3): 439–449.

Beaumont, J. (2010). A sequence of critical thinking tasks. *TESOL Journal*, *1*(4), 427–448. doi:10.5054/tj.2010.234763

Bednarowska, Z. (2015). *An investment, not an expense. Why marketing research and market analysis are essential for a company's success*. Retrieved from www.research-pmr.com/a44/an-investment-not-an-expense-why-marketing-research-and-market-analysis-are-essential-for-a-companys-success

Begg, D. (1997). *Microeconomics*. Warsaw: PWE. (in Polish)

Bell, R., & Loon, M. (2015). The impact of critical thinking disposition on learning using business simulations. *The International Journal of Management Education*, *13*(2), 119–127. doi:10.1016/j.ijme.2015.01.002

Ben-Chaim, D., Ron, S., & Zoller, U. (2000). The disposition of eleventh-grade science students toward critical thinking. *Journal of Science Education and Technology*, *9*(2), 149–159. doi:10.1023/A:1009474023364

Bengtson, P. (1994). Which Comes First, Internal Involvement or External? *Journal for Quality and Participation*, *17*(5).

Bensley, D. A., & Murtagh, M. P. (2012). Guidelines for a scientific approach to critical thinking assessment. *Teaching of Psychology*, *39*(1), 5–16. doi:10.1177/0098628311430642

Bensley, D. A., & Spero, R. A. (2014). Improving critical thinking skills and metacognitive monitoring through direct infusion. *Thinking Skills and Creativity*, *12*, 55–68. doi:10.1016/j.tsc.2014.02.001

Benyus, J. M. (n.d.). *Biomimicry: Innovation Inspired by Nature*. Retrieved from https://books.google.com.tr/books?id=mDHKVQyJ94gC&printsec=frontcover&dq=biomimicry&hl=tr&sa=X&ei=WOSvVKW7JYX3UKOFg6gH&redir_esc=y#v=snippet&q=3.8&f=false

Bergvall-Kåreborn, B., Ihlström Eriksson, C., Ståhlbröst, A., & Svensson, J. (2009). A Milieu for Innovation: Defining Living Labs. In K. R. E. Huizingh, S. Conn, M. Torkkeli, & I. Bitran (Eds.), *Proceedings of the 2nd ISPIM Innovation Symposium: Simulating Recovery – The Role of Innovation Management*.

Bernstein, D., & Crowley, K. (2008). Searching for signs of intelligent life: An investigation of young children's beliefs about robot intelligence. *Journal of the Learning Sciences*, *17*(2), 225–247. doi:10.1080/10508400801986116

Berthon, P., Leyland, P., & Campbell, C. (2008). Ad Lib: When Consumers Create the Ad. *California Management Review*, *50*(4), 6–30.

Bertman, S. (1998). *Hyperculture: The human cost of speed*. Westport, CT: Praeger Publishers.

Bertsch, V., Treitz, M., Geldermann, J., & Rentz, O. (2007). Sensitivity analyses in multi-attribute decision support for off-site nuclear emergency and recovery management. *International Journal of Energy Sector Management*, *1*(4), 342–365. doi:10.1108/17506220710836075

Bevan, R. (2009). Expanding rationality: The relation between epistemic virtue and critical thinking. *Educational Theory*, *59*(2), 167–179. doi:10.1111/j.1741-5446.2009.00312.x

Bigo, D. (2008). Globalized (In)Security: The field and the Ban-Opticon. In D. Bigo & A. Tsoukala (Eds.), *Terror, Insecurity and Liberty: Illiberal Practices of Liberal Regimes after 9/11* (pp. 10–48). London: Routledge. doi:10.4324/9780203926765.ch2

Bijker, W. E. (1997). *Of bicycles, bakelites, and bulbs: Toward a theory of socio-technical change*. Cambridge, MA: MIT Press.

Billings, D., & Halstead, J. (2012). *Teaching in nursing: A guide for faculty*. St. Louis, MO: Elsevier Saunders.

Biomimicry Enstitute Web Site. (n.d.). Retrieved 15 October 2015 from http://biomimicry.org/what-is-biomimicry/

Birkinshaw, J., & Gupta, K. (2013). Clarifying The Distincitve Contribution of Ambidexterity to the Field of Organization Studies. *The Academy of Management Perspectives*, *27*(4), 287–298. doi:10.5465/amp.2012.0167

Blackmore, S. (2009). Dangerous Memes; or, What the Pandorans let loose. In S. Dick & M. Lupisella (Eds.), *Cosmos and Culture: Cultural Evolution in a Cosmic Context* (pp. 297–318). NASA.

Boctor, L. (2013). Active-learning strategies: The use of a game to reinforce learning in nursing education. A case study. *Nurse Education in Practice*, *13*(2), 96–100. doi:10.1016/j.nepr.2012.07.010 PMID:22910398

Boekaerts, M. (1988). Emotion, motivation and learning. *International Journal of Educational Research*, *12*(3), 229–234. doi:10.1016/0883-0355(88)90001-8

Boekaerts, M. (2010). Motivation and self-regulation: Two close friends. In T. Urdan & S. Karabenick (Eds.), *The decade ahead: Applications and contexts of motivation and achievement* (pp. 69–108). Bingley, United Kingdom: Emerald Group Publishing. doi:10.1108/S0749-7423(2010)000016B006

Boje, D. M., Fedor, D. B., & Rowland, K. M. (1982). Myth making: A qualitative step in OD interventions. *The Journal of Applied Behavioral Science*, *18*(1), 17–28. doi:10.1177/002188638201800104

Bolman, L. G., & Deal, T. E. (1987). *Modern approaches to understanding and managing organizations*. San Francisco: Jossey-Bass.

Borg, M. O., & Stranahan, H. A. (2010). Evidence on the relationship between economics and critical thinking skills. *Contemporary Economic Policy*, *28*(1), 80–93. doi:10.1111/j.1465-7287.2008.00134.x

Bostrom, N. (2005). A history of transhumanist thought. *Journal of Evolution and Technology*.

Bostrom, N. (2014). *Superintelligence: Paths, dangers, strategies*. Oxford, UK: Oxford University Press.

Bowman, S., & Willis, C. (2003). *We media. How audiences are shaping the future of news and information*. Academic Press.

Boyd, D. M., & Ellison, N. B. (2007). Social Network Sites: Definition, History, and Scholarship. *Journal of Computer-Mediated Communication*, *13*(1), 210–230.

Brabham, D. (2008). Crowdsourcing as a model for problem solving. an introduction and cases. *Convergence (London)*, *14*(1), 75–90. doi:10.1177/1354856507084420

Brabham, D. (2013). *Crowdsourcing*. Boston, MA: The MIT Press.

Brachman, R., Cohen, W., & Dietterich, T. (2011). *Synthesis Lectures on Artificial Intelligence and Machine Learning*. Morgan Claypool Publishers.

Brin, S., & Page, L. (1998). The Anatomy of a Large-scale Hypertextual Web Search Engine. In *Proc. of the Seventh World Wide Web Conference*. doi:10.1016/S0169-7552(98)00110-X

Briones, R., Kuch, B., Liu, B. F., & Jin, Y. (2011). Keeping up with the digital age: How the American Red Cross Uses Social Media to Build Relationships. *Public Relations Review*, *37*(1), 37–43.

Broadbear, J. T., & Keyser, B. B. (2000). An approach to teaching for critical thinking in health education. *The Journal of School Health*, *70*(8), 322–326. doi:10.1111/j.1746-1561.2000.tb07266.x PMID:11044963

Broadfoot, P. M. (1996). *Education Assessment and Society*. Buckingham, UK: Open University Pres.

Brodie, R., Ilic, A., Juric, B., & Hollebeek, L. (2013). Consumer Engagement in a Virtual Brand Community: An Exploratory Analysis. *Journal of Business Research*, *66*(1), 105–114.

Brookfield, S. D. (2005). Overcoming impostorship, cultural suicide, and lost innocence: Implications for teaching critical thinking in the community college. *New Directions for Community Colleges*, *2005*(130), 49–57. doi:10.1002/cc.195

Brooks, F. P. (1995). *The Mythical Man-Month: Essays on Software Engineering*. Addison-Wesley Pub. Co.

Bruce, M., & Solomon, M. R. (2013). Managing for Media Anarchy: A Corporate Marketing Perspective. *Journal of Marketing Theory and Practice*, *21*(3), 307–318.

Brynjolfsson, E., & McAfee, A. (2014). *Second machine age: Work, progress, and prosperity in the time of brilliant technologies*. New York: W. W. Norton.

Brynjolfsson, E., & McAfee, A. (2014). *The Second Machine Age: Work, Progress and Prosperity in a Time of Brilliant Technologies*. New York: W.W. Norton & Company.

Buchanan, M. (2015, August 13). *Get Ready for the Robot Explosion*. Retrieved November 13, 2015, from http://www.bloombergview.com/articles/2015-08-13/get-ready-for-the-robot-explosion

Bulik, B. S. (2015). Worldwide PC Leader Lenovo Rebrands For Post-PC World. *Advertising Age*. Available at http://adage.com/article/digital/lenovo-rebrands-logo-renewed-post-pc-push/298773/

Burrell, L. A. (2014). Integrating critical thinking strategies into nursing curricula. *Teaching and Learning in Nursing*, *9*(2), 53–58. doi:10.1016/j.teln.2013.12.005

Butler, H. A. (2012). Halpern critical thinking assessment predicts real-world outcomes of critical thinking. *Applied Cognitive Psychology*, *26*(5), 721–729. doi:10.1002/acp.2851

Butler, H. A., Dwyer, C. P., Hogan, M. J., Franco, A., Rivas, S. F., Saiz, C., & Almeida, L. S. (2012). The Halpern critical thinking assessments and real-world outcomes: Cross-national applications. *Thinking Skills and Creativity*, *7*(2), 112–121. doi:10.1016/j.tsc.2012.04.001

Callison-Burch, C. (2009). *Fast, Cheap, and Creative: Evaluating Translation Quality Using Amazon's Mechanical Turk. In Proceedings of the 2009 Conference on Empirical Methods in Natural Language Processing* (pp. 286–295). Singapore: Association for Computational Linguistics. doi:10.3115/1699510.1699548

Cambria, E., Schuller, B., Xia, Y., & Havasi, C. (2013). New Avenues in Opinion Ming and Sentiment Analysis. *IEEE Intelligent Systems*, *28*(2), 15–21.

Campbell, C., Pitt, L. F., Parent, M., & Berthon, P. R. (2011). Understanding Consumer Conversations around Ads in a Web 2.0 World. *Journal of Advertising*, *40*(1), 87–102.

Carlgren, T. (2013). Communication, critical thinking, problem solving: A suggested course for all high school students in the 21st century. *Interchange*, *44*(1), 63–81. doi:10.1007/s10780-013-9197-8

Carmel-Gilfilen, C., & Portillo, M. (2010). Creating mature thinkers in interior design: Pathways of intellectual development. *Journal of Interior Design*, *35*(3), 1–20. doi:10.1111/j.1939-1668.2010.01039.x

Carter, L., & Rukholm, E. (2008). A study of critical thinking, teacher-student interaction, and discipline-specific writing in an online educational setting for registered nurses. *Journal of Continuing Education in Nursing*, *39*(3), 133–138. doi:10.3928/00220124-20080301-03 PMID:18386701

Caseau, Y. (2011). *Processus et Entreprise 2.0*. Paris: Dunod.

Casner-Lotto, J., & Barrington, L. (2006), *Are they really ready to work?Employers' Perspectives on the Knowledge and Applied Skills of New Entrants to the 21st Century U.S. Workforce*. The Conference Board, Corporate Voices for Working Families, Partnership for the 21st Century Skills and Society for Human Resource Management.

Cellan-Jones, R. (2015, January 29). *Office puts chips under staff's skin*. Retrieved November 20, 2015, from http://www.bbc.com/news/technology-31042477

Chandrasekar, R., Chi, E., Chickering, M., Ipeirotis, P. G., Mason, W., Provost, F., & Tam, J. (2010). Front matter.*Proc. SIGKDD HCOMP*.

Chang, M. J., Chang, Y. J., Kuo, S. H., Yang, Y. H., & Chou, F. H. (2011). Relationships between critical thinking ability and nursing competence in clinical nurses. *Journal of Clinical Nursing*, *20*(21/22), 3224–3232. doi:10.1111/j.1365-2702.2010.03593.x PMID:21414054

Chan, Z. C. (2013). Exploring creativity and critical thinking in traditional and innovative problem-based learning groups. *Journal of Clinical Nursing*, *22*(15/16), 2298–2307. doi:10.1111/jocn.12186 PMID:23452036

Chen, W. (2001). Description of an expert teacher's constructivist-oriented teaching: Engaging students' critical thinking in learning creative dance. *Research Quarterly for Exercise and Sport*, *72*(4), 366–375. doi:10.1080/02701367.2001.10608973 PMID:11770786

Chesbrough, H. (2011). Bringing open innovation to services. *MIT Sloan Manag. Rev.*, *52*(2), 85–90.

Chiou, L., & Tucker, C. E. (2010). *News, Copyright and Online Aggregators*. Discussion paper, NET Institute Working Paper No. 11-18.

Chircu, A. M., & Kauffman, R. J. (1999). Strategies for Internet middlemen in the intermediation/disintermediation/reintermediation cycle. *Electronic Markets*, *9*(1-2), 109–117. doi:10.1080/101967899359337

Choy, S., & Oo, P. (2012). Reflective thinking and teaching practices: A precursor for incorporating critical thinking into the classroom. *International Journal of Instruction*, *5*(1), 167–182.

Christopher, L. (2009). *Intuitive and rational approaches to decision making in education.* (Doctoral Dissertation). University of Hull.

Chun, W. H. K. (2011). *Programmed visions: Software and memory*. Cambridge, MA: MIT Press. doi:10.7551/mitpress/9780262015424.001.0001

Churchill, G. A. (2002). *Badania marketingowe: podstawy metodologiczne*. Warszawa: Wydaw. Naukowe PWN.

Clifford, J. S., Boufal, M. M., & Kurtz, J. E. (2004). Personality traits and critical thinking skills in college students: Empirical tests of a two-factor theory. *Assessment*, *11*(2), 169–176. doi:10.1177/1073191104263250 PMID:15171465

Corney, J., Torres-Sanchez, C., Jagadeesan, A., & Regli, W. (2009). Outsourcing labour to the cloud. *International Journal of Innovation and Sustainable Development, 4*(4), 294–301. doi:10.1504/IJISD.2009.033083

Cottrell, S. (2011). *Critical thinking skills: Developing effective analysis and argument.* New York, NY: Palgrave MacMillan.

Czakon, W. (Ed.). (2011). *Methodology bases of research in management sciences.* Warsaw: Wolter Kluwer. (in Polish)

Czarniawska, B. (2005). Don Quixote and Capitalism in Poland. On the Cultural Context of Organising. In B. Czarniawska-Joerges & P. de Monthoux (Eds.), *Good Novels, Better Management: Reading Organizational Realities in Fiction.* Routledge.

Czarniawska, B. (2011). *Cyberfactories: How News Agencies Produce News.* Edgar Publishing. doi:10.4337/9780857939135

Czarniawska, B. (2013). *Cyberfactories: How news agencies produce news.* Cheltenham, UK: Edward Elgar.

Dahlgaard-Park, S. M. (2013). Seven-S Framework. In E. H. Kessler (Ed.), *Encyclopedia of Management Theory.* Thousand Oaks, CA: Sage Pub.,Inc. doi:10.4135/9781452276090.n217

Daly, W. M. (2001). The development of an alternative method in the assessment of critical thinking as an outcome of nursing education. *Journal of Advanced Nursing, 36*(1), 120–130. doi:10.1046/j.1365-2648.2001.01949.x PMID:11555056

Daniel, M. F. (2001). Philosophical dialogue among peers: A study of manifestations of critical thinking in pre-service teachers. *Advances in Health Sciences Education: Theory and Practice, 6*(1), 49–67. doi:10.1023/A:1009827102022 PMID:11486139

Daniel, M. F., & Auriac, E. (2011). Philosophy, critical thinking and philosophy for children. *Educational Philosophy and Theory, 43*(5), 415–435. doi:10.1111/j.1469-5812.2008.00483.x

Davis, N. (2015, November 10). *5 ways of understanding the Fourth Industrial Revolution.* Retrieved November 12, 2015, from https://agenda.weforum.org/2015/11/5-ways-of-understanding-the-fourth-industrial-revolution/

Davis, D. F., Golicic, S. L., Boerstler, C. N., Choi, S., & Oh, H. (2013). Does marketing research suffer from methods myopia? *Journal of Business Research, 66*(9), 1245–1250. doi:10.1016/j.jbusres.2012.02.020

Dawar, N. (2013). When Marketing is Strategy. *Harvard Business Review, 91*(12), 101–108.

Dawkins, R. (1990). *The Selfish Gene*. New York: Oxford University Press, Inc.

De Bauw, H., & Vandenweghe, V. (2011). Brussels Court of Appeal Upholds Judgment Against Google News and Google Cache. *Eubelius*. Retrieved January 5, 2016 from http://www.eubelius.com/en/spotlight/brussels-court-appeal-upholds-judgment-against-google-news-and-google-cache

De Jong, J. P. J., Parker, S. K., Wennekers, S., & Wu, C. (2011). Corporate entrepreneurship at the individual level: measurement and determinants. In *EIM research reports*. Zoetermeer: EIM.

Deighton, J. (1996). The Future of Interactive Marketing. *Harvard Business Review*, *74*(6), 151–160.

Dell'Era, C., & Landoni, P. (2014). Living Lab: A Methodology between User-Centred Design and Participatory Design. *Creativity and Innovation Management, 23*(2).

Dellarocas, C., Sutanto, J., Calin, M., & Palme, E. (2015). Attention allocation in information-rich environments: The case of news aggregators. *Management Science*, mnsc.2015.2237. doi:10.1287/mnsc.2015.2237

Deloitte. (2015). *Liderzy na dziś – Liderzy na jutro. Jakie kompetencje przywódcze mają członkowie zarządów i rad nadzorczych?* Retrieved from http://www2.deloitte.com/content/dam/Deloitte/pl/Documents/Reports/pl_Rady-Nadzorcze-Liderzy-na-dzis.pdf

den Ouden, E., Valkenburg, R., & Blok, S. (2016). *Exploring the future of living labs. Research report - February 2016*. Eindhoven: TU/e LightHouse.

Dennett, D. (1991). *Consciousness Imagined. In Consciousness Explained* (pp. 431–455). Back Bay Books.

Deuze, M. (1999). Journalism and the Web An Analysis of Skills and Standards in an Online Environment. *International Communication Gazette*, *61*(5), 373–390. doi:10.1177/0016549299061005002

Dewey, J. (2004). *How we think*. Whitefish, MT: Kessinger Publishing.

Dijstelbloem, H., & Meijer, A. (Eds.). (2011). *Migration and the New Technological Borders of Europe*. Houndmills, UK: Palgrave Macmillan. doi:10.1057/9780230299382

Doan, A., Ramakrishnan, R., & Halevy, A. (2011). Crowdsourcing systems on the world-wide web. *Communications of the ACM*, *54*(4), 86–96. doi:10.1145/1924421.1924442

Doctor, K. (2009). News Users 2009. Outsell. Retrieved from http://www.outsellinc.com/store/products/886?refid=pr886

Doddington, C. (2007). Critical thinking as a source of respect for persons: A critique. *Educational Philosophy and Theory*, *39*(4), 449–459. doi:10.1111/j.1469-5812.2007.00350.x

Driscoll, M. P. (2005). *Psychology of learning for instruction*. Boston, MA: Pearson Education.

Drucker, P. (1954). *The Practice of Management*. New York: Harper & Row.

Duncan, R. (1976). The Ambidextrous Organization: Designing Dual Structures for Innovation. In R. H. Killman, L. R. Pondy, & D. Sleven (Eds.), *The Management of Organization, C.1* (pp. 167–188). New York: North Holland.

Duncan, T., & McKeachie, W. (2005). The making of the motivated strategies for learning questionnaire. *Educational Psychologist*, *40*(2), 117–128. doi:10.1207/s15326985ep4002_6

Dutilleul, B., Birrer, F. A. J., & Mensink, W. (2010). Unpacking European Living Labs: Analysing Innovation's Social Dimensions. *Central European Journal of Public Policy*, *4*(1).

Dwyer, C. P., Hogan, M. J., Harney, O. M., & O'Reilly, J. (2014). Using interactive management to facilitate a student-centred conceptualisation of critical thinking: A case study. *Educational Technology Research and Development*, *62*(6), 687–709. doi:10.1007/s11423-014-9360-7

Dwyer, C. P., Hogan, M. J., & Stewart, I. (2012). An evaluation of argument mapping as a method of enhancing critical thinking performance in e-learning environments. *Metacognition and Learning*, *7*(3), 219–244. doi:10.1007/s11409-012-9092-1

Dwyer, C. P., Hogan, M. J., & Stewart, I. (2014). An integrated critical thinking framework for the 21st century. *Thinking Skills and Creativity*, *12*, 43–52. doi:10.1016/j.tsc.2013.12.004

Dwyer, C. P., Hogan, M. J., & Stewart, I. (2015). The effects of argument mapping-infused critical thinking instruction on reflective judgement performance. *Thinking Skills and Creativity*, *16*, 11–26. doi:10.1016/j.tsc.2014.12.002

Edvardsson, B., Kristensson, P., Magnusson, P., & Sundström, E. (2012). Customer Integration within Service Development – A Review of Methods and an Analysis of Insitu and Exsitu Contributions. *Technovation*, *32*(7-8), 419–429. doi:10.1016/j.technovation.2011.04.006

Eguchi, A. (2016). Educational robotics as a learning tool for promoting rich environments for active learning (REALs). In *Human-computer interaction: Concepts, methodologies, tools, and applications* (pp. 740–767). Hershey, PA: IGI Global. doi:10.4018/978-1-4666-8789-9.ch033

Elder, L. (2005). Critical thinking as the key to the learning college: A professional development model. *New Directions for Community Colleges*, *2005*(130), 39–48. doi:10.1002/cc.194

Elizabeth, A. S. (2001). The Role of Tacit and Explicit Knowledge in the Workplace. *Journal of Knowledge Management*, *5*(4).

Ellermann, C. R., Kataoka-Yahior, M. R., & Wong, L. (2006). Logic models used to enhance critical thinking. *The Journal of Nursing Education*, *45*(6), 220–227. PMID:16780010

Ennis, R. H. (1996). Critical thinking dispositions: Their nature and assessability. *Informal Logic*, *18*(2/3), 165–182.

Ericson, J. (2013). *Information overload: how the Internet inhibits short-term memory*. Retrieved 21 January, 2016 from http://www.medicaldaily.com/information-overload-how-internet-inhibits-short-term-memory-257580

Ericson, R. V., & Haggerty, K. D. (2006). The New Politics of Surveillance and Visibility. In R. V. Ericson & K. D. Haggerty (Eds.), *The New Politics of Surveillance and Visibility*. Toronto: University of Toronto Press.

Eriksson, M., Niitamo, V.-P., Kulkki, S., & Hribernik, K. A. (2006). *Living Labs as aMulti-Contextual R&D Methodology.* The 12th International Conference on Concurrent Enterprising: Innovative Products and Services through Collaborative Networks, ICE 2006, Milan, Italy.

Ernst, J., & Monroe, M. (2006). The effects of environment-based education on students' critical thinking skills and disposition toward critical thinking. *Environmental Education Research*, *12*(3/4), 429–443. doi:10.1080/13504620600942998

ESOMAR. (2012). *Global Market Research 2012. An ESOMAR Industry Report in cooperation with KPMG Advisory*. Amsterdam: ESOMAR.

ESOMAR. (2015). *Global Market Research 2015. An ESOMAR Industry Report in cooperation with BDO Accountants & Advisors*. Amsterdam: ESOMAR.

Europe 2020. (n.d.). *European Strategy for smart, sustainable and inclusive growth*. European Commission.

Europe versus Facebook. (2015, November 24). Retrieved November 24, 2015, from http://www.europe-v-facebook.org/

European Commission Information Society and Media. (2009). *Unit F4 New Infrastructure Paradigms and Experimental Facilities. Living Labs for user-driven open innovation. An overview of the Living Labs methodology, activities and achievements*. Author.

European Communities. (2006). Evaluation Methods for the European Union's External Assistance. Guidelines for Project and Programme Evaluation. *Reproduction (Cambridge, England)*, *3*.

Evans, D. S. (2008). The Economics of the Online Advertising Industry. *Review of Network Economics*, *7*(3), 359–391.

Evans, D. S. (2009). The Online Advertising Industry: Economics, Evolution, and Privacy. *The Journal of Economic Perspectives*, *23*(3), 37–60.

Facione, P. A., & Facione, N. C. (2001). Analyzing explanations for seemingly irrational choices: Linking argument analysis and cognitive science. *The International Journal of Applied Philosophy*, *15*(2), 267–286. doi:10.5840/ijap200115217

Facione, P. A., Facione, N. C., & Giancarlo, C. A. F. (1998). *The California critical thinking disposition inventory*. Millbrae, CA: California Academic Press.

Facione, P. A., Sanchez, C. A., Facione, N. C., & Gainen, J. (1995). The dispositions towards critical thinking. *The Journal of General Education*, *44*(1), 1–25.

Fahim, M., Bagherkazemi, M., & Alemi, M. (2010). The relationship between test takers critical thinking ability and their performance on the reading section of TOEFL. *Journal of Language Teaching and Research*, *1*(6), 830–837. doi:10.4304/jltr.1.6.830-837

Fan, S., Lau, R. Y. K., & Leon Zhao, J. (2015). Demystifying Big Data Analytics for Business Intelligence through the Lens of Marketing Mix. *Big Data Research*, *2*(1), 28–32.

Fawkes, D., O'meara, B., Weber, D., & Flage, D. (2005). Examining the exam: A critical look at the California critical thinking skills test. *Science & Education*, *14*(2), 117–135. doi:10.1007/s11191-005-6181-4

Feigenbaum, L., Herman, I., Hongsermeier, T., Neuman, E., & Stephens, S. (2007). The semantic web in action. *Scientific American*, *297*(6), 90–97. doi:10.1038/scientificamerican1207-90 PMID:18237102

Ferrari, V., Mion, L., & Molinari, R. F. (2011). Innovating ICT Innovation: Trentino as a Lab. In *Proceedings of ICEGOV2011*. doi:10.1145/2072069.2072130

Finch, D., Nadeau, J., & O'Reilly, N. (2013). The Future of Marketing Education: A Practitioner's Perspective. *Journal of Marketing Education*, *35*(1), 54–67.

Finsen, S., & Finsen, L. (2009). Animal rights movement. In M. Bekoff (Ed.), *The Encyclopedia of Animal Rights and Animal Welfare*. Greenwood.

Flores, K. L., Matkin, G. S., Burbach, M. E., Quinn, C. E., & Harding, H. (2012). Deficient critical thinking skills among college graduates: Implications for leadership. *Educational Philosophy and Theory*, *44*(2), 212–230. doi:10.1111/j.1469-5812.2010.00672.x

Flueler, T. (2006). *Decision Making for Complex Socio-Technical Systems: Robustness from Lessons Learned in Long-Term Radioactive Waste Governance (Environment & Policy)*. Springer. doi:10.1007/1-4020-3529-2

Følstad, A. (2008). Living labs for innovation and development of information and communication technology: a literature review. *Electronic Journal of Virtual Organizations, 10*. Retrieved from http://graphene-flagship.eu/material/graphene/Pages/What-is-graphene.aspx

Ford, M. (2009). *The lights in the tunnel: Automation, accelerating technology and the economy of the future*. Acculant Publishing.

Forneris, S. G., & Peden-McAlpine, C. (2007). Evaluation of a reflective learning intervention to improve critical thinking in novice nurses. *Journal of Advanced Nursing*, *57*(4), 410–421. doi:10.1111/j.1365-2648.2007.04120.x PMID:17291205

Forneris, S. G., & Peden-McAlpine, C. (2009). Creating context for critical thinking in practice: The role of the preceptor. *Journal of Advanced Nursing*, *65*(8), 1715–1724. doi:10.1111/j.1365-2648.2009.05031.x PMID:19602012

Forrest, E., & Hoanca, B. (2015). Artificial Intelligence: Marketing's game changer. In T. Tsiakis (Ed.), Trends and Innovations in Marketing Information Systems, (pp. 45-64). Hersey, PA: IGI Global.

Foster, T. (2015). Industrial Marketing Communication: A (R)evolutionary Journey from Marketplace to Marketspace. In H. Spotts (Ed.), Revolution in Marketing: Market Driving Changes, (pp. 111-118). Springer.

Foucault, M. (1995). *Discipline & Punish: The Birth of the Prison*. New York: Vintage Books.

Frambach, R., Roest, H., & Krishnan, T. (2007). The Impact of Conusmer Internet Experience on Channel Preference and Usage Intentions Across the Different Stages of the Buying Process. *Journal of Interactive Marketing*, *21*(2), 26–41.

Frey, C., & Osborne, M. (2013, September 17). *The Future of Employment: How Susceptible are Jobs to Computerisation?* Retrieved from http://www.oxfordmartin.ox.ac.uk/downloads/academic/The_Future_of_Employment.pdf

Fung, D. (2014). Promoting critical thinking through effective group work: A teaching intervention for Hong Kong primary school students. *International Journal of Educational Research*, *66*, 45–62. doi:10.1016/j.ijer.2014.02.002

Fura, M. (2010). *Chcemy rozmawiać z wydawcami, jak rozliczać się za treści w sieci*. Gazeta Prawna 2010. Retrieved January 3, 2016 from http://www.gazetaprawna.pl/blogi/michal_fura/archiwa/P3.html

Gadamer, H.-G. (2004). Truth and Method (2nd ed.). London: Bloomsbury Academic.

Gagne, R. M. (1985). *The conditions of learning*. New York, NY: Holt, Rinehart and Winston.

Gambrill, E. (2006). Evidence-based practice and policy: Choices ahead. *Research on Social Work Practice*, *16*(3), 338–357. doi:10.1177/1049731505284205

Gandy, O. H. (1993). *The Panoptic Sort: A Political Economy of Personal Information*. Boulder, CO: Westview Press.

Gandy, O. H. Jr. (2010). Engaging rational discrimination: Exploring reasons for placing regulatory constraints on decision support systems. *Ethics and Information Technology*, *12*(1), 29–42. doi:10.1007/s10676-009-9198-6

Garrigos, F., Gil, I., & Narangajavana, Y. (2011). The impact of social networks in the competitiveness of the firms. In A. M. Beckford & J. P. Larsen (Eds.), *Competitiveness: Psychology, Production, Impact and Global Trends*. Hauppauge, NY: Nova Science Publishers.

Geiger, D., Seedorf, T., Schulze, C., & Nickerson, M. (2011). Managing the crowd: towards a taxonomy of crowdsourcing processes. *Proceedings of the Seventeenth Americas Conference on Information Systems*.

Gera, D. (2003). *Ancient Greek ideas on speech, language, and civilization*. Oxford, UK: Oxford University Press. doi:10.1093/acprof:oso/9780199256167.001.0001

Gerber, S., & Scott, L. (2011). Gamers and gaming context: Relationships to critical thinking. *British Journal of Educational Technology*, *42*(5), 842–849. doi:10.1111/j.1467-8535.2010.01106.x

Ghabanchi, Z., & Behrooznia, S. (2014). The impact of brainstorming on reading comprehension and critical thinking ability of EFL learners. *Procedia: Social and Behavioral Sciences*, *98*, 513–521. doi:10.1016/j.sbspro.2014.03.447

Giamanco, B., & Gregoire, K. (2012). Tweet Me, Friend Me, Make Me Buy. *Harvard Business Review*, *90*(7), 88–93. PMID:22299507

Gibbs, S., Steel, G., & Kuiper, A. (2011). Expectations of Competency: The Mismatch Between Employers' and Graduates' Views of End-User Computing Skills Requirements in the Workplace. *Journal of Information Technology Education*, *10*, 371–382.

Giglietto, F., & Selva, D. (2014). Second Screen and Participation: A Content Analysis on a Full Season Dataset of Tweets. *Journal of Communication*, *64*(2), 260–277.

Gillespie, T. (2014). 9 The Relevance of Algorithms. *Media technologies: Essays on communication, materiality, and society*, 167.

Glaser. (2010). Google News to Publishers: Let's Make Love Not War. *PBS*. Retrieved 2015 November, 2015 from http://www.pbs.org/mediashift/2010/02/google-news-to-publishers-lets-make-love-not-war035 Historia Onet.pl

Goldsmith, J. L., & Wu, T. (2006). *Who controls the Internet?: illusions of a borderless world* (Vol. 89). New York: Oxford University Press.

Gomez, K., Bernstein, D., Zywica, J., Hamner, E., Lee, U., & Cunningham, J. (2016). Robotics community experiences: Leveraging informal design and learning experiences to motivate urban youth in STEM. In Human-computer interaction: Concepts, methodologies, tools, and applications (pp. 120–147). Hershey, PA: IGI Global. doi:10.4018/978-1-4666-8789-9.ch006

Graham, S. (2004). The software-sorted city: Rethinking the 'digital divide'. *The Cybercities Reader*, 324-332.

Grandgenett, N., Ostler, E., Topp, N., & Goeman, R. (2014). Robotics and problem-based learning in STEM formal educational environments. In *K-12 education: Concepts, methodologies, tools, and applications* (pp. 463–488). Hershey, PA: IGI Global. doi:10.4018/978-1-4666-4502-8.ch028

Gredler, M. E. (2009). *Learning and instruction: Theory into practice*. Upper Saddle River, NJ: Pearson/Merrill Prentice Hall.

Grover, R., & Vriens, M. (2006). The Handbook of Marketing Research: Uses, Misuses, and Future Advances. Thousand Oaks, CA: Sage Publications, Inc.

Grunwald, A. (2014). The hermeneutic side of responsible research and innovation. *Journal of Responsible Innovation*, *1*(3), 274–291. doi:10.1080/23299460.2014.968437

Grzegorczyk, W. (2012). *Marketing Activities of Polish Banks*. Advanced Research in Scientific Areas.

Halpern, D. F. (1998). Teaching critical thinking for transfer across domains: Dispositions, skills, structure training, and metacognitive monitoring. *The American Psychologist*, *53*(4), 449–455. doi:10.1037/0003-066X.53.4.449 PMID:9572008

Halpern, D. F. (2006). Is intelligence critical thinking? Why we need a new definition of intelligence. In P. Kyllonen, R. Roberts, & L. Stankov (Eds.), *Extending intelligence: Enhancement and new constructs* (pp. 293–310). New York, NY: Taylor & Francis Group.

Halpern, D. F. (2014). *Thought and knowledge: An introduction to critical thinking*. New York, NY: Psychology Press.

Hamm, D. (2015). *Investing in soft skills pays off for workers, employers: create a work environment that values communication, decision making*. Retrieved 1 October 2015 from http://www.insightonmfg.com/ArticleText.aspx?articleId=3854&magzId=2

Hammond, K. R. (1996). *Human Judgment and Social Policy: Irreducible Uncertainty, Inevitable Error, Unavoidable Injustice*. New York: Oxford University Press.

Hannah, R., Rohm, A., & Crittenden, V. L. (2011). We're all connected: The Power of the Social Media Ecosystem. *Business Horizons*, *54*(3), 265–273.

Hansen, M. T. (2010). *T-Shaped Stars: The Backbone of IDEO's Collaborative Culture: An Interview with IDEO CEO Tim Brown*. Retrieved 5 October from http://chiefexecutive.net/ideo-ceo-tim-brown-t-shaped-stars-the-backbone-of-ideoae%E2%84%A2s-collaborative-culture/

Harvey, D. M., Bosco, S. M., & Emanuele, G. (2010). *The impact of "green-collar workers" on Integrative Decision Making*. Retrieved 5 October 2015 from http://www.businessdictionary.com/ definition/integrative-thinking.html#ixzz3si4xw9uY

Hastie, R., & Dawes, R. M. (2001). *Rational Choice in an Uncertain World: The psychology of judgment and decision making*. Thousand Oaks, CA: Sage Publications.

Hatch, M. J. (2002). *Teoria organizacji, przekł. Łuków, P.* Warszawa: Wydawnictwo Naukowe PWN.

Hatch, M. J., & Cunliffe, A. L. (1997). *Organization Theory: Modern. Symbolic-Interpretive and Postmodern Perspectives*. Oxford, UK: Oxford University Press.

Hayles, N. K. (2006). Unfinished work from cyborg to cognisphere. *Theory, Culture & Society*, *23*(7-8), 159–166. doi:10.1177/0263276406069229

Heidegger, M. (2008). *Being and Time (Reprint edition)*. New York: Harper Perennial Modern Classics.

Heijltjes, A., van Gog, T., Leppink, J., & Paas, F. (2015). Unraveling the effects of critical thinking instructions, practice, and self-explanation on students' reasoning performance. *Instructional Science*, *43*(4), 487–506. doi:10.1007/s11251-015-9347-8

Heijltjes, A., van Gog, T., & Paas, F. (2014). Improving students' critical thinking: Empirical support for explicit instructions combined with practice. *Applied Cognitive Psychology*, *28*(4), 518–530. doi:10.1002/acp.3025

Hendler, J. (2001). Agents and the Semantic Web. *IEEE Intelligent Systems*, *16*(2), 30–37. doi:10.1109/5254.920597

Hendler, J. (2009). Web 3.0 Emerging. *Computer*, *42*(1), 111–113. doi:10.1109/MC.2009.30

Heylighen, F. (2013). Return to Eden? Promises and Perils on the Road to a Global Superintelligence. In B. Goertzel & T. Goertzel (Eds.), *The End of the Beginning: Life*. Society and Economy on the Brink of the Singularity.

Hindman, M. (2008). *The myth of digital democracy*. Princeton University Press.

Hirsch, F. (2005). *Social limits to growth*. Routledge.

History. (n.d.). *ITI holdings*. Retrieved December 27, 2016 from http://www.itiholdings.com/index.php/History/History/

Hofer, B. K. (2004). Epistemological understanding as a metacognitive process: Thinking aloud during online searching. *Educational Psychologist*, *39*(1), 43–55. doi:10.1207/s15326985ep3901_5

Hoffman, D., & Fodor, M. (2010). Can You Measure the ROI of your Social Media Marketing? *MIT Sloan Management Review*, *52*(1), 41–49.

Hofmann, W., Schmeichel, B. J., & Baddeley, A. D. (2012). Executive functions and self-regulation. *Trends in Cognitive Sciences*, *16*(3), 174–180. doi:10.1016/j.tics.2012.01.006 PMID:22336729

Hofstede, G. (2000). *Kultury i organizacje: zaprogramowanie umysłu. Tłum. Durska, Małgorzata*. Warszawa: Polskie Wydawnictwo Ekonomiczne.

Ho, J. Y. C., & Dempsey, M. (2010). Viral Marketing: Motivations to Forward Online Content. *Journal of Business Research*, *63*(9-10), 1000–1006.

Hollebeek, L., Glynn, M., & Brodie, R. (2014). Consumer Brand Engagement in Social Media: Conceptualization, Scale Development and Validation. *Journal of Interactive Marketing*, *28*(2), 149–165.

Hong, S. (2012). Online news on Twitter: Newspapers' social media adoption and their online readership. *Information Economics and Policy*, *24*(1), 69–74. doi:10.1016/j.infoecopol.2012.01.004

Howard, J. (2014, December). *The wonderful and terrifying implications of computers that can learn.* Lecture presented at TED.

Howe, J. (2006). The rise of crowdsourcing. *Wired*, *14*(6), 176–180.

Howe, J. (2009). *Crowdsourcing: Why the power of the crowd is driving the future of business.* New York: Crown Business.

Huang, E., Zhang, H., Parkes, D. C., Gajos, K. Z., & Chen, Y. (2010). Toward automatic task design: A progress report. In *Proceedings of the ACM SIGKDD Workshop on Human Computation.* New York: ACM. doi:10.1145/1837885.1837908

Hudson, A. (2012). *The age of information overload.* Retrieved 28 December, 2015 from http://news.bbc.co.uk/2/hi/programmes/click_online/9742180.stm

Huffaker, D. A., & Calvert, S. L. (2003). The new science of learning: Active learning, metacognition and transfer of knowledge in e-learning applications. *Journal of Educational Computing Research*, *29*(3), 325–334. doi:10.2190/4T89-30W2-DHTM-RTQ2

Human Factors Guidelines for Safety Audits Manual . (2002). First Edition. Doc. 9806-AN/763. Canada, Montreal: International Civil Aviation Organization.

Humphries, L. (2000). *Extras-itlist and other bookmark managers.* Retrieved from http://www.llrx.com/extras/itlist.htm

Hwang, S. Y., Yen, M., Lee, B. O., Huang, M. C., & Tseng, H. F. (2010). A critical thinking disposition scale for nurses: Short form. *Journal of Clinical Nursing*, *19*(21/22), 3171–3176. doi:10.1111/j.1365-2702.2010.03343.x PMID:21040020

Ikuenobe, P. (2001). Questioning as an epistemic process of critical thinking. *Educational Philosophy and Theory*, *33*(3/4), 325–341. doi:10.1111/j.1469-5812.2001.tb00274.x

Ikuenobe, P. (2002). Epistemic foundation for teaching critical thinking in group discussion. *Interchange*, *33*(4), 371–393. doi:10.1023/A:1021564120979

IMF. (2016). *Too Slow for Too Long Cataloging-in-Publication Data*. Retrieved from http://www.imf.org/external/pubs/ft/weo/2016/01/pdf/text.pdf

Innovation Portal of the Polish Agency for Enterprise Development. (n.d.). Retrieved from http://www.pi.gov.pl/parp

Intille, S. S., Larson, K., Munguia Tapia, E., Beaudin, J., Kaushik, P., Nawyn, J., & Rockinson, R. R. (2006). Using a Live-In Laboratory for Ubiquitous Computing Research. In K. P. Fishkin, B. Schiele, P. Nixon, & A. Quigley (Eds.), *Proceedings of PERVASIVE 2006*, (LNCS), (vol. 3968, pp. 349-365). Berlin: Springer-Verlag.

Isbell, K. (2010). *The rise of the news aggregator: Legal implications and best practices*. Berkman Center Research Publication.

Iwaoka, W. T., Li, Y., & Rhee, W. Y. (2010). Research in food science education: Measuring gains in critical thinking in food science and human nutrition courses: The Cornell critical thinking test, problem-based learning activities, and student journal entries. *Journal of Food Science Education*, *9*(3), 68–75. doi:10.1111/j.1541-4329.2010.00100.x

Jenkins, D. (2012). Global critical leadership: Educating global leaders with critical leadership competencies. *The Journal of Leadership Studies*, *6*(2), 95–101. doi:10.1002/jls.21241

Jenkins, H., Ford, S., & Green, J. (2013). *Spreadable media: Creating value and meaning in a networked culture*. NYU Press.

Jeon, D. S., & Esfahani, N. N. (2012). *News Aggregators and Competition among Newspapers in the Internet*. Academic Press.

Johann, M. (2013). *Strategie marketingowe stosowane w praktyce polskich przedsiębiorstw*. Katedra Zarządzania Strategicznego Kolegium Nauk o Przedsiębiorstwie. Retrieved from http://www.sgh.waw.pl/katedry/kzs/Konferencja/10.%20Maria%20Johann.pdf

Jukes, I., & McCain, T. (2002). *Minds in play: Computer game design as a context of children's learning*. Mahwah, NJ: Lawrence Erlbaum Associates.

Juujärvi, S., & Pesso, K. (2013). Actor Roles in an Urban Living Lab: What Can We Learn from Suurpelto, Finland? *Technology Innovation Management Review*, *3*(12).

Kaddoura, M. A. (2010). New graduate nurses' perceptions of the effects of clinical simulation on their critical thinking, learning, and confidence. *Journal of Continuing Education in Nursing*, *41*(11), 506–516. doi:10.3928/00220124-20100701-02 PMID:20672760

Kahneman, D. (1994). New Challenges to the Rationality Assumption. *Journal of Institutional and Theoretical Economics*, (150), 18–36.

Kalpic, B., & Bernus, P. (2006). Business process modeling through the knowledge management perspective. *Journal of Knowledge Management*, *10*(3), 40–56. doi:10.1108/13673270610670849

Kaminski, J. (2014). *Cyborganization*. Retrieved 15 October 2015 from http:// jermainkaminski.com/ cyborganization/

Kaniewska-Sęba, A., Leszczyński, G., & Pilarczyk, B. (2006). Badania marketingowe na rynku business-to-business. Kraków: Oficyna Ekonomiczna. Oddział Polskich Wydawnictw Profesjonalnych.

Kapera, K., & Kuziak, M. (2007). *Pozyskiwanie i wykorzystywanie informacji rynkowych. Raport z badania ankietowego przedsiębiorstw.* Katedra Marketingu UEK Kraków. Retrieved from http://ankieta.uek.krakow.pl/report/Raport_Badanie_AE_Krakow_2006.pdf

Kaplan, D. (2010). *Mark Cuban: Google, Content Aggregators Are Vampires; Newspapers Are Zombies*. PaidContent.org. Retrieved 12 December, 2015, from https:// gigaom.com/2010/02/02/419-onmedia-mark-cuban-google-content-aggregators-are-vampires-newspapers/

Kasemsap, K. (2016b). Mastering big data in the digital age. In M. Singh & D. G. (Eds.), Effective big data management and opportunities for implementation (pp. 104–129). Hershey, PA: IGI Global. doi:10.4018/978-1-5225-0182-4.ch008

Kasemsap, K. (2016a). Analyzing the roles of human capital and competency in global business. In S. Sen, A. Bhattacharya, & R. Sen (Eds.), *International perspectives on socio-economic development in the era of globalization* (pp. 1–29). Hershey, PA: IGI Global. doi:10.4018/978-1-4666-9908-3.ch001

Kasemsap, K. (2016c). The fundamentals of neuroeconomics. In B. Christiansen & E. Lechman (Eds.), *Neuroeconomics and the decision-making process* (pp. 1–32). Hershey, PA: IGI Global. doi:10.4018/978-1-4666-9989-2.ch001

Kasemsap, K. (2016d). Promoting leadership development and talent management in modern organizations. In U. Aung & P. Ordoñez de Pablos (Eds.), *Managerial strategies and practice in the Asian business sector* (pp. 238–266). Hershey, PA: IGI Global. doi:10.4018/978-1-4666-9758-4.ch013

Katz, D., & Kahn, R. L. (1966). *The social psychology of organizations*. New York: John Wiley & Sons.

Keating, C. B., Fernandez, A. A., Jacobs, D. A., & Kauffmann, P. (2001). A methodology for analysis of complex sociotechnical processes. *Business Process Management Journal, 7*(1), 33–50. doi:10.1108/14637150110383926

Kennaly, M. (2012). Pinterest now the THIRD most popular social media network beating LinkedIn and Tumblr. *Daily Mail.* Retrieved 23 December, 2015 from http://www.dailymail.co.uk/news/article-2127078/Pinterest-popular-social-media-network-beating-LinkedIn-Tumblr.html?ito=feeds-newsxml

Keynes, J. M. (1931). Economic possibilities for our grandchildren (1930). In Essays in persuasion, (pp. 358–73). Academic Press.

Kharchenko, V. P. (2011). Methodology for Analysis of Decision Making in Air Navigation System. *Proceedings of the National Aviation University.* doi:10.18372/2306-1472.48.51

Kharchenko, V. P. (2012). Modeling of behavioral activity of air navigation system's human-operator in flight emergencies. *Proceedings of the National Aviation University.* doi:10.18372/2306-1472.51.2150

Kharchenko, V. P., Shmelova, Y. V., & Sikirda, N. N. (2011). The Human Operator's Decision-Making graphic analytical models. *Proceedings of the National Aviation University.*

Kharchenko, V. P. (2012) A methodology for analysis of flight situation development using GERT's and Markov's networks. *World Congress "Aviation in the XXIst century. Safety in Aviation And Space Technologies.*

Kilgour, M., Sasser, S. L., & Larke, R. (2015). The Social Media Transformation Process: Curating Content into Strategy. *Corporate Communications: An International Journal, 20*(3), 326–343.

Kim, K., Sharma, P., Land, S. M., & Furlong, K. P. (2013). Effects of active learning on enhancing student critical thinking in an undergraduate general science course. *Innovative Higher Education, 38*(3), 223–235. doi:10.1007/s10755-012-9236-x

Kinaret, J., Ferrari, A., Fal'ko, V., & Kivioja, J. (2011). Graphene-Driven Revolutions in ICT and Beyond. *Procedia Computer Science, 7*, 30–33. doi:10.1016/j.procs.2011.12.013

King, P. M., & Kitchener, K. S. (1994). *Developing reflective judgment: Understanding and promoting intellectual growth and critical thinking in adolescents and adults.* San Francisco, CA: Jossey–Bass.

King, P. M., Wood, P. K., & Mines, R. A. (1990). Critical thinking among college and graduate students. *The Review of Higher Education, 13*(2), 167–186.

Kırbaşlar, M., & Özsoy-Güneş, Z. (2015). The effect of critical thinking disposition on entrepreneurship levels: A study on future teachers. *Procedia: Social and Behavioral Sciences, 174*, 199–207. doi:10.1016/j.sbspro.2015.01.647

Kirsch, I., & Thorn, W. (2013). *Technical Report of the Survey of Adult Skills*. PIAAC.

Kleemann, F., Voß, G., & Rieder, K. (2008). Un(der)paid innovators: The commercial utilization of consumer work through crowdsourcing. *Science, Technology & Innovation Studies, 4*(1), 5–26.

Knight, T. (1749). *Reflections upon catholicons, or universal medicines*. Printed for T. Osborne in Gray's-Inn.

Kodama, M. (2005). New knowledge creation through leadership-based strategic community –a case of new product development in IT and multimedia business fields. *Technovation, 25*(8), 895–908. doi:10.1016/j.technovation.2004.02.016

Kormyshev, V.M. (2011). *Definition of Competence in Information Systems*. Academic Press.

Kostera, M., Aggestam, M., & Eriksson-Zetterquist, U. (2008). *Introduction to the Trilogy: Mythologies of Organizational Everyday Life. In Organizational Olympians: Heroes, heroines and villains of organizational myths*. London: Palgrave Macmillan.

Kotler, P., & Caslione, J. A. (2009). *Chaotics: the business of managing and marketing in the age of turbulence*. AMACOM Div American Mgmt Assn.

Kovalev, Yu. N. (2012). Modeling of personality and activities of air navigation system operator.*World Congress "Aviation in the XXIst century. Safety in Aviation and Space Technologies*.

Kovalik, D. L., & Kovalik, L. M. (2007). Language simulations: The blending space for writing and critical thinking. *Simulation & Gaming, 38*(3), 310–322. doi:10.1177/1046878106298271

Kreft, J. (2016). From Readers to Users and Creators. Migration of Printed Media Readers to the Internet: Research Results of 2007-2015. *Social Sciences, 90*(4), 32–42. doi:10.5755/j01.ss.90.4.14250

Krugman, P. (2013, June 13). *Sympathy for the Luddites*. Retrieved November 13, 2015, from http://www.nytimes.com/2013/06/14/opinion/krugman-sympathy-for-the-luddites.html

Kuiper, R. A., & Pesut, D. J. (2004). Promoting cognitive and metacognitive reflective reasoning skills in nursing practice: Self-regulated learning theory. *Journal of Advanced Nursing*, *45*(4), 381–391. doi:10.1046/j.1365-2648.2003.02921.x PMID:14756832

Ku, K. Y. L. (2009). Assessing students' critical thinking performance: Urging for measurements using multi-response format. *Thinking Skills and Creativity*, *4*(1), 70–76. doi:10.1016/j.tsc.2009.02.001

Ku, K. Y. L., & Ho, I. T. (2010). Metacognitive strategies that enhance critical thinking. *Metacognition and Learning*, *5*(3), 251–267. doi:10.1007/s11409-010-9060-6

Ku, K. Y. L., Ho, I. T., Hau, K. T., & Lai, E. C. M. (2014). Integrating direct and inquiry-based instruction in the teaching of critical thinking: An intervention study. *Instructional Science*, *42*(2), 251–269. doi:10.1007/s11251-013-9279-0

Kumar, V. (2015). Evolution of Marketing as a Discipline: What Has Happened and What to Look Out For. *Journal of Marketing*, *79*(1), 1–9.

Kusiak, A. (2007). Innovation: The Living Laboratory Perspective. *Computer-Aided Design and Applications*, *4*(6), 863–876. doi:10.1080/16864360.2007.10738518

Kwan, Y. W., & Wong, A. F. L. (2014). The constructivist classroom learning environment and its associations with critical thinking ability of secondary school students in Liberal Studies. *Learning Environments Research*, *17*(2), 191–207. doi:10.1007/s10984-014-9158-x

Kwan, Y. W., & Wong, A. F. L. (2015). Effects of the constructivist learning environment on students' critical thinking ability: Cognitive and motivational variables as mediators. *International Journal of Educational Research*, *70*, 68–79. doi:10.1016/j.ijer.2015.02.006

Lamb, C. J., & Lachow, I. (2006). *Reforming Pentagon Strategic Decision-making*. Institute for National Strategic Studies.

Lander, S. (2014). *Conversations with History*. Hay House, Inc.

Langlois, G. (2013). Participatory Culture and the New Governance of Communication The Paradox of Participatory Media. *Television & New Media*, *14*(2), 91–105. doi:10.1177/1527476411433519

Lasher, D. R., Ives, B., & Järvenpää, S. L. (1991). USAA-IBM Partnerships in Information Technology: Managing the Image Project. *Management Information Systems Quarterly*, *15*(4), 551. doi:10.2307/249458

Lash, S. (2007). Power after Hegemony Cultural Studies in Mutation? *Theory, Culture & Society*, *24*(3), 55–78. doi:10.1177/0263276407075956

Law, E., & von Ahn, L. (2009). Input-agreement: a new mechanism for collecting data using human computation games.*Proc. CHI*, (pp. 1197-1206). doi:10.1145/1518701.1518881

Law, E., West, K., Mandel, M., Bay, M., & Downie, J. S. (2009). Evaluation of algorithms using games: the case of music tagging.*Proc. ISMIR*2009, (pp. 387-392).

Leavy, B. (2014). Strategy, organization and leadership in a new transient-advantage world. *Strategy and Leadership*, *42*(4), 3–13. doi:10.1108/SL-05-2014-0038

Lefebvre, B. A. (2001). *Functions of Fast Reflexion in Bipolar Choice. In Reflexing processes and operations* (Vol. 1, pp. 34–46).

Lefrancois, G. R. (2000). *Psychology for teaching*. Belmont, CA: Wadsworth/ Cengage Learning.

Leminen, S. (2013). Coordination and Participation in Living Lab Networks. *Technology Innovation Management Review, 3*(11).

Leminen, S. (2015). What Are Living Labs? *Technology Innovation Management Review*, *5*(9), 29–35.

Leminen, S., Nyström, A.-G., & Westerlund, M. (2015a). A Typology of Creative Consumers in Living Labs. *Journal of Engineering and Technology Management*, *37*, 6–20. doi:10.1016/j.jengtecman.2015.08.008

Leminen, S., & Westerlund, M. (2012). Towards Innovation in Living Labs Network. *International Journal of Product Development*, *17*(1/2), 43. doi:10.1504/ IJPD.2012.051161

Lemon, & Verhoef. (2016). Understanding Customer Experience throughout the Costumer Journey. *Journal of Marketing*.

Lepik, K.-L., Krigul, M., & Terk, E. (2010). Introducing Living Lab's Method as Knowledge Transfer from One Socio-Institutional Context to Another: Evidence from Helsinki-Tallinn Cross-Border Region. *Journal of Universal Computer Science*, *16*(8).

Lévinas, E. (2008). *Bóg, śmierć i czas*. Kraków: Wydawnictwo Znak.

Levy, P. (1997). *Collective intelligence: Mankind's emerging world in cyberspace*. Cambridge, MA: Perseus.

Leychenko, S.D., Malishevskiy, A.V., & Mikhalic N.F. (2006). *Human Factors in Aviation*. Kirovograd: YMEKS.

Licklider, J. C. R. (1960). Man-computer symbiosis.IRE Transactions on Human Factors in Electronics, 1, 4-11.

Lievens, B., Schaffers, H., Turkama, P., Ståhlbröst, A., & Ballon, P. (2011). Cross Border Living Labs Networks to Support SMEs Accessing New Markets. In P. Cunningham & M. Cunningham (Eds.), *Proceedings of the eChallenges e-2011 Conference: 8*. International Information Management Corporation (IIMC).

Lim, L. A. Y. L. (2011). A comparison of students' reflective thinking across different years in a problem-based learning environment. *Instructional Science*, *39*(2), 171–188. doi:10.1007/s11251-009-9123-8

Lippitt, M. (2007). Fix the disconnect between strategy and execution. *T&D, 61*(8), 54-57.

Liu, O. L., Frankel, L., & Roohr, K. C. (2014). Assessing critical thinking in higher education: Current state and directions for next-generation assessment. *ETS Research Report Series, 2014*(1), 1–23.

Liu, B. (2012). Sentiment Analysis and Opinion Mining. *Synthesis Lectures on Human Language Technologies*, *5*(1), 1–167.

Liu, X., Guo, C., & Zhang, L. (2014). Scholar metadata and knowledge generation with human and artificial intelligence. *Journal of the Association for Information Science and Technology*, *65*(6), 1187–1201. doi:10.1002/asi.23013

Livingstone, S. (2008). Internet Literacy: Young People's Negotiation of New Online Opportunities. In T. McPherson (Ed.), Unexpected Outcomes and Innovative Uses of Digital Media by Youth, (pp. 101-121). Cambridge, MA: The MIT Press.

Lochrie, M., & Coulton, P. (2011). Mobile Phones as Second Screen for TV, Enabling Inter-audience Interaction. In *Proceedings of the 8th International Conference on Advances in Computer Entertainment Technology*.

Loes, C. N., Salisbury, M. H., & Pascarella, E. T. (2015). Student perceptions of effective instruction and the development of critical thinking: A replication and extension. *Higher Education*, *69*(5), 823–838. doi:10.1007/s10734-014-9807-0

Lohr, S. (2012). Big data's impact in the world. *The New York Times*.

Lordan, E. J. (2006). Interactivity: The Latest Trend in Effective Communications. *Public Relations Quarterly*, *51*(3), 27–29.

Lovejoy, K., & Saxton, G. D. (2012). Information, Community, and Action: How Nonprofit Organizations Use Social Media. *Journal of Computer-Mediated Communication, 17*(3), 337–353.

Luger, G., & Stubblefield, W. (2004). *Artificial Intelligence: Structures and Strategies for Complex Problem Solving* (5th ed.). Benjamin/Cummings.

Lunenburg, F. C. (2010). The Decision Making Process. *National Forum of Educational a MD Administration and Supervision Journal, 27*(4).

Luo, A., & Kumar, V. (2013). Recovering Hidden Buyer-Seller Relationship States to Measure the Return on Marketing Investment in Business-to-Business Markets. *Journal of Marketing Research, 50*, 143-160.

Lyon, D. (2013). *The Electronic Eye: The Rise of Surveillance Society-Computers and Social Control in Context*. Minneapolis: John Wiley & Sons.

Madsen, P. (2011). *Aggregating agendas: Online news aggregators as agenda setters*. Paper presented at 36th Annual Conference of the Midwest Association for Public Opinion Research, Chicago, IL.

Mager, A. (2012). Algorithmic ideology. *Information Communication and Society, 15*(5), 269–787. doi:10.1080/1369118X.2012.676056

Magno, C. (2010). The role of metacognitive skills in developing critical thinking. *Metacognition and Learning, 5*(2), 137–156. doi:10.1007/s11409-010-9054-4

Makarov, R. N., Nidziy, N. A., & Shishkin, G. K. (2000). *Psychological Foundations of Didactics in Flight*. MAPCHAK.

Malone, R., Laubacher, C., & Dellarocas. (2009). *Harnessing Crowds: Mapping the Genome of Collective Intelligence*. MIT Sloan Research Paper, 4732-09.

Malone, T. W., Laubacher, R., & Dellarocas, C. N. (2009). *Harnessing crowds: Mapping the genome of collective intelligence*. MIT Sloan Research Paper 4732-09.

Malone, T. W., Laubacher, R., & Chrysanthos, D. (2010). The collective ıntelligence genome. *MIT Sloan Management Review, 51*(3), 21–22.

Manalo, E., Kusumi, T., Koyasu, M., Michita, Y., & Tanaka, Y. (2013). To what extent do culture-related factors influence university students' critical thinking use? *Thinking Skills and Creativity, 10*, 121–132. doi:10.1016/j.tsc.2013.08.003

Mangold, W. G., & Faulds, D. J. (2009). Social Media: The New Hybrid Element of the Promotion Mix. *Business Horizons, 52*(4), 357–365.

Manovich, L. (1999). Database as symbolic form. *Convergence (London), 5*(2), 80–99. doi:10.1177/135485659900500206

Manovich, L. (2001). *The language of new media*. Cambridge, MA: MIT Press.

March, J. G. (1991). Exploration and Exploitation in Organization Learning. *Organization Science, 2*.

Marchigiano, G., Eduljee, N., & Harvey, K. (2011). Developing critical thinking skills from clinical assignments: A pilot study on nursing students' self-reported perceptions. *Journal of Nursing Management, 19*(1), 143–152. doi:10.1111/j.1365-2834.2010.01191.x PMID:21223414

March, J. G. (1991). Exploration and Exploitation in Organizational Learning. *Organization Science, 2*(1), 71–87. doi:10.1287/orsc.2.1.71

Markkula, M., & Kune, H. (2015). Smart Specialization and the Role of Universities in Regional Innovation Ecosystems.Technology Innovation Management Review, 5(10).

Marshak, R. J. (2004). Morphing: The leading edge of organizational change in the 21st century. *Organization Development Journal, 22*(3), 8–21.

Martincová, J., & Lukešová, M. (2015). Critical thinking as a tool for managing intercultural conflicts. *Procedia: Social and Behavioral Sciences, 171*, 1255–1264. doi:10.1016/j.sbspro.2015.01.239

Mathews, S. R., & Lowe, K. (2011). Classroom environments that foster a disposition for critical thinking. *Learning Environments Research, 14*(1), 59–73. doi:10.1007/s10984-011-9082-2

Mathiesen, T. (1997). The Viewer Society Michel Foucault's `Panopticon' Revisited. *Theoretical Criminology, 1*(2), 215–234. doi:10.1177/1362480697001002003

Mathuros, F. (2015, November 10). *What is the theme of Davos 2016?* Retrieved November 12, 2015, from https://agenda.weforum.org/2015/11/what-is-the-theme-of-davos-2016/

Mazurek-Łopacińska, K., & Sobocińska, M. (2002). Badania marketingowe. Podstawowe metody i obszary zastosowań. Wrocław: Wydawnictwo Akademii Ekonomicznej.

Mazurek-Łopacińska, K., & Sobocińska, M. (2010). Kierunki rozwoju badań marketingowych w dobie społeczeństwa i gospodarki opartych na wiedzy. In *Badania marketingowe: nowe wyzwania*. Wrocław: Wydawnictwo Uniwersytetu Ekonomicznego. Prace Naukowe Uniwersytetu Ekonomicznego we Wrocławiu.

McCorduck, P. (2004). *FAQ answered by Pamela McCorduck*. Retrieved November 7, 2015, from http://www.pamelamc.com/html/machines_who_think.html

McCorduck, P. (1979). *Machines who think: A personal inquiry into the history and prospects of artificial intelligence*. San Francisco: W.H. Freeman.

McGrath, R. M. (2013, June). Transient advantage. *Harvard Business Review*, 64–70.

McNichols, T. J. (1999). Deconstructing constructivism: The Kantian connection. *Journal of Philosophy and History of Education*, *49*, 141–146.

McQuail, D. (2010). *McQuail's mass communication theory*. Sage Publications.

McWilliam, G. (2000). Building Stronger Brands through Online Communities. *Sloan Management Review*, *41*(3), 43–54.

Meece, J. L., Blumenfeld, P. C., & Hoyle, R. H. (1988). Students' goal orientations and cognitive engagement in classroom activities. *Journal of Educational Psychology*, *80*(4), 514–523. doi:10.1037/0022-0663.80.4.514

Megapanel PBI/Gemius, IV.2014 – IV.2015.

Melaia, S., Abratt, R., & Bick, G. (2008). Competencies of Marketing Managers in South Africa. *Journal of Marketing Theory and Practice*, *16*(3), 233–246.

Michelucci, P. (2011). *Handbook on human computation*. Springer.

Miller, E. K., & Cohen, J. D. (2001). An integrative theory of prefrontal cortex function. *Annual Review of Neuroscience*, *24*(1), 167–202. doi:10.1146/annurev.neuro.24.1.167 PMID:11283309

Mintzberg, H. (1980). Structure in 5's: A synthesis of the research on organizational design. *Management Science*, *26*(3), 322–341. doi:10.1287/mnsc.26.3.322

Mirijamdotter, A., Ståhlbröst, A., Sällström, A., Niitamo, V.-P., & Kulkki, S. (2006). *The European Network of Living Labs for CWE - user-centric co-creation and innovation*. E-Challenges 2006, Barcelona, Spain.

Mohamud, A. (2012). Nearly 50 Percent of Internet Users in Europe Visit Newspaper Sites. *comScore*. Retrieved 25 December, 2015 from http://www.comscore.com/por/Insights/Press-Releases/2012/1/Nearly-50-Percent-of-Internet-Users-in-Europe-Visit-Newspaper-Sites

Molinari, F. (2011). Living Labs as Multi-Stakeholder Platforms for the eGovernance of Innovation. In *Proceedings of the International Conference on Theory and Practice of Electronic Governance (ICEGOV2011)*. doi:10.1145/2072069.2072092

Moon, J. (2008). *Critical thinking: An exploration of theory and practice*. Abingdon, United Kingdom: Routledge.

Moorman, C., Deshpandé, R., & Zaltman, G. (1993). Factors Affecting Trust in Market Research Relationships. *Journal of Marketing*, *57*(1), 81-101.

Moran, R. (2013). *The Futures of Marketing Research*. Retrieved from http://www.sagepub.com/upm-data/43895_Epilogue.pdf

Morgan, G., Frost, P. J., & Pondy, L. R. (1983). Organizational symbolism. *Organizational Symbolism, 3*, 35.

Morgan, J. (2014). *The Future of Work: Attract New Talent, Build Better Leaders and Create a Competitive Organization*. Wiley.

Morlino, G., Gianelli, C., Borghi, A. M., & Nolfi, S. (2015). Learning to manipulate and categorize in human and artificial agents. *Cognitive Science*, *39*(1), 39–64. doi:10.1111/cogs.12130 PMID:25041751

Morozov, E. (2011). Don't be evil. *The New Republic*, 13-07.

Morozov, E. (2014, July 30). *Digital Surveillance. Like clueless guinea pigs*. Retrieved from http://www.faz.net/aktuell/feuilleton/debatten/the-digital-debate/digital-surveillance-like-clueless-guinea-pigs-13070758.html

Mulder, I., & Stappers, P. J. (2009). *Co-Creating in Practice: Results and Challenges*. Paper presented at the 15th International Conference on Concurrent Engineering (ICE 2009), Leiden, The Netherlands.

Mun, M. S. (2010). An analysis of narratives to identify critical thinking contexts in psychiatric clinical practice. *International Journal of Nursing Practice*, *16*(1), 75–80. doi:10.1111/j.1440-172X.2009.01803.x PMID:20158552

Musk, Wozniak and Hawking urge ban on warfare AI and autonomous weapons. (n.d.). Retrieved November 5, 2015, from http://www.msn.com/en-gb/news/other/musk-wozniak-and-hawking-urge-ban-on-warfare-ai-and-autonomous-weapons/ar-AAdxy1e

Nam, H., & Kannan, P. K. (2014). The Informational Value of Social Tagging Networks. *Journal of Marketing*, *78*(4), 21–40.

Napoli, P. M. (2013). *The algorithm as institution: Toward a theoretical framework for automated media production and consumption*. Fordham University Schools of Business Research Paper.

Nielsen, H. (2011). Bounded Rationality in Decision Making. Dobbs Ferry, NY: Manchester University Press.

Niitamo, V.-P., Kulkki, S., Eriksson, M., & Hribernik, K. A. (2006). State-of-the-Art and Good Practice in the Field of Living Labs. In *Proceedings of the 12th International Conference on Concurrent Enterprising: Innovative Products and Services through Collaborative Networks* (pp. 349–357). Milan, Italy: Nottingham University Business School.

Nilsson, N. (1984). Artificial Intelligence, Employment, and Income. *AI Magazine*, 5–14.

Niu, L., Behar-Horenstein, L. S., & Garvan, C. W. (2013). Do instructional interventions influence college students' critical thinking skills? A meta-analysis. *Educational Research Review*, *9*, 114–128. doi:10.1016/j.edurev.2012.12.002

Nyström, A. G., Leminen, S., Westerlund, M., & Kortelainen, M. (2014). Actor roles and role patterns influencing innovation in living labs. *Industrial Marketing Management*, *43*(3), 483–495. doi:10.1016/j.indmarman.2013.12.016

Oczachowski, D. (2011). Kultura kognitywna organizacji a prowadzenie badań marketingowych. In *Badania marketingowe: metody, nowe podejścia i konteksty badawcze*. Wrocław: Wydawnictwo Uniwersytetu Ekonomicznego. Prace Naukowe Uniwersytetu Ekonomicznego we Wrocławiu.

OECD. (2013). *OECD Skills Outlook 2013 - First results from the survey of adult skills*. http://doi.org/10.1787/9789264204256-en

Olejniczuk-Merta, A. (2010). Trzy generacje badań marketingowych. In *Badania marketingowe: nowe wyzwania*. Wrocław: Wydawnictwo Uniwersytetu Ekonomicznego. Prace Naukowe Uniwersytetu Ekonomicznego we Wrocławiu.

Olesiński, Z. (2010). *Managing Inter-Organizational Relations*. Warsaw: C.H. Beck. (in Polish)

Oliveira, A., Fradinho, E., Caires, R., Oliveira, J., & Barbosa, A. (2006). Successful Regional Information Society Strategy to an Advanced Living Lab in Mobile Technologies and Services System Sciences. In *HICSS '06: Proceedings of the 39th Annual Hawaii International Conference on System Sciences*.

Oliver, R., Rust, R., & Varki, S. (1998). Real-Time Marketing. *Marketing Management*, *7*(Fall), 28–37.

Ordinary Home Appliances Are About to Get Really Sexy. (2016, January 6). *Fortune*. Available at http://fortune.com/2016/01/06/home-appliances-ces-2016/

Orlikowski, W. J., & Gash, D. C. (1994). Technological frames: Making sense of information technology in organizations. *ACM Transactions on Information Systems*, *12*(2), 174–207. doi:10.1145/196734.196745

Orliński, W. (2013). *Internet. Czas się bać.* Warszawa: Wydawnictwo Agora.

Page, L., Brin, S., Motwani, R., & Winograd, T. (1999). *The PageRank Citation Ranking: Bringing order to the web. Technical report, Computer Science Department*. Stanford University.

Palis, C. (2012). *Pinterest traffic growth soars to new heights: Experian report.* The Huffington Post.

Parameswaran, M., & Whinston, A.B. (2007). Social Computing: An Overview. *CAIS, 19*(37), 762-780.

Paul, R. (2005). The state of critical thinking today. *New Directions for Community Colleges*, *2005*(130), 27–38. doi:10.1002/cc.193

Paul, S. A. (2014). Assessment of critical thinking: A Delphi study. *Nurse Education Today*, *34*(11), 1357–1360. doi:10.1016/j.nedt.2014.03.008 PMID:24713126

Pavlenko, M. A. (2007). The Method of Analysis of the Automated Control Systems of Air Traffic Operator. *Information Processing Systems, No.*, *1*(59), 78–81.

Pavlovskaia, O.O. (2005). Methods for Determining the Parameters of a Linear Mathematical Model of the Operator. *Computer Technology, Management, Electronics, 5*(7), 41-44.

Peters, T. (2011). *A Brief History of 7S (McKinsey's 7S) Model.* Retrieved 15 August 2015 from http://tompeters.com/docs/7SHistory.pdf

PEW. (n.d.). *Special Reports – Economic Attitudes*. Retrieved January 10, 2016, from http://www.stateofthemedia.org/2010/special-reports-economic-attitudes/nielsen-analysis

Picard, R. G. (2009). Why journalists deserve low pay. *The Christian Science Monitor*, 19.

Picard, R. G. (Ed.). (2014). *Media product portfolios: Issues in management of multiple products and services*. Routledge.

Plato. (1903), Platonis Opera. Oxford University Press.

Policastro, E. (1999). Intuition. In M. A. Runco & S. R. Pritzker (Eds.), *Encyclopedia of creativity* (Vol. 2, pp. 89–93). San Diego, CA: Academic Press.

Popper, K. (1973). Of Clouds and Clocks. In Objective Knowledge (pp. 206–55). Oxford, UK: Oxford University Press.

Porter, M. E. (1980). *Competitive strategy: techniques for analyzing industries and competitors*. New York: Free Press.

Porter, M. E. (1996). What is strategy? *Harvard Business Review*, *74*(6), 61–81. PMID:10158474

Potatoes, C. C. (2013, February 9). *The Economist*. Available at http://www.economist.com/news/business/21571471-measuring-tv-audiences-keeps-getting-harder-counting-couch-potatoes

Power, M. (1997). *The Audit Society: Rituals of Verification*. Oxford, UK: Oxford University Press.

Prahalad, C.K. & Ramaswamy, V. (2000). Co-opting Customer Competence. *Harvard Business Review*, *78*(January-February), 79–87.

Pratt, M. G., et al. (2007). Exploring Intuition and its Role in Managerial Decision Making. Academy of Management Review, 32(1).

Pratt, G. (2015). Is a Cambrian Explosion Coming for Robotics? *The Journal of Economic Perspectives*, *29*(3), 51–60. doi:10.1257/jep.29.3.51

Profetto-McGrath, J. (2003). The relationship of critical thinking skills and critical thinking dispositions of baccalaureate nursing students. *Journal of Advanced Nursing*, *43*(6), 569–577. doi:10.1046/j.1365-2648.2003.02755.x PMID:12950562

PTBRiO. (2009). *Wizerunek branży badawczej 2009*. Prezentacja wyników badania IX-X 2009. PTBRiO.

PTBRiO. (2011). *Rocznik PTBRiO 2010/2011*. Retrieved from http://ptbrio.pl/index.php/publikacje/rocznik-ptbrio

PTBRiO. (2012a). *Wizerunek branży badawczej 2011*. PTBRiO. Retrieved from www.ptbrio.pl/images/stories/publikacje/wizerunek_branzy_badawczej_raport_wersja_pelna.pdf

PTBRiO. (2012b). *Rocznik PTBRiO 2011/2012*. Retrieved from http://ptbrio.pl/index.php/publikacje/rocznik-ptbrio

PTBRiO. (2012c). *Raport idealny*. Retrieved from http://www.ptbrio.pl/index.php/publikacje/raporty

PTBRiO. (2013a). *Rocznik PTBRiO 2012/2013*. Retrieved from http://ptbrio.pl/index.php/publikacje/rocznik-ptbrio

PTBRiO. (2013b). *Badania marketingowe*. Rocznik Polskiego Towarzystwa Badaczy Rynku i Opinii. Edycja XVII, 2012/13.

PTBRiO. (2014). *Rocznik PTBRiO 2013/2014*. Retrieved from http://ptbrio.pl/index.php/publikacje/rocznik-ptbrio

PTBRiO. (2015). *Rocznik PTBRiO 2014/2015*. Retrieved from http://ptbrio.pl/index.php/publikacje/rocznik-ptbrio

PTBRiO. (2016). *Rocznik PTBRiO 2015/2016*. Retrieved from http://ptbrio.pl/index.php/publikacje/rocznik-ptbrio

Pucer, P., Trobec, I., & Žvanut, B. (2014). An information communication technology based approach for the acquisition of critical thinking skills. *Nurse Education Today*, *34*(6), 964–970. doi:10.1016/j.nedt.2014.01.011 PMID:24581890

Pulizzi, J. (2012). The Rise of Storytelling as the New Marketing. *Publishing Research Quarterly*, *28*(2), 116–123.

Pun, H., & Heese, H. S. (2015, August). A note on budget allocation for market research and advertising. *International Journal of Production Economics*, *166*, 85–89. doi:10.1016/j.ijpe.2015.04.013

Pupkoff, K. (1974). The Optimization of Connection Between Human and Techniques in Man-Machine Systems. *Preprints of JFAC-JFORS Symposium*, 419-426.

Qing, Z., Xiang, W., & Linna, Y. (2007). A preliminary investigation into critical thinking of urban Xi'an high school students. *Frontiers of Education in China*, *2*(3), 447–468. doi:10.1007/s11516-007-0036-5

Quinn, A., & Bederson, B. (2009). *A Taxonomy of Distributed Human Computation*. Tech. Rep. HCIL-2009-23. University of Maryland.

Quinn, S., & Quinn-Allan, D. (2005, January). The World-wide spread of journalism convergence. In *Proceedings of the 2005 Journalism Education Association Conference*. Griffith University, School of Arts.

Rahimi, A., & Sajed, M. A. (2014). The interplay between critical pedagogy and critical thinking: Theoretical ties and practicalities. *Procedia: Social and Behavioral Sciences*, *136*, 41–45. doi:10.1016/j.sbspro.2014.05.284

Raisch, S., & Birkinshaw, J. (2008). Organizational Ambidexterity: Antecedents, Outcomes, and Moderators. *Journal of Management*, *34*(3), 375–409. doi:10.1177/0149206308316058

Rajani, S. (2011). Artificial intelligence: Man or machıne? *International Journal of Information Technology and Knowledge Management*, *4*(1), 173–176.

Ramaswamy, V. (2010). The Power of Co-Creation: Build It with Them To Boost Growth. Productivity, and Profits, Gouillart, Francis.

Ramaswamy, V., & Gouillart, F. (2010). *The Power of Co-creation. Build It With Them to Boost Growth, Productivity, and Profits*. New York, NY: Free Press.

Rapanta, C., Garcia-Mila, M., & Gilabert, S. (2013). What is meant by argumentative competence? An integrative review of methods of analysis and assessment in education. *Review of Educational Research*, *83*(4), 483–520. doi:10.3102/0034654313487606

Rayport, J. F., & Sviokla, J. J. (1994). Managing in the Marketspace. *Harvard Business Review*, *72*, 141–150.

Reason, J. (1990). *Human Error*. Ashgate.

Reichheld, F. (2001). *Loyalty rules! How today's leaders build lasting relationships*. Boston: Harvard Business School Press.

Report, M. (2014). *15 economic facts about millennials*. Executive Office of the President of the United States: The Council of Economic Advisers. Retrieved from 10 October 2015 https://www.whitehouse.gov/sites/default/files/docs/millennials_report.pdf

Ricoeur, P. (1981). *Hermeneutics and the Human Sciences: Essays on Language, Action and Interpretation*. Cambridge, UK: Cambridge University Press. doi:10.1017/CBO9781316534984

Rieder, K., & Voß, G. (2010). The working customer- an emerging new type of consumer. *Psychology of Everyday Activity*, *3*(2), 2–10.

Rifkin, J. (n.d.). *Book Summary: Third industrial revolution: how lateral power is transforming energy, the economy, and the world*. Retrieved from http://www.wermutham.com/pdf/The%20Third%20Industrial %20Revolution.pdf

Rifkin, J. (2011). *Third industrial revolution: how lateral power is transforming energy, the economy, and the world*. London: Palgrave Macmillan.

Rifkin, J. (2014). *The zero marginal cost society: the internet of things, the collaborative commons, and the eclipse of capitalism*. London: Palgrave Macmillan.

Rimiene, V. (2002). Assessing and developing students' critical thinking. *Psychology Learning & Teaching*, *2*(1), 17–22. doi:10.2304/plat.2002.2.1.17

Ritola, J. (2011). Justificationist social epistemology and critical thinking. *Educational Theory, 61*(5), 565–585. doi:10.1111/j.1741-5446.2011.00421.x

Rizun, N. (2012). The Questions of the Testee Identification as PID-Controller in the Automated System of Professional Certification. *Vimiryuvalna obchislyuvalna tehnika tehnologichnih, 3*(41), 133-141.

Rizun, N. (2013). Development of Methods and Models of Inaccuracy Minimization of the Machine-to-Human Interaction in Automated Systems of Professional Readiness Level Diagnostics. *Nauchniy Visnyk NMU, 2*.

Rizun, N. (2014). Methodology of Intellectual Express-Evaluation of the Decision-Making Effectiveness. Economics & Business Communication Challenges. *International Week. Research Papers*, 209-229.

Rizun, N., & Taranenko, Y. (2014a). Simulation Models of Human Decision-Making Processes. *Management Dynamics in the Knowledge Economy, 2*(5), 241-264.

Rizun, N., & Taranenko, Y. (2014b). Simulation Modelling of the Decision-Making Activity of the Automated Control Systems Operators Using Nonlinear Elements. *Ambient Technology & Creativity Support Systems, 188*, 157-178.

Rizun, N. (2013). Computer Modelling of Cognitive Processes.*Proc. of the Federated Conference on Computer Science and Information Systems* (FedCSIS).

Roberts, N., & Stockport, G. J. (2014). Defining Strategic Flexibility. In The Flexible Enterprise, Flexible Systems Management (vol. 1, pp. 37-45). Springer Publishing. doi:10.1007/978-81-322-1560-8_3

Roberts, M. L., & Zahay, D. (2012). *Internet Marketing: Integrating Online and Offline Strategies*. Stamford, CT: Cengage Learning.

Robertson, B. J. (2007). *Organization at the Leading Edge: Introducing Holacracy*. Retrieved 15 October 2015 from http://www.integralesleben.org/fileadmin/user_upload/images/DIA/Flyer/ Organization_at_the_Leading_Edge_2007-06_01.pdf

Rodzalan, S. A., & Saat, M. M. (2015). The perception of critical thinking and problem solving skill among Malaysian undergraduate students. *Procedia: Social and Behavioral Sciences, 172*, 725–732. doi:10.1016/j.sbspro.2015.01.425

Ronit, R. (2011). Technophilia: a new model for technology adoption. *UK Academy for Information Systems Conference Proceedings*.

Roos, J. M., Mela, C. F., & Shachar, R. (2013). *Hyper-media search and consumption*. Academic Press.

Rosanvallon, P. (2011). Demokracja interakcyjna. *Res Publica Nowa, 16*(206), 18–29.

Rothenbuhler, E. W., & Coman, M. (Eds.). (2005). *Media anthropology*. Thousand Oaks, CA: Sage Publications.

Rowley, J. (2008). Understanding Digital Content Marketing. *Journal of Marketing Management, 24*(5-6), 517–540.

Royle, J., & Laing, A. (2014). The Digital Marketing Skills Gap: Developing a Digital Marketer Model for the Communication Industries. *International Journal of Information Management, 34*(2), 65–73.

Rud, O. P. (2009). *Business intelligence success factors: tools for aligning your business in the global economy*. John Wiley & Sons, Inc.

Rusbase. (2012). *Управляющий партнер Altair Capital Management инвестировал в Likehack 4,5 млн руб.* Retrieved 23 December, 2015 from http://rusbase.vc/news/upravlyayushij-partner-altair-capital-management-investiroval-v-likehack-45-mln-rub/

Russell, S., & Norvig, P. (1995). *Artificial intelligence: A modern approach*. Englewood Cliffs, NJ: Prentice Hall.

Russell, S., & Norvig, P. (2003). *Artificial Intelligence: A Modern Approach* (2nd ed.). Upper Saddle River, NJ: Prentice Hall.

Rust, R. T., & Huang, M.-H. (2014). The Service Revolution and the Transformation of Marketing Science. *Marketing Science, 33*(2), 206–221.

Ryu, K., & Jung, M. (2004). Fractal Approach to Managing Intelligent Enterprises. In Creating Knowledge Based Organizations. Idea Group. Retrieved from http://books.google.com.tr/books?id=7aHM3hjWsjUC&lpg=PA319&dq=fractal%20organization &hl=tr&pg=PA348#v=onepage&q=fractal%20organization&f=false

Saadé, R. G., Morin, D., & Thomas, J. D. E. (2012). Critical thinking in e-learning environments. *Computers in Human Behavior, 28*(5), 1608–1617. doi:10.1016/j.chb.2012.03.025

Salecl, R. (2011). *The tyranny of choice*. Profile Books.

Salleh, S. M., Tasir, Z., & Shukor, N. A. (2012). Web-based simulation learning framework to enhance students' critical thinking skills. *Procedia: Social and Behavioral Sciences, 64*, 372–381. doi:10.1016/j.sbspro.2012.11.044

Saroj. (2009). *Decision Making: Meaning and Definition*. Retrieved from http://www.excellentguru.com/index.php

Sashi, C. M. (2012). Customer Engagement, Buyer-Seller Relationships, and Social Media. *Management Decision*, *50*(2), 253–272.

Schaffers, H., & Kulkki, S. (2007). *Living Labs: An Open Innovation Concept Fostering Rural Development. Tech Monitor.*

Schaffers, H., & Turkama, P. (2012). Living Labs for Cross-Border Systemic Innovation. *Technology Innovation Management Review*, *2*(9), 25–30.

Schall, D., Truong, H., & Dustdar, S. (2008). The Human Provided Services Framework Science Platform. In *Zooniverse: Observing the World's Largest Citizen Science Platform*. Academic Press.

Schlee, R. P., & Harich, K. R. (2010). Knowledge and Skill Requirements for Marketing Jobs in the 21st Century. *Journal of Marketing Education*, *32*(3), 341–352.

Schmid, G. (2003). *Beyond Conventional Service Economics*. Discussion Paper FS I -2003.

Schmitt, B., & Zarantonello, L. (2013). Consumer Experience and Experiential Marketing: A Critical Review. *Review of Marketing Research*, *10*, 25–61.

Schumpeter, J. A. (1960). *Economic Theory Development*. Warsaw: PWN.

Schunk, D. H. (2012). *Learning theories: An educational perspective*. Upper Saddle River, NJ: Pearson/Merrill Prentice Hall.

Schuurman, D., & De Marez, L. (2009). User-Centred Innovation: Towards a Conceptual Integration of Lead Users and Living Labs. In *Proceedings of COST298: The Good, the Bad and the Challenging*.

Schuurman, D., De Marez, L., & Ballon, P. (2013). *Open Innovation Processes in Living Lab Innovation Systems: Insights from the LeYLab*. Carleton University.

Schwab, K. (2015, October 27). *Will the Fourth Industrial Revolution have a human heart?* Retrieved November 12, 2015, from https://agenda.weforum.org/2015/10/will-the-fourth-industrial-revolution-have-a-human-heart-and-soul/

Schwab, K. (2016). *The fourth industrial revolution*. World Economic Forum.

Scott, W. R. (2013). *Institutions and organizations: Ideas, interests, and identities*. Los Angeles, CA: Sage Publications.

Seker, H., & Komur, S. (2008). The relationship between critical thinking skills and in-class questioning behaviors of English language teaching students. *European Journal of Teacher Education*, *31*(4), 389–402. doi:10.1080/02619760802420784

Seymour, B., Kinn, S., & Sutherland, N. (2003). Valuing both critical and creative thinking in clinical practice: Narrowing the research–practice gap? *Journal of Advanced Nursing*, *42*(3), 288–296. doi:10.1046/j.1365-2648.2003.02618.x PMID:12680973

Shafer. (2002). *Automated News. The new Google News site, news untouched by human hands*. Retrieved 17 December, 2015 from http://www.slate.com/articles/news_and_politics/press_box/2002/09/automated_news.html

Shapiro, S. C. (1992). Artificial Intelligence. In S. C. Shapiro (Ed.), *Encyclopedia of Artificial Intelligence* (2nd ed.; pp. 54–57). New York: John Wiley & Sons, Inc.

Shenk, D. (1998). *Data smog: Surviving the information glut*. Harper San Francisco.

Sheth, J. N., Sisodia, R., & Sharma, A. (2000). The Antecedents and Consequences of Customer-Centric Marketing. *Journal of the Academy of Marketing Science*, *28*(1), 55–66.

Shiffman, D. (2008). *The Age of Engage: Reinventing Marketing for Today's Connected, Collaborative, and Hyper-Interactive Culture*. Ladera, CA: Hunt Street Press.

Shin, H., Ma, H., Park, J., Ji, E. S., & Kim, D. H. (2015). The effect of simulation courseware on critical thinking in undergraduate nursing students: Multi-site pre-post study. *Nurse Education Today*, *35*(4), 537–542. doi:10.1016/j.nedt.2014.12.004 PMID:25549985

Shin, K. R., Lee, J. H., Ha, J. Y., & Kim, K. H. (2006). Critical thinking dispositions in baccalaureate nursing students. *Journal of Advanced Nursing*, *56*(2), 182–189. doi:10.1111/j.1365-2648.2006.03995.x PMID:17018066

Shinnick, M. A., & Woo, W. A. (2013). The effect of human patient simulation on critical thinking and its predictors in prelicensure nursing students. *Nurse Education Today*, *33*(9), 1062–1067. doi:10.1016/j.nedt.2012.04.004 PMID:22564925

Shirky, C. (2010). *Cognitive surplus: how technology makes consumers into collaborators*. New York: Penguin Press.

Shmelova, T.F. (2012). Analysis of Human-Operator's Decision-Making in Air Navigation System. *Radio Electronic and Computer Systems*, *9*(52), 287-292.

Silk, A. J., Klein, L. R., & Berndt, E. R. (2001). The Emerging Position of the Internet as an Advertising Medium. *NETNOMICS: Economic Research and Electronic Networking*, *3*(2), 129–148.

Silvertown, J. (2009). A new dawn for citizen science. *Trends in Ecology & Evolution*, *24*(9), 467–471. doi:10.1016/j.tree.2009.03.017 PMID:19586682

Simon, H. A. (1997). *Models of bounded rationality: Empirically Grounded Economic Reason*. Cambridge, MA: MIT Press.

Simon, H. A. (2009). *Economics, Bounded Rationality, and the Cognitive Revolution*. Northampton, MA: Edward Elgar Publishing.

Simon, M., van den Driest, F., & Wilms, T. (2016). Driving Customer-centric Growth: A Practical Roadmap. *Journal of Advertising Research*, *56*(2), 159–168.

Simpson, R., Page, K., & De Roure, D. (2014). *Zooniverse: Observing the World's Largest Citizen*. Academic Press.

Simpson, E., & Courtney, M. (2002). Critical thinking in nursing education: Literature review. *International Journal of Nursing Practice*, *8*(2), 89–98. doi:10.1046/j.1440-172x.2002.00340.x PMID:11993582

Simpson, E., & Courtney, M. (2008). Implementation and evaluation of critical thinking strategies to enhance critical thinking skills in Middle Eastern nurses. *International Journal of Nursing Practice*, *14*(6), 449–454. doi:10.1111/j.1440-172X.2008.00719.x PMID:19126073

Simsek, Z., Heavey, C., Veiga, J. F., & Souder, D. (2009). A Typology for Aligning Organizational Ambidexterity's Conceptualizations, Antecedents, and Outcomes. *Journal of Management Studies*, *46*(5), 864–894. doi:10.1111/j.1467-6486.2009.00841.x

Singh, S. (2010). *Social Influence Marketing Trends for 2010*. Available at http://www.slideshare.net/shivsingh/social-influence-marketing-trends-2967561

Smetanová, V., Drbalová, A., & Vitáková, D. (2015). Implicit theories of critical thinking in teachers and future teachers. *Procedia: Social and Behavioral Sciences*, *171*, 724–732. doi:10.1016/j.sbspro.2015.01.184

Smith, A., Lynn, S., & Lintott, S. (2013). *Introduction to the Zooniverse*. AAAI Workshop - Technical Report WS-13-18.

Smith, E. (2011). Teaching critical reflection. *Teaching in Higher Education*, *16*(2), 211–223. doi:10.1080/13562517.2010.515022

Snow, R., O'Connor, B., Jurafsky, D., & Ng, A. (2008). Cheap and fast - but is it good? Evaluating non-expert annotations for natural language tasks.*Proceedings of the 2008 Conference on Empirical Methods in Natural Language Processing*.

Sosu, E. M. (2013). The development and psychometric validation of a critical thinking disposition scale. *Thinking Skills and Creativity*, *9*, 107–119. doi:10.1016/j.tsc.2012.09.002

Spring, T. (2011). *How Google, Facebook and Amazon run the internet*. PC Advisor.

Staib, S. (2003). Teaching and measuring critical thinking. *The Journal of Nursing Education*, *42*(11), 498–508. PMID:14626388

Stanovich, K. E., & West, R. F. (1999). Discrepancies Between Normative and Descriptive Models of Decision Making and the Understanding/Acceptance Principle. *Cognitive Psychology*, *38*(3), 349–385. doi:10.1006/cogp.1998.0700 PMID:10328857

Stapleton, P. (2010). A survey of attitudes towards critical thinking among Hong Kong secondary school teachers: Implications for policy change. *Thinking Skills and Creativity*, *6*(1), 14–23. doi:10.1016/j.tsc.2010.11.002

Stedman, N. L., & Andenoro, C. (2007). Identification of relationships between emotional intelligence and critical thinking dispositions in undergraduate leadership students. *Journal of Leadership Education*, *6*(1), 190–208. doi:10.12806/V6/I1/RF10

Steiner, C. (2012). *Automate This: How Algorithms Took Over Our Markets, Our Jobs, and the World*. New York: Penguin.

Stone, M., & Woodcock, N. (2014). Interactive, Direct and Digital Marketing. *Journal of Research in Interactive Marketing*, *8*(1), 4–17.

Strelkov, Y. K. (2001). *Engineering and Professional Psychology*. Academic Press.

Sullins, J. (2014). Information Technology and Moral Values. In *The Stanford Encyclopedia of Philosophy*. Retrieved from http://plato.stanford.edu/archives/spr2014/entries/it-moral-values/

Surowiecki, J. (2005). *The Wisdom of Crowds*. New York: Anchor Books.

Szahaj, A. (2003). Co to jest postmodernizm? In M. A. Potocka (Ed.), Postmodernizm: teksty polskich autorów (pp. 41–52). Kraków: "Bunkier Sztuki".

Tajvidi, M., Ghiyasvandian, S., & Salsali, M. (2014). Probing concept of critical thinking in nursing education in Iran: A concept analysis. *Asian Nursing Research*, *8*(2), 158–164. doi:10.1016/j.anr.2014.02.005 PMID:25030649

Tan, S. (2014). *The European Security Complex*. documentary film.

Taneja, H., Webster, J., Malthouse, E. C., & Ksiazek, T. B. (2012). Media consumption across platforms: Identifying user-defined repertoires. *New Media & Society*, *14*(6), 951–968.

Tang, T., & Hämäläinen, M. (2012). Living Lab Methods and Tools for Fostering Everyday Life Innovation. In B. Katzy, T. Holzmann, K. Sailer, & K. D. Thoben (Eds.), *Proceedings of the 2012 18th International Conference on Engineering, Technology and Innovation*. Academic Press.

Taranenko, Y. (2008). Formation of the Optimal Characteristics for Oscillating Systems vibro-frequently Data Densitometers. *Math. Electronics Universities*, *7*(51), 47–58.

Tarricone, P. (2011). *The taxonomy of metacognition*. New York, NY: Psychology Press.

Taylor, B. J. (2006). *Reflective practice: A guide for nurses and midwives*. Maidenhead, United Kingdom: Open University Press.

Taylor, C. (1985). The concept of a person. *Philosophical Papers Human Agency and Language*, *1*, 97–114. doi:10.1017/CBO9781139173483.005

Taylor, J., & Wros, P. (2007). Concept mapping: A nursing model for care planning. *The Journal of Nursing Education*, *46*(5), 211–216. PMID:17547344

Taylor, K. (2003). Marketing Yourself in the Competitive Job Market: An Innovative Course Preparing Undergraduates for Marketing Careers. *Journal of Marketing Education*, *25*(2), 97–107.

Taylor, S. S. (2015). Controls and Constraints. *Organizational Aesthetics*, *4*(1), 1–3.

Teece, D., & Pisano, G. (1994). The dynamic capabilities of firms: An introduction. *Industrial and Corporate Change*, *3*(3), 537–556. doi:10.1093/icc/3.3.537-a

The Nobel Prize in Physics 2010. (2011). Retrieved from http://nobelprize.org/nobelprizes/physics/laureates/2010/

Timochko, A.I. (2007). Modeling of the Decision Maker Activities in the Management Systems Networks. *Radiotekhnika*, *151*, 85-91.

Tiwari, A., Lai, P., So, M., & Yuen, K. (2006). A comparison of the effects of problem-based learning and lecturing on the development of students' critical thinking. *Medical Education*, *40*(6), 547–554. doi:10.1111/j.1365-2929.2006.02481.x PMID:16700770

Toffler, A. (1980). *Third wave*. New York: William Morrow and Co.,Inc.

Tomlinson, B., & Silberman, M. S. (2012). The cognitive surplus is made of fossil fuels. *First Monday*, *17*(11). doi:10.5210/fm.v17i11.4120

Towler, M. (2010). *Rational Decision Making: An introduction*. New York, NY: Wiley.

Tractenberg, R. E., Gushta, M. W., Mulroney, S. E., & Weissinger, P. A. (2013). Multiple choice questions can be designed or revised to challenge learners' critical thinking. *Advances in Health Sciences Education: Theory and Practice*, *18*(5), 945–961. doi:10.1007/s10459-012-9434-4 PMID:23288470

Tran-Thanh, L., Stein, S., Rogers, A., & Jennings, N. (2014). Efficient Crowdsourcing of Unknown Experts using Bounded Multi-Armed Bandits. *Artificial Intelligence*, *214*, 89–111. doi:10.1016/j.artint.2014.04.005

Tunc, O., & Aysegul, A. T. (2013). Businesses as Fractal Organisms in Chaotic Aura. *International Journal of Business and Management Studies*, 10-20.

Tuten, T. L., & Solomon, M. R. (2015). *Social Media Marketing*. London, UK: SAGE Publications.

Tversky, A., & Kahneman, D. (1974). Judgment under Uncertainty: Heuristics and Biases. *Science*, *185*(4157), 1124–1131. doi:10.1126/science.185.4157.1124 PMID:17835457

Tynjala, P. (2008). Perspectives into learning at the workplace. *Educational Research Review*, *3*(2), 130–154. doi:10.1016/j.edurev.2007.12.001

Tynjala, P., Slotte, V., Nieminen, J., Lonka, K., & Olkinuora, E. (2007). From university to working life: Graduates workplace skills in practice. In P. Tynjala, J. Valimaa, & G. Boulton-Lewis (Eds.), *Higher education and working life: Collaborations, confrontations and challenges* (pp. 73–88). Oxford, United Kingdom: Elsevier.

Tyreman, S. (2000). Promoting critical thinking in health care: Phronesis and criticality. *Medicine, Health Care, and Philosophy*, *3*(2), 117–124. doi:10.1023/A:1009973021449 PMID:11079339

Ushahidi. (2016). Available at https://www.ushahidi.com

Vahlberg, V., Shearer, E., Peer, L., Williams, V., & Figueroa, B. (2008). From "Too Much" to "Just Right". Engaging Millenials in Election News on the Web. Media Management Center, Northwestern University. *Tulostettu*, *5*, 2009.

Valenzuela, J., Nieto, A. M., & Saiz, C. (2011). Critical thinking motivational scale: A contribution to the study of relationship between critical thinking and motivation. *Electric Journal of Research in Educational Psychology*, *9*(2), 823–848.

Van Dijck, J. (2013). *The Culture of Connectivity: A Critical History of Social Media*. New York, NY: Oxford University Press.

Van Dijck, J., & Nieborg, D. (2009). Wikinomics and its discontents: A critical analysis of Web 2.0 business manifestos. *New Media & Society*, *11*(5), 855–874. doi:10.1177/1461444809105356

Van Otterlo, M. (2013). A Machine Learning View on Profiling. In M. Hildebrandt & K. de Vries (Eds.), *Privacy, Due Process and the Computational Turn. Philosophers of Law Meet Philosophers of Technology*. New York: Routledge.

Vandermerwe, S. (2014). *Breaking Through*. Palgrave Macmillan.

Vargo, S., & Lusch, R. (2004). Evolving to a New Dominant Logic for Marketing. *Journal of Marketing*, *68*(1), 1–17.

Vattimo, G. (1992). *The transparent society*. Cambridge, MA: Polity Press.

Veeckman, C., Schuurman, D., Leminen, S., & Westerlund, M. (2013). Linking Living Lab Characteristics and Their Outcomes: Towards a Conceptual Framework. *Technology Innovation Management Review*, *3*(12).

Veeckman, C., Schuurman, D., Leminen, S., & Westerlund, M. (2013, December). Linking Living Lab Characteristics and Their Outcomes: Towards a Conceptual Framework. Technology Innovation Management Review.

Verzera, S. (2014). *Liquid organizations: building the next evolutionary stage of anti-fragility*. Academic Press.

von Ahn, L. (2005). *Human Computation*. (Doctoral Thesis). CMU.

von Glasersfeld, E. (1990). An exposition of constructivism: Why some like it radical. In R. Davis, C. Maher, & N. Noddings (Eds.), *Constructivist views on the teaching and learning of mathematics* (pp. 19–29). Reston, VA: National Council of Teachers of Mathematics.

Von Hippel, E. (2005). *Democratizing innovation*. Cambridge, MA: MIT Press.

Vygotsky, L. S. (1978). *Mind in society: The development of higher psychological processes*. Cambridge, MA: Harvard University Press.

Wagner, R. K. (1986). *Tacit Knowledge and Intelligence in the Everyday World in Practical Intelligence: Nature and Origins of Competence in the Everyday World*. New York: Cambridge University Press.

Wallace, E., Buil, I., & de Chernatony, L. (2014). Consumer Engagement with Self-expressive Brands: Brand Love and WOM outcomes. *Journal of Product and Brand Management*, *23*(1), 33–42.

Wallach, W. (2009). *Moral machines: Teaching robots right from wrong*. Oxford, UK: Oxford University Press. doi:10.1093/acprof:oso/9780195374049.001.0001

Walthew, P. J. (2004). Conceptions of critical thinking held by nurse educators. *The Journal of Nursing Education*, *43*(9), 408–411. PMID:15478693

Wangensteen, S., Johansson, I. S., Björkström, M. E., & Nordström, G. (2011). Research utilisation and critical thinking among newly graduated nurses: Predictors for research use. A quantitative cross-sectional study. *Journal of Clinical Nursing*, *20*(17/18), 2436–2447. doi:10.1111/j.1365-2702.2010.03629.x PMID:21689180

Wang, Q., & Woo, H. L. (2010). Investigating students' critical thinking in weblogs: An exploratory study in a Singapore secondary school. *Asia Pacific Education Review*, *11*(4), 541–551. doi:10.1007/s12564-010-9101-5

Waterman, R. Jr, Peters, T., & Phillips, J. R. (1980). Structure is not Organization. *Business Horizons*, *23*(3), 14–26. doi:10.1016/0007-6813(80)90027-0

Watson, H. (2012). The Necessary Skills for Advanced Analytics. *Business Intelligence Journal*, *17*(4), 4–7.

Webber, R. (2013). The Evolution of Direct, Data and Digital Marketing. *Journal of Direct, Data and Digital Marketing Practice*, *14*(4), 291–309.

Webster, F., Malter, A., & Ganesan, S. (2005). The Decline and Dispersion of Marketing Competence. *MIT Sloan Management Review*, *46*(4), 35–43.

Webster, J. G., & Ksiazek, T. B. (2012). The Dynamics of Audience Fragmentation: Public Attention in an Age of Digital Media. *Journal of Communication*, *62*(1), 39–56.

Weick, K. E. (2001). *Making Sense of the Organization*. Oxford, UK: Blackwell Ltd.

Weissberg, R. (2013). Critically thinking about critical thinking. *Academic Questions*, *26*(3), 317–328. doi:10.1007/s12129-013-9375-2

Wellman, N. (2010). The Employability Attributes Required of New Marketing Graduates. *Marketing Intelligence & Planning*, *28*(7), 908–930.

Westerlund, M., & Leminen, S. (2011). Managing the Challenges of Becoming an Open Innovation Company: Experiences from Living Labs. *Technology Innovation Management Review*, *1*(1), 9–25.

Westley, B. (2011). How a Narrow Application of 'Hot News' Missapropriation Can Help Save Journalism. *The American University Law Review*, *60*(3), 4.

Wilgis, M., & McConnell, J. (2008). Concept mapping: An educational strategy to improve graduate nurses' critical thinking skills during a hospital orientation program. *Journal of Continuing Education in Nursing*, *39*(3), 119–126. doi:10.3928/00220124-20080301-12 PMID:18386699

Williams, L., & Lahman, M. (2011). Online discussion, student engagement, and critical thinking. *Journal of Political Science Education*, *7*(2), 143–162. doi:10.1080/15512169.2011.564919

Winter, M., & Siddharth, S. (2012). Conducting Behavioral Research on Amazon's Mechanical Turk. *Behavior Research Methods*, *44*(1), 1–23. doi:10.3758/s13428-011-0124-6 PMID:21717266

Wymbs, C. (2011). Digital Marketing: The Time for a New "Academic Major" Has Arrived. *Journal of Marketing Education*, *33*(1), 93–106.

Yadav, M. S., & Pavlou, P. A. (2014). Marketing in Computer-Mediated Environments: Research Synthesis and New Directions. *Journal of Marketing*, *78*(1), 20–40.

Yang, Y. T. C. (2008). A catalyst for teaching critical thinking in a large university class in Taiwan: Asynchronous online discussions with the facilitation of teaching assistants. *Educational Technology Research and Development*, *56*(3), 241–264. doi:10.1007/s11423-007-9054-5

Yang, Y. T. C., Gamble, J. H., Hung, Y. W., & Lin, T. Y. (2014). An online adaptive learning environment for critical-thinking-infused English literacy instruction. *British Journal of Educational Technology*, *45*(4), 723–747. doi:10.1111/bjet.12080

Yang, Y. T. C., Newby, T. J., & Bill, R. L. (2005). Using Socratic questioning to promote critical thinking skills through asynchronous discussion forums in distance learning environments. *American Journal of Distance Education*, *19*(3), 163–181. doi:10.1207/s15389286ajde1903_4

Yang, Y., Zhu, B. B., Guo, R., Yang, L., Li, S., & Yu, N. (2008). A comprehensive human computation framework: with application to image labeling.*Proc. MM*2008. doi:10.1145/1459359.1459423

Yin, R. K. (1994). *Case study research. Design and methods*. Thousand Oaks, CA: Sage.

Young, J.F. (1979). *Robotics*. Mechanical Engineering.

Yuen, M., Chen, L., & King, I. (2009). A Survey of Human Computation Systems. *Proc CSE*.

Yu, K. C., Lin, K. Y., & Fan, S. C. (2015). An exploratory study on the application of conceptual knowledge and critical thinking to technological issues. *International Journal of Technology and Design Education*, *25*(3), 339–361. doi:10.1007/s10798-014-9289-5

Zaidan, C. (2011). Crowdsourcing translation: professional quality from non-professionals. In *Proceedings of the 49th Annual Meeting of the Association for Computational Linguistics: Human Language Technologies.*

Żakowska, M. (2007). Kultura organizacyjna polskiego biznesu według kryteriów zaprogramowania kulturowego. In *Kultura i gospodarka: ku antropologii życia gospodarczego we współczesnej Polsce*. Tychy: Śląskie Wydawnictwo Naukowe.

Zhabreev, V.S. (2007). *Models and Evaluation of Human Factors in the Large Systems: a Tutorial*. Academic Press.

Zooniverse. (2016). Available at https://www.zooniverse.org

Zori, S., Nosek, L. J., & Musil, C. M. (2010). Critical thinking of nurse managers related to staff RNs' perceptions of the practice environment. *Journal of Nursing Scholarship*, *42*(3), 305–313. doi:10.1111/j.1547-5069.2010.01354.x PMID:20738741

Zuboff, S. (1988). *In the Age of the Smart Machine: The Future of Work and Power*. New York: Basic Books.

About the Contributors

Bionote Roman Batko is Associate Professor of Management in the Faculty of Management and Social Communication at the Jagiellonian University Krakow, Poland and director of the Institute of Culture. He holds an MA in Polish Literature and a PhD in Management from the Jagiellonian University. He is the author and editor of numerous articles and expert analyses on process and quality management, as well as management in public institutions, especially in the context of liquid modernity; he has written four books, *Quality in Public Administration* (2009), *Modern Museum* (2010), *A new paradigm for the public administration* (2012) and *Golem, Avatar, Midas, Golden Calf: Public Organization in Liquid Modernity* (2013). His most recent interests concentrate on humanistic approach to management, philosophical, ethical and aesthetic aspects of the organization as well as issues related to the new media and organization of the future.

Anna Szopa, PhD, designs and teaches undergraduate and graduate courses in entrepreneurship and technology commercialization. She holds an MS in Quality Management from Silesian University of Technology and earned her doctoral degree from Jagiellonian University examining innovation strategies of spin-off companies. She has twelve years' experience in managing companies including research and consultancy. She awarded the fellowship of the The Ryoichi Sasakawa Young Leaders Fellowship Fund (Sylff). She was a visiting researcher at Maryland Technology Enterprise Institute and University of Central Florida. Her primary research interests include university-industry relationships and university spin-offs. She is and editor of books about innovations, and published several articles in management journals and books.

* * *

Dennis Anderson is Chairman and Professor of Management and Information Technology at St. Francis College, New York City. He also serves as Founding Executive Director of the Institute of E-government and Global Sustainability and the Center for Entrepreneurship. Prior to this appointment, he was a professor of information systems and Associate Dean at Pace University. He also served as Founding Director of the University's Center for Advanced Media. He is a strong advocate of technology-enhanced learning, emerging technologies, sustainable technologies, and knowledge entrepreneurship. He has taught various business, information systems, and computer science courses at NYU Courant Institute, City University of New York, and Pace University.

Zofia Bednarowska is a PhD visiting student at Centre for Mobilities Research, Department of Sociology, Lancaster University in 2016. A PhD student in the Department of Sociology, Economy, Education and Methods of Social Research at the Institute of Sociology at the Jagiellonian University in Kraków. Researcher in Centre for Evaluation and Analysis of Public Policies of the Jagiellonian University. BA in Interfaculty Studies in Humanities, MA in Political Science and MA in Sociology from the Jagiellonian University in Kraków. Participant of the interdisciplinary Ph.D. studies "Society – Environment – Technologies". Graduate of the Essex Summer School in Social Science Data Analysis in 2014. In 2015 completed research stay as a Visiting scholar at University at Buffalo, NY. Worked 5 years in a market research agency as a research executive, responsible for the management and execution of research projects, as well as for research design and data analysis. Research interests are focused on the consumer behaviour theory, sociology of space, spatial analysis, in particular spatial statistics, geostatistics, data visualisation, and also test pricing, methodology of market research in social and consumer behaviour analysis. Selected publications: Bednarowska, Zofia. (2015). A spatial regression model of retail chains development in Poland. pp. 45-54. "Ekonometria=Econometrics", 3(49). Bednarowska, Zofia. (2015). Desk research - exploiting the potential of secondary data in market and social research. pp. 18-26. "Marketing i Rynek", Nr 7/2015.

Michał Andrzej Chrzanowski is a Ph.D. student at Institute of Sociology of the Jagiellonian University.

Anna Katarzyna Florek-Paszkowska is a lecturer and trainer; she realizes projects and writing scientific texts from the following areas: management, multi-criteria decision-making methods (AHP/ANP), quality management and sustainable food production and consumption.

Mario Gonzalez-Fuentes is an Assistant Professor of Marketing in the Department of Business Administration. For over a decade, he has taught at several European and American business schools and universities. His research interests focus on consumption identity and acculturation as well as the effects of technology in the way individuals interact with each other and with institutions.

Kijpokin Kasemsap received his BEng degree in Mechanical Engineering from King Mongkut's University of Technology, Thonburi, his MBA degree from Ramkhamhaeng University, and his DBA degree in Human Resource Management from Suan Sunandha Rajabhat University. He is a Special Lecturer in the Faculty of Management Sciences, Suan Sunandha Rajabhat University, based in Bangkok, Thailand. He is a Member of the International Association of Engineers (IAENG), the International Association of Engineers and Scientists (IAEST), the International Economics Development and Research Center (IEDRC), the International Association of Computer Science and Information Technology (IACSIT), the International Foundation for Research and Development (IFRD), and the International Innovative Scientific and Research Organization (IISRO). He also serves on the International Advisory Committee (IAC) for International Association of Academicians and Researchers (INAAR). He has had numerous original research articles in top international journals, conference proceedings, and books on the topics of business management, human resource management, and knowledge management, published internationally.

Jan Kreft is an associate professor at Jagiellonian University in Cracow, Faculty of Management and Social Communication. CEO and member board of directors in media enterprises (2001-2016).

Robert Niewiadomski, MA is an educator and a member of NABU, (a think tank with the UN's Economic and Social Council consultative status, focused on empowering communities through sharing of knowledge and technology). Robert graduated from Columbia University and Fordham University with degrees in Philosophy and Education, respectively. He is also an alumnus of Teach for America. Robert's areas of in inquiry include the societal impact of emerging technologies and cyberculture.

Mariana Petrova, comes from Saint-Petersburg, where she graduated from The Saint-Petersburg State University of Economics and Finance with a degree in Logistics and Supply Chain Management. In 2015 she graduated with honors from The Jagiellonian University with a Master degree in Culture and Media Management. Currently, Mariana is a doctoral student at The Jagiellonian University in Cracow, the media studies.

Nina Rizun is an Associate Professor and Candidate of Technical Sciences. She is working at the Department of Applied Informatics in Management, Gdansk University of Technology, Poland. She provides the courses "Mathematical Methods in Marketing and Management", "Forecasting of the of Social and Economic Processes", "Mathematical Methods in Applied Economics", "Data Science", "Methods and Tools of Economic

Analysis" . In 2014 and 2015 she provided thecourse "Social Network Analysis And Social Media: Theory And Applications" for students of the Master program "Cooperative Systems and Knowledge Engineering" of the faculty "Business Informatics" of Brandenburg University of Applied Sciences, Germany on the basis of individual Partnership Program within the framework of the Organization of International Academic Exchanges DAAD. Since 2010 till present time Nina Rizun also doing the postdoctoral research in the area of development the Simulation Models of Human Decision-Making Processes and Methodologies of Decision-Making Effectiveness Evaluation. Nina Rizun is the Member of: Editorial Board of the COMM. Scientific Journal; Program Committee of the Federated Conference on Computer Science and Information Systems, Poland, 2014 /2015. Workshop: 9-th Conference on Information Systems; Program Committee of the 7th International Conference on Information Technology (ICIT 2015), Prague etc.

Anna Maria Sabat is an expert in strategic management, creating business models, trainer, coach business with 13 years experience.

Shmelova Tatyana, 55 years, professor of Department of Air Navigation Systems in National Aviation University Author of more than 200 scientific articles, guides, monograph in fields of aviation, economics, mathematics, theory of system. Doctor of Science, thesis in the field of Scientific and Methodological Basis of Decision-Making in the Air Navigation System. Teaching courses: Theory of decision-making, Air Navigation, Operations Research, Mathematical Programming. Areas of Scientific Interests: Mathematical Models of Decision-Making by human operator, especially in Emergency Situation; Decision Support System; Research of air-navigation system as socio-technical system; Methodology of scientific research; Aviation Sociometry and Socionics; Informatics of Decision Making. She has got two children: daughter and son. Hobbies: Science, traveling, music and winter swimming.

Ayşegül Özbebek Tunç is an assistant professor of Management and Organization at Faculty of Political Sciences, Istanbul University. She holds a PhD degree with the dissertation examines the relationship between Organizational Ambidexterity, Top Management Team Behavioral Integration and Organizational Performance. Her major research interests and ongoing studies are about Sustainability, Strategic Management, Top Management Teams and Organizational Theory.

Oya Zincir is a research assistant of Management and Organization in Department of Business Administration at Faculty of Political Sciences, Istanbul University since November 2012. She is studying on her PhD thesis. Her major research interests and ongoing studies are about Knowledge Management, Digital Entrepreneurship, Open Innovation and Organizational Theory.

Index

P

R

S

T

W

www.ingramcontent.com/pod-product-compliance
Lightning Source LLC
LaVergne TN
LVHW061544290926
849162LV00004B/903